TEACHER'S EDITION

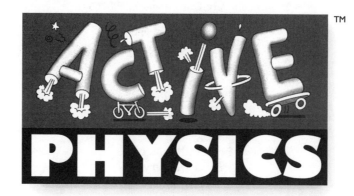

Arthur Eisenkraft, Ph.D.

Active Physics has been developed in association with the
American Association of Physics Teachers (AAPT)
and the
American Institute of Physics (AIP)

SPORTS

IT's ABOUT TIME ®

Published by
IT'S ABOUT TIME, Inc.
Armonk, NY

Published in 1999 by

It's About Time, Inc.

84 Business Park Drive, Armonk, NY 10504

Phone (914) 273-2233 Fax (914) 273-2227

Toll Free (888) 698-TIME

http://Its-About-Time.com

Publisher

Laurie Kreindler

Project Director

Dr. Arthur Eisenkraft

Project Manager

Ruta Demery

Design

John Nordland

Production Manager

Barbara Zahm

Studio Manager

Leslie Jander

Cover Illustration

Steven Belcher

Student's Edition Illustrations and Photos

Chapter 1: Tomas Bunk pages 13, 26, 27, 29, 35, 38; David Madison/Tony Stone Images pages 1, 2; Ed Mahan Photos page 7; Rueters/Gary Hershorn/Archive Photos page 18. **Chapter 2:** Tomas Bunk pages 54, 61, 68, 73, 81, 86, 92, 97, 103; NBC pages 53, 60, 67, 72, 80, 85, 91, 96, 102, 108, 110. **Chapter 3:** Tomas Bunk pages 114, 118, 125, 129, 134, 140, 145, 150, 156; NAS pages 113(2), 115, 116, 117, 122, 128, 133, 139, 144, 149, 154, 155, 159, 161, 163(2).

Teacher's Edition Illustrations

Kathleen Bowen, Robert Hansmann, Cathy Vidal

This project was supported, in part,
by the

National Science Foundation

Opinions expressed are those of the authors
and not necessarily those of the Foundation

TABLE OF CONTENTS

Acknowledgements iv

Meeting *Active Physics* for the First Time viii

Features of *Active Physics* xiv

Active Physics and the National Science Education Standards xx

Active Physics Addresses Key NSES Recommendations xxi

Cooperative Learning xxiii

Chapter 1 THE TRACK AND FIELD CHAMPIONSHIP .1

NSE Content Standards 2

Key Physics Concepts and Skills 3

Equipment List 4

Chapter Organizer 5

Chapter and Challenge Overview 7

Assessment Rubrics 9

Activities One through Ten 12

Alternative Chapter Assessment and Answers 112

Chapter 2 PHYSICS IN ACTION 117

NSE Content Standards 118

Key Physics Concepts and Skills 119

Equipment List 120

Chapter Organizer 121

Chapter and Challenge Overview 123

Assessment Rubrics 125

Activities One through Nine 128

Alternative Chapter Assessment and Answers 225

Chapter 3 SPORTS ON THE MOON 231

NSE Content Standards 232

Key Physics Concepts and Skills 233

Equipment List 234

Chapter Organizer 235

Chapter and Challenge Overview 237

Assessment Rubrics 239

Activities One through Nine 243

Alternative Chapter Assessment and Answers 326

BLACKLINE MASTERS: Reproducible pages for Additional Materials and Alternative/Optional Activities follow each activity.

All pages bound in the TEACHER'S ADDITION are perforated and three-hole punched to provide you with the flexibility of organizing your materials

Acknowledgments

Project Director

Arthur Eisenkraft teaches physics and serves as science coordinator in the Bedford Public Schools in N.Y. Dr. Eisenkraft is the author of numerous science and educational publications. He holds a US Patent for a laser vision testing system and was featured in *Scientific American*.

Dr. Eisenkraft is chair of the Duracell Science Scholarship Competition; chair of the Toyota TAPESTRY program giving grants to science teachers; and chair of the Toshiba/NSTA ExploraVisions Awards competition for grades K-12. He is co-author of a contest column and serves on the advisory board of *Quantum* magazine, a collaborative effort of the US and Russia. In 1993, he served as Executive Director for the XXIV International Physics Olympiad after being Academic Director for the United States Team for six years. He served on the content committee and helped write the National Science Education Standards of the NRC (National Research Council).

Dr. Eisenkraft received the Presidential Award for Excellence in Science Teaching at the White House in 1986, and the AAPT Distinguished Service Citation for "excellent contributions to the teaching of physics" in 1989. In 1991 he was recognized by the Disney Corporation as Science Teacher of the Year in their American Teacher Awards program. In 1993 he received an Honorary Doctor of Science degree from Rensselaer Polytechnic Institute.

Primary and Contributing Authors

Communication

Richard Berg
University of Maryland
College Park, MD

Ron DeFronzo
Eastbay Ed. Collaborative
Attleboro, MA

Harry Rheam
Eastern Senior High School
Atco, NJ

John Roeder
The Calhoun School
New York, NY

Patty Rourke
Potomac School
McLean, VA

Larry Weathers
The Bromfield School
Harvard, MA

Home

Jon L. Harkness
Active Physics Regional Coordinator
Wausau, WI

Douglas A. Johnson
Madison West High School
Madison, WI

John J. Rusch
University of Wisconsin, Superior
Superior, WI

Ruta Demery
Blue Ink Editing
Stayner, ON

Medicine

Russell Hobbie
University of Minnesota
St. Paul, MN

Terry Goerke
Hill-Murray High School
St. Paul, MN

John Koser
Wayzata High School
Plymouth, MN

Ed Lee
WonderScience, Associate Editor
Silver Spring, MD

Predictions

Ruth Howes
Ball State University
Muncie, IN

Chris Chiaverina
New Trier Township High School
Crystal Lake, IL

Charles Payne
Ball State University
Muncie, IN

Ceanne Tzimopoulos
Omega Publishing
Medford, MA

Sports

Howard Brody
University of Pennsylvania
Philadelphia, PA

Mary Quinlan
Radnor High School
Radnor, PA

Carl Duzen
Lower Merion High School
Havertown, PA

Jon L. Harkness
Active Physics Regional Coordinator
Wausau, WI

David Wright
Tidewater Comm. College
Virginia Beach, VA

Transportation

Ernest Kuehl
Lawrence High School
Cedarhurst, NY

Robert L. Lehrman
Bayside, NY

Salvatore Levy
Roslyn High School
Roslyn, NY

Tom Liao
SUNY Stony Brook
Stony Brook, NY

Bob Ritter
University of Alberta
Edmonton, AB, CA

Principal Investigators

Bernard V. Khoury
American Association of Physics
Teachers

Dwight Edward Neuenschwander
American Institute of Physics

Consultants

Peter Brancazio
Brooklyn College of CUNY
Brooklyn, NY

Robert Capen
Canyon del Oro High School
Tucson, AZ

Carole Escobar

Earl Graf
SUNY Stony Brook
Stony Brook, NY

Jack Hehn
American Association of
Physics Teachers
College Park, MD

Donald F. Kirwan
Louisiana State University
Baton Rouge, LA

Gayle Kirwan
Louisiana State University
Baton Rouge, LA

James La Porte
Virginia Tech
Blacksburg, VA

Charles Misner
University of Maryland
College Park, MD

Robert F. Neff
Suffern, NY

Ingrid Novodvorsky
Mountain View High School
Tucson, AZ

John Robson
University of Arizona
Tucson, AZ

Mark Sanders
Virginia Tech
Blacksburg, VA

Brian Schwartz
Brooklyn College of CUNY
New York, NY

Bruce Seiger
Wellesley High School
Newburyport, MA

Clifford Swartz
SUNY Stony Brook
Setauket, NY

Barbara Tinker
The Concord Consortium
Concord, MA

Robert E. Tinker
The Concord Consortium
Concord, MA

Joyce Weiskopf
Herndon, VA

Donna Willis
American Association of
Physics Teachers
College Park, MD

Safety Reviewer

Gregory Puskar
University of West Virginia
Morgantown, WV

Equity Reviewer

Leo Edwards
Fayetteville State University
Fayetteville, NC

Spreadsheet and MBL

Ken Appel
Yorktown High School
Peekskill, NY

Physics at Work

Barbara Zahm
Zahm Productions
New York, NY

Physics InfoMall

Brian Adrian
Bethany College
Lindsborg, KS

Unit Reviewers

George A. Amann
F.D. Roosevelt High School
Rhinebeck, NY

Patrick Callahan
Catasaugua High School
Center Valley, PA

Beverly Cannon
Science and Engineering
Magnet High School
Dallas, TX

Barbara Chauvin

Elizabeth Chesick
The Baldwin School
Haverford, PA 19041

Chris Chiaverina
New Trier Township High School
Crystal Lake, IL

Andria Erzberger
Palo Alto Senior High School
Los Altos Hills, CA

Elizabeth Farrell Ramseyer
Niles West High School
Skokie, IL

Mary Gromko
President of Council of State Science
Supervisors
Denver, CO

Thomas Guetzloff

Jon L. Harkness
Active Physics Regional Coordinator
Wausau, WI

Dawn Harman
Moon Valley High School
Phoenix, AZ

James Hill
Piner High School
Sonoma, CA

Bob Kearney

Claudia Khourey-Bowers
McKinley Senior High School

Steve Kliewer
Bullard High School
Fresno, CA

Ernest Kuehl
Roslyn High School
Cedarhurst, NY

Jane Nelson
University High School
Orlando, FL

John Roeder
The Calhoun School
New York, NY

Patty Rourke
Potomac School
McLean, VA

Gerhard Salinger
Fairfax, VA

Irene Slater
La Pietra School for Girls

Pilot Test Teachers

John Agosta

Donald Campbell
Portage Central High School
Portage, MI

John Carlson
Norwalk Community
Technical College
Norwalk, CT

Veanna Crawford
Alamo Heights High School
New Braunfels

Janie Edmonds
West Milford High School
Randolph, NJ

Eddie Edwards
Amarillo Area Center for
Advanced Learning
Amarillo, TX

Arthur Eisenkraft
Fox Lane High School
Bedford, NY

Tom Ford

Bill Franklin

Roger Goerke
St. Paul, MN

Tom Gordon
Greenwich High School
Greenwich, CT

Ariel Hepp

John Herrman
College of Steubenville
Steubenville, OH

Linda Hodges

Ernest Kuehl
Lawrence High School
Cedarhurst, NY

Fran Leary
Troy High School
Schenectady, NY

Harold Lefcourt

Cherie Lehman
West Lafayette High School
West Lafayette, IN

Kathy Malone
Shady Side Academy
Pittsburgh, PA

Bill Metzler
Westlake High School
Thornwood, NY

Elizabeth Farrell Ramseyer
Niles West High School
Skokie, IL

Daniel Repogle
Central Noble High School
Albion, IN

Evelyn Restivo
Maypearl High School
Maypearl, TX

Doug Rich
Fox Lane High School
Bedford, NY

John Roeder
The Calhoun School
New York, NY

Tom Senior
New Trier Township High School
Highland Park, IL

John Thayer
District of Columbia Public Schools
Silver Spring, MD

Carol-Ann Tripp
Providence Country Day
East Providence, RI

Yvette Van Hise
High Tech High School
Freehold, NJ

Jan Waarvick

Sandra Walton
Dubuque Senior High School
Dubuque, IA

Larry Wood
Fox Lane High School
Bedford, NY

Field Test Coordinator

Marilyn Decker
Northeastern University
Acton, MA

Field Test Workshop Staff

John Carlson

Marilyn Decker

Arthur Eisenkraft

Douglas Johnson

John Koser

Ernest Kuehl

Mary Quinlan

Elizabeth Farrell Ramseyer

John Roeder

Field Test Evaluators

Susan Baker-Cohen

Susan Cloutier

George Hein

Judith Kelley

all from Lesley College,
Cambridge, MA

Field Test Teachers and Schools

Rob Adams
Polytech High School
Woodside, DE

Benjamin Allen
Falls Church High School
Falls Church, VA

Robert Applebaum
New Trier High School
Winnetka, IL

Joe Arnett
Plano Sr. High School
Plano, TX

Bix Baker
GFW High School
Winthrop, MN

Debra Beightol
Fremont High School
Fremont, NE

Patrick Callahan
Catasaugua High School
Catasaugua, PA

George Coker
Bowling Green High School
Bowling Green, KY

Janice Costabile
South Brunswick High School
Monmouth Junction, NJ

Stanley Crum
Homestead High School
Fort Wayne, IN

Russel Davison
Brandon High School
Brandon, FL

Christine K. Deyo
Rochester Adams High School
Rochester Hills, MI

Jim Doller
Fox Lane High School
Bedford, NY

Jessica Downing
Esparto High School
Esparto, CA

Douglas Fackelman
Brighton High School
Brighton, CO

Rick Forrest
Rochester High School
Rochester Hills, MI

Mark Freeman
Blacksburg High School
Blacksburg, VA

Jonathan Gillis
Enloe High School
Raleigh, NC

Karen Gruner
Holton Arms School
Bethesda, MD

Larry Harrison
DuPont Manual High School
Louisville, KY

Alan Haught
Weaver High School
Hartford, CT

Steven Iona
Horizon High School
Thornton, CO

Phil Jowell
Oak Ridge High School
Conroe, TX

Deborah Knight
Windsor Forest High School
Savannah, GA

Thomas Kobilarcik
Marist High School
Chicago, IL

Sheila Kolb
Plano Senior High School
Plano, TX

Todd Lindsay
Park Hill High School
Kansas City, MO

Malinda Mann
South Putnam High School
Greencastle, IN

Steve Martin
Maricopa High School
Maricopa, AZ

Nancy McGrory
North Quincy High School
N. Quincy, MA

David Morton
Mountain Valley High School
Rumford, ME

Charles Muller
Highland Park High School
Highland Park, NJ

Fred Muller
Mercy High School
Burlingame, CA

Vivian O'Brien
Plymouth Regional High School
Plymouth, NH

Robin Parkinson
Northridge High School
Layton, UT

Donald Perry
Newport High School
Bellevue, WA

Francis Poodry
Lincoln High School
Philadelphia, PA

John Potts
Custer County District High School
Miles City, MT

Doug Rich
Fox Lane High School
Bedford, NY

John Roeder
The Calhoun School
New York, NY

Consuelo Rogers
Maryknoll Schools
Honolulu, HI

Lee Rossmaessler, Ph.D
Mott Middle College High School
Flint, MI

John Rowe
Hughes Alternative Center
Cincinnati, OH

Rebecca Bonner Sanders
South Brunswick High School
Monmouth Junction, NJ

David Schlipp
Narbonne High School
Harbor City, CA

Eric Shackelford
Notre Dame High School
Sherman Oaks, CA

Robert Sorensen
Springville-Griffith Institute and
Central School
Springville, NY

Teresa Stalions
Crittenden County High School
Marion, KY

Roberta Tanner
Loveland High School
Loveland, CO

Anthony Umelo
Anacostia Sr. High School
Washington, D.C.

Judy Vondruska
Mitchell High School
Mitchell, SD

Deborah Waldron
Yorktown High School
Arlington, VA

Ken Wester
The Mississippi School for
Mathematics and Science
Columbus, MS

Susan Willis
Conroe High School
Conroe, TX

Meeting Active Physics for the First Time

Welcome! A Five-Minute Introduction

Active Physics is a different species of physics course. It has the mechanics, optics, and electricity you anticipate, but not where you expect to find them. In a traditional physics course, we teach forces in the fall, waves in the winter, and solenoids in the spring. In *Active Physics*, students are introduced to physics concepts on a need-to-know basis as they explore issues in Communication, Home, Medicine, Predictions, Sports, and Transportation.

Every chapter is independent of any other chapter. You can begin the year with any one of three chapters in any one of the six thematic units. As an example, let's start the year with Chapter 3 of the Sports unit.

On Day One, students are introduced to the chapter challenge. NASA, recognizing that residents of a future moon colony will need physical exercise, has commissioned our physics class to develop, adapt, or create a sport for the moon.

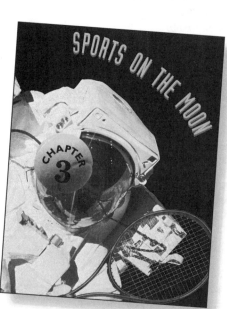

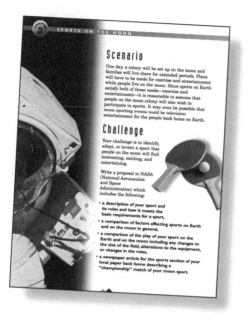

Our proposal to NASA will have to include the following:

a) a description of a sport and its rules;

b) a comparison of factors affecting sports on Earth and the moon in general;

c) a comparison of play of the sport on Earth and the moon including any changes to field, rules, or equipment;

d) a newspaper article for the people back 'home' describing a championship match of the moon sport.

How can students get started? How can students complete such a challenge without the requisite physics knowledge? Before the chapter activities begin, a discussion takes place about the criteria for success. The class discusses what is expected in an excellent proposal. How will this proposal be graded? For instance, the rubric for grading will describe whether "a comparison of factors affecting sports on Earth

and the moon" implies two factors or four factors and whether an equation or graph or description should be a part of the comparison. Similarly, do the factors and newspaper article carry equal weight, or does one have a greater impact on the final grade? Students will have a sense of what is required for an excellent proposal before they begin. This will be revisited before work on the project begins.

Day Two begins with the first of nine activities. Each successive day begins with another activity. *Active Physics* is an activity-based curriculum. Let's look at Activity Seven: **Friction on the Moon**.

The activity begins by mentioning, "The Lunar Rover proved that there is enough frictional force on the moon to operate a passenger-carrying wheeled vehicle." The students are then asked, "How do frictional forces on Earth and the moon compare?"

This **What Do You Think?** question is intended to find out what students know about friction—to get into the 'friction part' of their brains. Formally, we say that this question is to elicit the student's prior understanding and is part of the constructivist approach. Students write a response for one minute and discuss for another two minutes. But we don't reach closure. The question opens the conversation.

Students then begin the **For You To Do** activity.

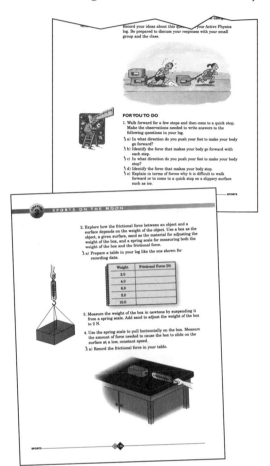

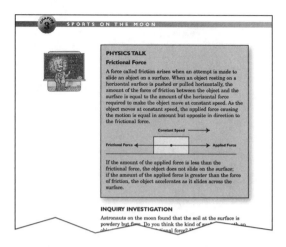

An **Inquiry Investigation** is presented for specific students or classes who wish to go further independently. In this case, they can investigate the effect of friction of different surfaces.

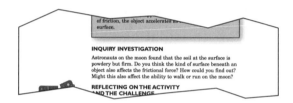

A **Reflecting On The Activity And The Challenge** relates the activity to the larger challenge of developing the moon sport.

In this activity, students weigh a box with a spring scale and measure the force required to pull it across a table at constant speed. By adding sand to the box, they take repeated measurements of weight and frictional force. A graph then shows them that the frictional force is directly proportional to the weight—more weight, more friction. An earlier activity convinced students that all objects weigh less on the moon. And so they can now conclude that friction must be less on the moon.

A **Physics Talk** summarizes the physics principle and includes equations where appropriate.

"Friction is involved somehow in most if not all sports. . . One thing is certain, your proposal to NASA won't 'slide through' if you don't demonstrate that you understand frictional forces on the moon." Students have been given another piece of the jigsaw puzzle. How is the sport that they are developing going to be modified because of the decreased friction on the moon?

The activity concludes with a **Physics To Go** homework assignment.

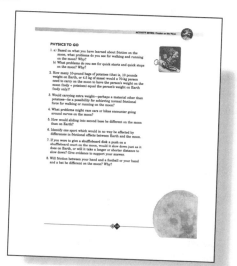

Here students are asked about the specifics of the activity and required to explain how sliding into second base would be different on the moon; how shuffleboard play would be different on the moon; and whether the friction between your hand and a football would be different on the moon.

The chapter also has activities which help students discover that projectiles travel differently on the moon, how mass and weight relationships change sports, how running and jumping are different and how collisions could be changed to limit the range of a golf ball.

The chapter concludes with a **Physics At Work** profile where students are introduced to someone whose job is related to the chapter challenge. In this chapter, astronaut Linda Godwin describes adapting to zero gravity during flight and space walks.

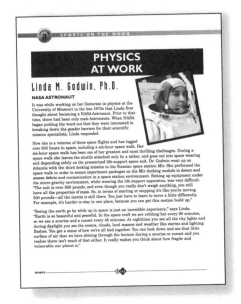

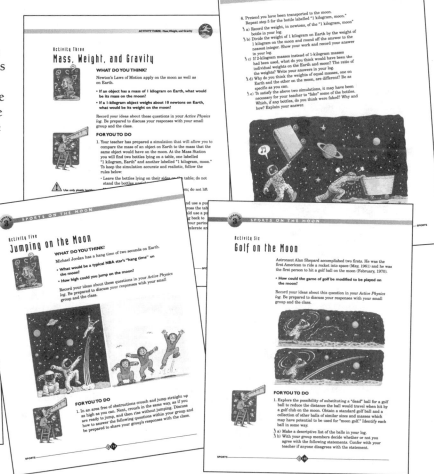

With the results of all of the activities before them, student teams now complete the challenge. They put the jigsaw pieces of friction, trajectories, collisions, running and jumping all together to construct their sport. Each team creates their own sport, reflecting the interests and creativity of the team members. The teams share their work with the rest of the class and the Sports chapter concludes.

CHAPTER 1 · HEARING

Scenario

A 23-year-old rock musician has difficulty understanding speech. The musician goes to an audiologist, a medical person who helps people with hearing loss. A hearing test shows that the musician has a loss of hearing at high frequencies. The audiologist says that loud noise can cause a hearing loss, but the loss may be only temporary. Hoping the problem will go away, the musician stops playing music for a month. Unfortunately, the hearing loss remains. The audiologist suggests a hearing aid. Another musician, who plays in a symphony orchestra, has no hearing loss. What is the difference between the two situations?

Challenge

Your committee has been put in charge of the school dance. You enjoyed the local band that played last year, but the principal of your school objects to inviting them back. He explains that, after leaving the last dance, his ears were "ringing" for the rest of the evening. You try to calm him down by explaining that this is normal at rock concerts and the famous bands are actually much louder than the local band. The principal decides that there will be no school dances where hearing loss or damage could occur.

M 2

One strength of *Active Physics* is the independence of the chapters. After finishing Sports, we begin anew. Let's choose Chapter 1 of Medicine as the next adventure. In this chapter, students are challenged to write a position paper to the school principal convincing him that a school dance can be held and guaranteeing that nobody's hearing will be damaged. It is for this purpose that students will learn about sound travel, decibels and frequency response, or human hearing. Or perhaps Chapter 2 of Transportation should be initiated. In this chapter, students are required to design and build an improved safety device for cars or bicycles. And it is in this context that students will learn about impulse, momentum, forces, and acceleration.

The beginning of a new chapter has two distinct advantages. For the students who did not do well on the Sports unit, they have a fresh start. Maybe they didn't do well because Sports didn't interest them and rock concerts or car collisions will. Or maybe they didn't do well because they missed school due to illness or a suspension. It's time to start over. The horizon for success is only four weeks. *Active Physics* does not ask students to worry about a final exam that will be given eight months from now, but rather to focus on one challenge that will be completed within a month.

A second advantage is apparent when one considers the transient nature of our school populations. In most courses, when that new student arrives in November, we do our best as teachers to greet the student and help them make the transition to the class. But we are also keenly aware of how much the student has missed and how difficult the learning situation really is. In an *Active Physics* course, that new student in November is asked to hang in for a week, get used to the class, work with the group over there and is reassured that we will soon be beginning a brand new chapter where they will be full participants irrespective of their late arrival. This removes one of the large hurdles which some students must face as they transfer programs, schools or communities.

Active Physics offers 18 chapters: six units with three chapters each. In a one-year physics or physical science course, students can be expected to complete 12 of the chapters at the most. This provides the teacher and students with a wide selection of content that meets local interests and course objectives.

Students in *Active Physics* never ask, "Why am I learning this?" Teachers of *Active Physics* never have to respond, "Because one day it will be useful to you." *Active Physics* is relevant physics. Students know that they have a challenge and they know that the activities will help them to be successful.

Please take a more careful, leisurely look at *Active Physics*. It's probably just what you and your students have been looking for.

CHAPTER 2 SAFETY

Scenario

Probably the most dangerous thing you will do today is travel to your destination. Transportation is necessary, but the need to get there in a hurry, and the large number of people and vehicles, have made transportation very risky. There is a greater chance of being killed or injured traveling than in any other common activity. Realizing this, people and governments have begun to take action to alter the statistics. New safety systems have been designed and put into use in automobiles and airplanes. New laws and a new awareness are working together with these systems to reduce the danger in traveling.

What are these new safety systems? You are probably familiar with many of them. In this chapter, you will become more familiar with most of these designs. Could you design or even build a better safety device for a car or a plane? Many students around the country have been doing just that, and with great success!

Challenge

Your design team will develop a safety system protecting automobile, airplane, bicycle, motorcycle, or train passengers. As you study existing safety systems, you and your design team should develop ideas for improving an existing system or building a new system for preventing accidents. You might consider a system that will minimize the injuries caused by accidents.

Your final product will be a working model or prototype of a safety system. On the day that you bring the final product to class, the teams will display them around the room while class members informally view them and discuss them with members of the design team. During this time, class members will ask questions about each others products. The questions will be placed in envelopes provided to each team by the teacher. The teacher will use some of these questions during the oral presentations on the next day.

The product will be judged according to the following three parts:

1. The quality of your safety feature enhancement and the working model or prototype.

2. The quality of a 5-minute oral report that should include:

   - the need for the system;
   - the method used to develop the working model;
   - the demonstration of the working model;
   - the discussion of the physics concepts involved;
   - the description of the next-generation model as envisioned by the team.

3. The quality of a written and/or multimedia report including:

   - the information from the oral report;
   - the documentation of the sources of expert information;
   - the discussion of consumer acceptance and market potential;
   - the discussion of the physics concepts applied in the design of the safety system.

Criteria

You and your classmates will work with your teacher to define the criteria for determining grades. You will also be asked to evaluate your own work. Discuss as a class the performance task and the points that should be allocated for each part. A starting point for your discussions may be:

- Part 1 = 40 points
- Part 2 = 30 points
- Part 3 = 30 points

Since group work is made up of individual work, your teacher will assign some points to each individual's contribution to the project. If individual points total 30 points, then parts 1, 2 and 3 must be changed so that the total remains at 100.

MEDICINE

As a member of the committee in charge of the school dance, you must take a stand on whether or not the dance should be held. You then have the opportunity to write a position paper to the principal explaining your position and giving the reasons for it. To make your paper as effective as possible, include as much science in it as you can.

Your position paper should have at least four paragraphs. The first paragraph should state your position—should the school hold the dance or not? The remaining paragraphs should present the arguments for your position.

The paper should demonstrate an understanding of hearing and hearing loss. Here are important topics to include:

- measurement of sound levels
- how loud sounds can contribute to hearing loss
- the role of frequency and overtones in hearing and hearing loss
- how the ear works
- how the ear is such a remarkable organ

To receive full credit, you should support your position with your experimental results and any research or information presented in this chapter. Also, you should demonstrate an understanding of the science concepts involved in the hearing process.

Criteria

How will I be graded?

What quality standards must be met to successfully complete the above challenge?

You and your classmates will work with your teacher to define the criteria for determining grades. You will also be asked to evaluate the quality of your own work—both by how much effort you put in and by how well you met the standards set by your class.

Features of Active Physics

1. Scenario

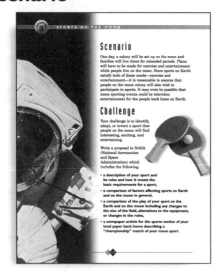

Each *Active Physics* chapter opens with an engaging scenario. Students from diverse backgrounds and localities have been interviewed in order to find situations which are not only realistic but meaningful to the high school population. The scenarios (only a paragraph or two in length) set the stage for the chapter challenge which immediately follows. Many teachers choose to read the scenario aloud to the class as a way of introducing the new chapter.

2. Challenge

The chapter challenge is the heart and soul of *Active Physics*. It provides a purpose for all of the work that will follow. The challenges provide the rationale for learning. One of the common complaints teachers hear from students is, "Why am I learning this?" In *Active*

Physics, no students raise this criticism. Similarly, no teacher has to answer, "Because one day it will be useful to you." The complaint is avoided because on Day One of the chapter students are presented with a challenge that, in essence, becomes their job for the next few weeks.

In Medicine, Chapter 1, students are challenged with a situation where the school principal is not going to permit a school dance because his ears were ringing after the last band performed. Students must write a position paper either agreeing with the principal or convincing him that nobody's hearing will be damaged if another dance is held. This is why the students then learn about decibels, frequencies and human response to sound.

In Transportation, Chapter 2, students are challenged to design and build an improved safety device for an automobile. The study of momentum, forces and Newton's Laws will be integral to their understanding of the required features in a safety device.

In Home, Chapter 2, students must create an appliance package that can be used in developing nations. The appliance package is limited by the wind generator available to the households. Students must also supply a rationale for how each suggested appliance will enhance the well-being of the family using it. This requires students to be able to differentiate between power and energy. It also provides a basis for students to reflect on quality of life issues in parts of the globe that they learn about in their social studies classes.

The beauty of the challenges lies in the variety of tasks and opportunities for students of different talents and skills to excel. Students who express themselves artistically will have an opportunity to shine in some challenges, while the student who can design and build may be the group leader in another challenge. Some challenges have a major component devoted to writing while others require oral or visual presentations. All challenges require the demonstration of solid physics understanding.

The challenges are not contrived situations for high school students. Professional engineers also design and build improved safety devices. Medical writers and illustrators design posters and pamphlets. The challenge in Chapter 3 of Sports requires students to create, invent, or adapt a sport that can be played on the moon. This challenge has been successfully completed by 9th grade high school students, 12th grade *Active Physics* students, and by NASA engineers. The expectation may be different for each of these audiences, but the challenge is consistent.

3. Criteria

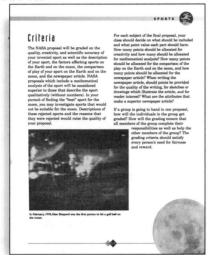

4. What Do You Think?

In creating *Active Physics*, we had thought that the generation of the challenge was good enough. Upon reflection, we soon realized that criteria for success must also be included. When students agree to the matrix by which they will be measured, the research has shown that the students will perform better and achieve more. It makes sense. In the simplest situation of cleaning a lab room, the teacher may simply state, "Please clean up the lab." The results are often a minimal cleanup. If the teacher begins by asking, "What does a clean lab room look like?" and students and teacher jointly list the attributes of a clean lab room (i.e., no paper on the floor, all beakers put away, all materials on the back of the lab tables, all power supplies unplugged and all water removed), the students respond differently and the cleanup is better. When students are asked to include physics principles in an explanation, the students should know whether the expectation is for three physics principles or five.

The discussion of grading criteria and the creation of a grading rubric is a crucial ingredient for student success. *Active Physics* requires a class discussion, after the introduction of the challenge, about the grading criteria. How much is required? What does an "A" presentation look like? Should creativity be weighed more than delivery? The criteria can be visited again at the end of the chapter, but at this point it provides a clarity to the challenge and the expectation level that the students should set for themselves.

During the past few years much has been written about a constructivist approach to learning. Videos of Harvard graduates, in caps and gowns, show that the students are not able to explain correctly why it is colder in the winter than it is in the summer. These students have previously answered these questions correctly in 4th grade, in Middle School, and then again in High School. How else would they have gotten into Harvard? We believe that they never internalized the logic and understanding of the seasons. One reason for this problem is that they were never confronted by what they did believe, and were never adequately shown why they should give up that belief system. Certainly, it is worth writing down a "book's perfect answer" on a test to secure a good grade, but to actually believe requires a more thorough examination of competing explanations.

The best way to ascertain a student's prior understanding is through extensive interviewing. Much of the research literature in this area includes the results of these interviews. In a classroom, this one-on-one dialogue is rarely possible. The **What Do You Think?** question introduces each activity in a way in which to elicit prior understandings. It gives students an opportunity to verbalize what they think about friction, or energy, or light, before they embark on an activity. The brief discussion of the range of answers brings the student a little closer in touch with that part of his/her brain which understands friction, energy, or light. The **What Do You Think?** question is not intended to produce a correct answer or a discussion of the features of the questions. It is not intended to bring closure. The activity which follows will provide that discussion as experimental results are analyzed. The **What Do You Think?** question should take no more than a few minutes of class time. It is the lead into the physics investigation. Students should be strongly encouraged to write their responses to the questions in their logs, to ensure that they have in fact addressed their prior conceptions. After students have discussed their responses in their small groups, activate a class discussion. Ask students to volunteer other students' answers which they found interesting. This may encourage students to exchange ideas without the fear of personally giving a "wrong" answer.

5. For You To Do

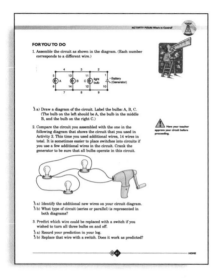

Active Physics is a hands-on, minds-on curriculum. Students *do* physics; they do not *read* about doing physics. Each activity has instructions for each part of the investigation. The pencil icons are provided to remind students that data, hypotheses, or conclusions should be recorded in their log or laboratory manual.

Activities are the opportunity for students to garner the knowledge that they will need to complete the chapter challenge. Students will understand the physics principle involved because they have investigated it. In *Active Physics*, if a student is asked, "How do you know?" the response is, "Because I did an experiment!"

Recognizing that many students know how to read, but do not like reading, background information is provided within the context of the activity. Students have demonstrated that they will read when the information is required for them to continue with their exploration.

Occasionally, the activity will require the entire class to participate in a large, single demonstration simultaneously. The teacher, on other occasions, may decide that a specific activity is best done as a demonstration. This would be appropriate if there is limited equipment for that one activity, or the facilities are not available.

Viewing demonstrations on an ongoing basis, though, is not what *Active Physics* is about.

There are specific **For You To Do** activities where computer spreadsheets, force transducers, or specific electronic equipment is required. Most of these activities have 'low-tech' alternatives provided in the Teacher's Edition. In the initial teaching of *Active Physics*, the low-tech alternative may be the only reasonable approach. As the course becomes a staple of the school offerings, it is hoped that funds can be set aside to improve the students' access to equipment.

Most of the **For You To Do** activities require between one and two class periods. With the present trend toward block scheduling, there are so many time structures that it is difficult to predict how *Active Physics* will best fit with your schedule. The other impact on time is the achievement and preparation level of the students. In a given activity, students may be required to complete a graph of their data. This is considered one small part of the activity. If the students have never been exposed to graphing, this could require a two-period lesson to teach the rudiments of graphing with suitable practice in interpretation. *Active Physics* is accessible to all students. The teacher is in the best position to make accommodations in time reflecting the needs of the students.

6. Physics Talk

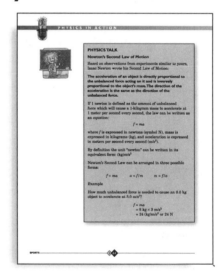

Equations are often the simplest, most straightforward, most concise, and clearest way of expressing physics principles. *Active Physics* limits the mathematics to the ninth grade curriculum. Students who have shied away from studying physics because of the mathematics prerequisites find that they are welcomed into *Active Physics*. **Physics Talk** is a means by which specific attention can be given to the mathematical equations. It also provides an opportunity to illustrate a problem solution or to derive a complex equation. For some students, there is a need to guide them through the algebraic manipulation which shows the equivalence of F=ma and a = F/m. Where appropriate, this manipulation is explicitly shown. Finally, sample problems required for the chapter challenge will also be in **Physics Talk**.

7. For You To Read

The **For You To Read** inserts provide students with some reading at the ninth grade level. This section may be used to tie together concepts from the present activity or a set of activities. It may also be used to provide a glimpse into the history of the physics principle being investigated. Finally, **For You To Read** may provide background information which will help clarify the meaning of the physics principle investigated in **For You To Do**.

8. Reflecting On The Activity And The Challenge

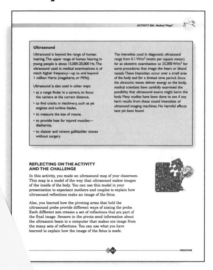

each activity, the student is often so the completion of the single experiment ntext of the investigation is lost. he **Activity And The Challenge** is students to place the new insights the context of the chapter and allenge. If the chapter challenge is ed a completed picture, each activity is a jigsaw piece. By completing enough of the **For You To Do** activities, the students will be able to fit the jigsaw pieces together and complete the challenge. This summary section ensures that the students do not forget about the larger context and continue their personal momentum toward completion of the challenge.

9. Physics To Go

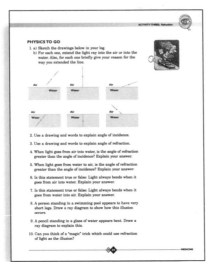

This section provides additional questions and problems that can be completed outside of class. Some of the problems are applications of the principles involved in the preceding activity. Others are replication of the work in the **For You To Do** activity. Still others provide an opportunity to transfer the results of the investigation to the context of the chapter challenge. **Physics To Go** provides a means by which students can be working on the larger chapter challenge in smaller chunks during the chapter.

10. Inquiry Investigation

The outcome of good science instruction should be the ability of students to conceive of an experiment, design that experiment, complete the data collection, interpret the data and draw suitable conclusions based on the experiment. The nature of the daily immersion in activities in *Active Physics* often, by necessity, provides for detailed instructions in how to proceed. Inquiry is an opportunity to provide the right stimulus for students to try their

hands at designing a specific experiment to answer a specific question. It affords students the chance to mirror the techniques and approaches that they have experienced in *Active Physics* and to expand the approach to secure new information. The Inquiry can be assigned as independent study or as a class extension to the lab.

11. Stretching Exercises

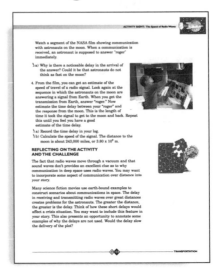

Some students express additional interest in a specific topic or an extension to a topic. The **Stretching Exercises** provide an avenue in which to pursue that interest. **Stretching Exercises** often require additional readings or interviews. They may be given for extra credit to students who wish to attempt a more in-depth problem or a tougher exercise.

12. Chapter Assessment

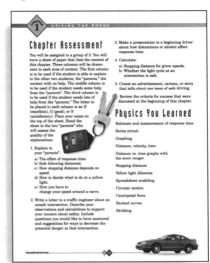

The **Chapter Assessment** is the return to the **Chapter Challenge and Criteria**. The students are ready to complete the challenge. They are able to view the challenge with a clarity that has emerged from the completion of the **For You To Do** activities of the chapter. Students are able to review the chapter as they discuss the synthesis of the information into the required context of the challenge. The students should have some class time to work together to complete the challenge and to present their project. In many physics courses, all students are expected to converge on the same solution. In *Active Physics*, each group is expected to have a unique solution. All solutions must have correct physics, but there is ample room for creativity on the students' part. This is one of the features that captures the imagination of students who have often previously chosen not to enroll in physics classes.

13. Physics You Learned

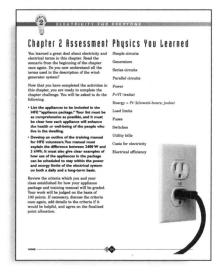

This small section at the end of the chapter provides a list of physics concepts and equations which were studied in the context of the **For You To Do** activities. It provides students with a sense of accomplishment and serves as a quick review of all that was learned during the preceding weeks.

14. Physics At Work

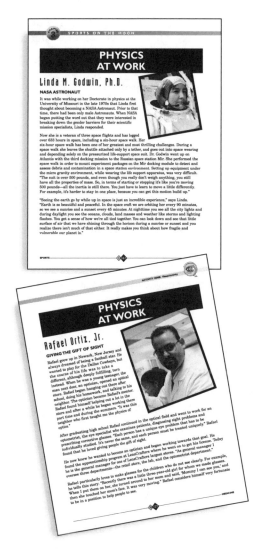

These sections highlight an individual whose work or hobby is illustrative of the **Chapter Challenge.** **Physics At Work** speaks to the authenticity of the **Chapter Challenges.** The profiles illustrate how knowledge of physics is important and valuable in different walks of life. The choice of profiles span the ethnic, racial, and gender diversity that we find in our nation.

Active Physics & National Science Education Standards

Active Physics was designed and developed to provide teachers with instructional strategies that model the following from *The Standards*:

Guide and Facilitate Learning

• Focus and support inquiries while interacting with students.

• Orchestrate discourse among students about scientific ideas.

• Challenge students to accept and share responsibility for their own learning.

• Recognize & respond to student diversity; encourage all to participate fully in science learning.

• Encourage and model the skills of scientific inquiry as well as the curiosity openness to new ideas and data and skepticism that characterize science.

Engage in ongoing assessment of their teaching and student learning

• Use multiple methods & systematically gather data about student understanding & ability.

• Analyze assessment data to guide teaching.

• Guide students in self-assessment.

Design and manage learning environments that provide students with time, space and resources needed for learning science

• Structure the time available so students are able to engage in extended investigations.

• Create a setting for student work that is flexible and supportive of science inquiry.

• Make available tools, materials, media, & technological resources accessible to students.

• Identify and use resources outside of school.

Develop communities of science learners that reflect the intellectual rigor of scientific attitudes and social values conducive to science learning

• Display and demand respect for diverse ideas, skills, & experiences of students.

• Enable students to have significant voice in decisions about content & context of work & require students to take responsibility for the learning of all members of the community.

• Nurture collaboration among students.

• Structure and facilitate ongoing formal and informal discussion based on shared understanding of rules.

• Model and emphasize the skills, attitudes and values of scientific inquiry.

Assessment Standards

• Features claimed to be measured are actually measured.

• Students have adequate opportunity to demonstrate their achievement and understanding.

• Assessment tasks are authentic and developmentally appropriate, set in familiar context, and engaging to students with different interests and experiences.

• Assesses student understanding as well as knowledge.

• Improve classroom practice and plan curricula.

• Develop self-directed learners.

Active Physics Addresses Key NSES Recommendations

Active Physics addresses the following science curriculum recommendations:

Scenario-Driven

In each thematic unit there are three chapters, each requiring approximately two to three weeks of class time. Each chapter begins with an engaging scenario or project assignment that challenges the students and sets the stage for the learning activities and chapter assessments to follow. Chapter contents and activities are selectively aimed at providing the students with the knowledge and skills needed to address the introductory challenge, thus providing a natural content filter in the "less is more" curriculum.

Flexibly Formatted

Units are designed to stand alone, so teachers have the flexibility of changing the sequence of presentation of the units, omitting the entire unit, or not finishing all of the chapters within a unit. Although intended to serve as a full-year physics course, the units of *Active Physics* could be adapted to spread across a four-year period in an integrated high school curriculum.

Multiple Exposure Curriculum

The thematic nature of the course requires students to continually revisit fundamental physics principles throughout the year, extending and deepening their understanding of these principles as they apply them in new contexts. This repeated exposure fosters the retention and transferability of learning, and promotes the development of critical thinking skills.

Constructivist Approach

Students are continually asked to explore how they think about certain situations. As they investigate new situations, they are challenged to either explain observed phenomena using an existing paradigm or to develop a more consistent one. This approach can be helpful in including situations to abandon previously held notions in favor of the more powerful ideas and explanations offered by scientists.
.

Authentic Assessment

For the culmination of each chapter, students are required to demonstrate the usefulness of their newly acquired knowledge by adequately meeting the challenge posed in the chapter introduction. Students are then evaluated on the degree to which they accomplish this performance task. The curriculum also includes other methods and instruments for authentic assessments as well as non-traditional procedures for evaluating and rewarding desirable behaviors and skills.

Cooperative Grouping Strategies

Use of cooperative groups is integral to the course as students work together in small groups to acquire the knowledge and information needed to address the series of challenges presented through the chapter scenarios. Ample teacher guidance is provided to assure that effective strategies are used in group formation, function, and evaluation.

Math Skills Development/Graphing Calculators and Computer Spreadsheets

The presentation and use of math in *Active Physics* varies substantially from traditional high school physics courses. Math, primarily algebraic expressions, equations, and graphs is approached as a way of representing ideas symbolically. Students begin to recognize the usefulness of math as an aid in exploring and understanding the world about them. Finally, since many of the students in the target audience are insecure about their math backgrounds, the course engages and provides instruction for the use of graphing calculators and computer spreadsheets to provide math assistance.

Minimal Reading Required

Because it is assumed that the target audience reads only what is absolutely necessary, the entire course is activity-driven. Reading passages are presented mainly within the context of the activities, and are written at the ninth grade level.

Use of Educational Technologies

Videos which capture students' attention explore a variety of the *Active Physics* topics. Opportunities are also provided for students to produce their own videos in order to record and analyze events. Computer software programs make use of various interfacing devices.

Problem Solving

For the curriculum to be both meaningful and relevant to the target population, problem solving related to technological applications and related issues is an essential component of the course. Problem solving ranges from simple numerical solutions where one result is expected, to more involved decision-making situations where multiple alternatives must be compared.

Challenging Learning Extensions

Throughout the text, a variety of **Stretching Exercises** are provided for more motivated students. These extensions range from more challenging design tasks, to enrichment readings, to intriguing and unusual problems. Many of the extensions take advantage of the frequent opportunities the curriculum provides for oral and written expression of student ideas.

Cooperative Learning
Benefits of Cooperative Learning

Cooperative learning requires you to organize and structure a lesson so that students work with other students to jointly accomplish a task. Group learning is an essential part of balanced methodology. It should be blended with whole-class instruction and individual study to meet a variety of learning styles and expectations as well as maintain a high level of student involvement.

Cooperative learning has been thoroughly researched and agreement has been reached on a number of results. Cooperative learning:

- promotes trust and risk-taking

- elevates self-esteem

- encourages acceptance of individual differences

- develops social skills

- permits a combination of a wide range of backgrounds and abilities

- provides an inviting atmosphere

- promotes a sense of community

- develops group and individual responsibility

- reduces the time on a task

- results in better attendance

- produces a positive effect on student achievement

- develops key employability skills

As with any learning approach, some students will benefit more than others from cooperative learning. Therefore, you may question as to what extent you should use cooperative learning strategies. It is important to involve the student in helping decide which type of learning approaches they prefer, and to what extent each is used in the classroom. When students have a say in their learning, they will accept to a greater extent any method which you choose to use.

Phases of Cooperative Learning Lessons

Organizational Pre-lesson Decisions

What academic and social objectives will be emphasized? In other words, what content and skills are to be learned and what interaction skills are to be emphasized or practiced?

What will be the group size? Or, what is the most appropriate group size to facilitate the achievement of the academic and social objectives? This will depend on the amount of individual involvement expected (small groups promote more individual involvement), the task (diverse thinking is promoted by larger groups), nature of the task or materials available and the time available (shorter time demands smaller groupings to promote involvement).

Who will make up the different groups? Teacher-selected groups usually have the best mix, but this can only happen after the teacher gets to know his/her students well enough to know who works well together. Heterogeneous groupings are most successful in that all can learn through active participation. The duration of the groups' existence may have some bearing on deciding the membership of groups.

How should the room be arranged? Practicing routines where students move into their groups quickly and quietly is an important aspect. Having students face-to-face is important. The teacher should still be able to move freely among the groups.

What Materials and/or Rewards Might be Prepared in Advance?

Setting the Lesson

Structure for Positive Interdependence: When students feel they need one another, they are more likely to work together--goal interdependence becomes important. Class interdependence can be promoted by setting class goals which all teams must achieve in order for class success.

Explanation of the Academic Task: Clear explanations and sometimes the use of models can help the students. An explanation of the relevance of the activity is importance. Checks for clear understanding can be done either before the groups form or after, but they are necessary for delimiting frustrations.

Explanation of Criteria for Success: Groups should know how their level of success will be determined.

Structure for Individual Accountability: The use of individual follow-up activities for tasks or social skills will provide for individual accountability.

Specification of Desired Social Behaviors: Definition and explanations of the importance of values of social skills will promote student practice and achievement of the different skills.

Monitoring/Intervening During Group Work

Through monitoring students' behaviors, intervention can be used more appropriately. Students can be involved in the monitoring by being "a team observer," but only when the students have a very clear understanding of the behavior being monitored.

Interventions to increase chances for success in completing the task or activity and for the teaching of collaborative skills should be used as necessary-- they should not be interruptions. This means that the facilitating teacher should be moving among the groups as much as possible. During interventions, the problem should be turned back to the students as often as possible, taking care not to frustrate them.

Evaluating the Content and Process of Cooperative Group Work

Assessment of the achievement of content objectives should be completed by both the teacher and the students. Students can go back to their groups after an assignment to review the aspects in which they experienced difficulties.

When assessing the accomplishment of social objectives, two aspects are important: how well things proceeded and where/how improvements might be attempted. Student involvement in this evaluation is a very basic aspect of successful cooperative learning programs.

Organizing and Monitoring Groups

An optimum size of group for most activities appears to be four; however, for some tasks, two may be more efficient. Heterogeneous groups organized by the teacher are usually the most sucessful. The teacher will need to decide what factors should be considered in forming the heterogeneous groups. Factors which can be considered are: academic achievement, cultural background, language proficiency, sex, age, learning style, and even personality type.

Level of academic achievement is probably the simplest and initially the best way to form groups. Sort the students on the basis of marks on a particular task or on previous year's achievement. Then choose a student from each quartile to form a group. Once formed, groups should be flexible. Continually monitor groups for compatability and make adjustments as required.

Students should develop an appreciation that it is a privilege to belong to a group. Remove from group work any student who is a poor participant or one who is repeatedly absent. These individuals can then be assigned the same tasks to be completed in the same time line as a group. You may also wish to place a ten percent reduction on all group work that is completed individually.

The chart on the next page presents some possible group structures and their functions.

What Does Cooperative Learning Look Like?

During a cooperative learning situation, students should be assigned a variety of roles related to the particular task at hand. Following is a list of possible roles that students may be given. It is important that students are given the opportunity of assuming a number of different roles over the course of a semester.

Leader:

Assigns roles for the group. Gets the group started and keeps the group on task.

Organizer:

Helps focus discussion and ensures that all members of the group contribute to the discussion. The organizer ensures that all of the equipment has been gathered and that the group completes all parts of the activity.

Recorder:

Provides written procedures when required, diagrams where appropriate and records data. The recorder must work closely with the organizer to ensure that all group members contribute.

Researcher:

Seeks written and electronic information to support the findings of the group. In addition, where appropriate, the researcher will develop and test prototypes. The researcher will also exchange information gathered among different groups.

Encourager:

Encourages all group members to participate. Values contributions and supports involvement.

Checker:

Checks that the group has answered all the questions and the group members agree upon and understand the answers.

Diverger:

Seeks alternative explanations and approaches. The task of the diverger is to keep the discussion open. "Are other explanations possible?"

Some Possible Group Structures and Their Functions*

	Structure	Brief Description	Academic and Social Functions
Team Building	Round-robin	Each student in turn shares something with his/her teammates.	Expressing ideas and opinions, creating stories. Equal participation, getting acquainted with each other.
Class Building	Corners	Each student moves to a group in a corner or location as determined by the teacher through specified alternatives. Students discuss within groups, then listen to and paraphrase ideas from other groups.	Seeing alternative hypotheses, values, and problem solving approaches. Knowing and respecting differing points of view.
Mastery	Numbered heads together	The teacher asks a question, students consult within their groups to make sure that each member knows the answer. Then one student answers for the group in response to the number called out by the teacher.	Review, checking for knowledge comprehension, analysis, and divergent thinking. Tutoring.
	Color coded co-op cards	Students memorize facts using a flash card game or an adaption. The game is structured so that there is a maximum probability for success at each step, moving from short to long-term memory. Scoring is based on improvement.	Memorizing facts. Helping, praising.
	Pairs check	Students work in pairs within groups of four. Within pairs students alternate-one solves a problem while the other coaches. After every problem or so, the pair checks to see if they have the same answer as the other pair.	Practicing skills. Helping, praising.
Concept Development	Three-step interview	Students interview each other in pairs, first one way, then the other. Each student shares information learned during interviews with the group.	Sharing personal information such as hypotheses, views on an issue, or conclusions from a unit. Participation, involvement.
	Think-pair-share	Students think to themselves on a topic provided by the teacher; they pair up with another student to discuss it; and then share their thoughts with the class.	Generating and revising hypotheses, inductive and deductive reasoning, and application. Participation and involvement.
	Team word-webbing	Students write simultaneously on a piece of paper, drawing main concepts, supporting elements, and bridges representing the relation of concepts/ideas.	Analysis of concept into components, understanding multiple relations among ideas, and differentiating concepts. Role-taking.
Multifunctional	Roundtable	Each student in turn writes one answer as a paper and a pencil are passed around the group. With simultaneous roundtable, more than one pencil and paper are used.	Assessing print knowledge, practicing skills, recalling information, and creating designs. Team building, participation of all.
	Partners	Students work in pairs to create or master content. They consult with partners from other teams. Then they share their products or understandings with the other partner pair in their team.	Mastery and presentation of new material, concept development. Presentation and communication skills.
	Jigsaw	Each student from each team becomes an "expert" on one topic area by working with members from other teams assigned to the same topic area. On returning to their own teams, each one teaches the other members of the group and students are assessed on all aspects of the topic.	Acquisition and presentation of new material review and informed debate. Independence, status equalization.

* Adapted from Spencer Kagan (1990), "*The Structural Approach to Cooperative Learning*," Educational Leadership, December 1989/January 1990.

Active Listener:

Repeats or paraphrases what has been said by the different members of the group.

Idea Giver:

Contributes ideas, information, and opinions.

Materials Manager:

Collects and distributes all necessary material for the group.

Observer:

Completes checklists for the group.

Questioner:

Seeks information, opinions, explanations, and justifications from other members of the group.

Reader:

Reads any textual material to the group.

Reporter:

Prepares and/or makes a report on behalf of the group.

Summarizer:

Summarizes the work, conclusions, or results of the group so that they can be presented coherently.

Timekeeper:

Keeps the group members focused on the task and keeps time.

Safety Manager:

Reponsible for ensuring that safety measures are being followed, and the equipment is clean prior to and at the end of the activity.

Group Assessment

Assessment should not end with a group mark. Students and their parents have a right to expect marks to reflect the students' individual contributions to the task. It is impossible for you as the instructor to continuously monitor and record the contribution of each individual student. Therefore, you will need to rely on the students in the group to assign individual marks as merited.

There are a number of ways that this can be accomplished. The group mark can be multiplied by the number of students in the group, and then the total mark can be divided among the students, as shown in the graphics that follow.

Activity:_____

Group Mark: 8/10

Number in Group: 4

Total Marks: 32/40

Distribution of Marks

Student's Name	Mark	Signature
Ahmed	8/10	_____
Jasmin	8/10	_____
Mike	7/10	_____
Tabitha	9/10	_____

Another way to share group marks is to assign a factor to each student. The mark factors must total the number of students in the group. The group mark is then multiplied by this factor to arrive at each student's individual mark which best represents their contribution to the task, as shown below.

Activity:_____

Group Mark: 8/10

Number in Group: 4

Mark Factors and Individual Marks

Student's Name	Mark Factor	Individual Mark	Signature
Ahmed	1.0	8/10	_____
Jasmin	1.0	8/10	_____
Mike	0.9	7.2/10	_____
Tabitha	1.1	8.8/10	_____
Total Mark Factor	4		

In any case, students must sign to show that they are in agreement with the way the individual marks were assigned.

You may also wish to provide students with an assessment rubric similar to the one shown which they can use to assess the manner in which the group worked together.

Assessment Rubric for Group Work:
Individual Assessment of the Group

Individual's name: _____

Names of group members: _____

Name of activity: _____

Circle the appropriate number: #1 is excellent, #2 is good, #3 is average, and #4 is poor.

1. The group worked cooperatively. Everyone assumed a role and carried it out.	1	2	3	4
2. Everyone contributed to the discussion. Everyone's opinion was valued.	1	2	3	4
3. Everyone assumed the roles assigned to them.	1	2	3	4
4. The group was organized. Materials were gathered, distributed, and collected.	1	2	3	4
5. Problems were addressed as a group.	1	2	3	4
6. All parts of the task were completed within the time assigned.	1	2	3	4

Comments:

If you were to repeat the activity, what things would you change?

CHAPTER
1

THE
TRACK
& FIELD
CHAMPIONSHIP

Sports Chapter 1- The Track & Field Championship
National Science Education Standards

Chapter Summary

The scenario and challenge for this chapter is for the physics class to write a physics manual about track and field training for your high school team to help improve its performance.

To gain knowledge and understanding of physics principles necessary to meet this challenge, students work collaboratively on activities in which they apply concepts of potential and kinetic energy as they collect and analyze data collected in investigations of speed, acceleration, velocity, projectile motion, and gravity. These experiences engage students in the following content identified in the *National Science Education Standards*.

Content Standards

Unifying Concepts

- Evidence, models and explanations
- Constancy, change and measurement

Science as Inquiry

- Identify questions and concepts that guide scientific investigations
- Use technology and mathematics to improve investigations
- Formulate and revise scientific explanations and models using logic and evidence

Physical Science

- Motions and Forces: Objects change their motion when a net force is applied
- Laws of Motion are used to calculate the effects of forces in motion

Key Physics Concepts and Skills

Activity Summaries	Physics Principles

Activity One: Running the Race

Students time classmates to calculate the average speed as they run set distances. They then compare and analyze differences in speed of one runner at different distances and among different runners for the same distance.

- **Relationship of speed, distance and time**
- **Kinetic energy and motion**

Activity Two: Analysis of Trends

Students examine graphs representing results from track meets to explore trends in average speed. From this, they learn about extrapolation of data and the use of data in making predictions.

- **Average Speed**
- **Using data as basis for predictions**

Activity Three: Who Wins the Race?

Students use a ticker-tape timer to investigate speed of cars on sloped tracks. Analysis of the tapes introduces the concept that the average speed is not the same as actual speed at all points.

- **Acceleration**
- **Average Speed**
- **Friction**

Activity Four: Understanding the Sprint

Students create and analyze graphs of split time vs. distance as an introduction to instantaneous speed and how a runner changes speed during a race.

- **Acceleration**
- **Instantaneous Speed**
- **Average Speed**

Activity Five: Acceleration

Using a simple accelerometer to monitor changes in motion, students investigate acceleration and deceleration while walking, then with physics lab carts.

- **Acceleration**
- **Instantaneous Speed**
- **Average Speed**

Activity Six: Running a Smart Race

Focusing on split times and the changes in speed over the total distance, students explore how a runner can apply concepts of constant speed and acceleration to plan a strategy for winning a race.

- **Acceleration**
- **Velocity and Speed**

Activity Seven: Increasing Top Speed

Students measure their own stride length as an introduction to wavelengths. They then investigate the relationship of speed in a race to stride length and stride frequency of the runner.

- **Wavelength**
- **Velocity = Frequency x Wavelength**

Activity Eight: Projectile Motion

To develop understanding of the shot put, students explore the differences between the motion and landing position of objects dropped straight down to those with projected motion.

- **Projectile Motion**
- **Gravity**
- **Free Fall**

Activity Nine: The Shot Put

Students compare mathematical and physical models of projectile motion to that of a shot put. They apply this to describe the vertical and horizontal motion of the projected object, and predict its trajectory.

- **Projectile Motion**
- **Gravity**
- **Trajectories**

Activity Ten: Energy in the Pole Vault

Students use a penny launched from a ruler to model motion during the pole vault. They connect their observations to the concept of energy conservation.

- **Gravitational Potential Energy**
- **Transfer of Mechanical Energy**
- **Conservation of Energy**

Equipment List For Chapter One

QTY	TO SERVE	ACTIVITY	ITEM	COMMENT
1	Class	7	*Active Physics Sports* content video	Segment: Runner on football field.
1	Class	9	Apparatus for measuring "g"	See suggestions in Teacher's Edition for Activity Nine.
1	Group	5	Attachment device, accelerometer-to-cart	Kind depends on qualities of cart, accelerometer.
1	Class	8,9	Ball	Tennis ball (or similar) size.
1	Group	10	Ball	Marble or ball bearing which will deflect ruler.
1	Individual	All	Calculator, basic	One per student best; one per group minimum.
1	Group	5	Cart pulled by falling weights	See equipment arrangement on *Sports* text p.S23.
1	Class	8	Chair with wheels	Substitute: Wagon or cart capable of carrying person.
1	Class	1,7	Chalk stick or tape	For marking running surface.
1	Class	9	Chalk, bright	To mark projectile positions on chalkboard.
1	Class	9	Chalkboard	4-m length preferred.
1	Group	10	Clamp(s) for holding ruler	Kind depends on furniture
4	Class	3	Clamps for holding car tracks	Substitute: Stacks of books to elevate ends of tracks.
2	Group	8	Coins, matched	Metal washers, matched
1	Group	5	Cork accelerometer	Substitute: Liquid accelerometer
1	Group	10	Flexible plastic ruler	Substitute: Strip of any flexible solid material.
1	Individual	4,6	Graph paper, any grid size	May use page of students' logs, if grid ruled.
1	Individual	2	Graph, "Speed vs. Distance, Men and..."	Blackline Master provided in Background Material.
1	Individual	2	Graph, "Speed vs. Year, Men's Olympic..."	Blackline Master provided in Background Material.
1	Group	3	Hot Wheels™ car	Substitute: Equivalent toy car.
1	Individual	2	Local track records (running events)	Obtain from local track team.
1	Group	10	Marking pen or pencil	
1	Group	10	Marking surface, rigid	Cardboard or notebook cover.
1	Group	1,3,8,9,10	Meter stick	
1	Group	10	Metric ruler, mm markings	
1	Class	1,7	Metric tape measure	Substitute: Meter stick(s) placed end-to-end.
1	Group	10	Penny coin	Substitute: Small metal washer.
1	Class	9	Pins (9) along 4-meter row on wall	See Background Material for specifications.
1	Group	9	Post-It-Note	To label string and weight assembly.
1	Class	9	Protractor, large	Substitute: Cardboard templates for 30, 45, 60 degree angles.
1	Group	5	Pulley, clamped	To pull cart via falling weight.
1	Group	10	Rubber band	Holds pen on ruler; tape could be substituted.
1	Class	3	Set of 4 tracks	Arranged as shown on *Sports* text p.S15.
1	Group	10	Starting ramp, directional	Sloped ruler having groove on surface would serve.
1	Class	9	Stepladder or stool	Use when launching ball.
4	Class	1,7	Stopwatch	Substitute: Wristwatch having stopwatch function.
1	Class	9	String	Several meters needed.
1	Group	5	String, 1-meter length	Suspend over pulley to pull cart.
1	Group	3	Tabletop surface, 1-m minimum length	For horizontal run of toy car.
1	Class	9	Tacking strip on wall, 4-m long	Substitute: Paper strip.
1	Group	3	Tape	For attaching paper strip of timer to car.
1	Group	3	Ticker-tape timer & supplies	Calibrated (known frequency) version preferred.
9	Class	9	Time labels for pin positions	See Background Material for specifications.
1	Class	9	Trajectory model, 2.4-meter	See Background Material for specifications.
1	Class	6	Transparency, "Final Lap: Kicker, You..."	Blackline Master provided in Background Materials.
1	Class	7	VCR & TV monitor	
8	Class	9	Weight, lead fishing	Substitute: Metal washers.
2	Group	5	Weight, suspended to pull cart	To produce two reasonable, contrasting accelerations.

Organizer for Materials Available in Teacher's Edition

Activity in Student Text	Additional Material	Alternative / Optional Activities
ACTIVITY ONE: Running the Race p. S4	Sample Data of Records From High School Track Meets pgs. 22-23	
ACTIVITY TWO: Analysis of Trends p. S9	Speed vs. Year, Men's Olympic 400-m Dash p. 31 Speed vs. Distance, Men and Women, Penn Relays p. 32 Performance-Based Assessment Rubric: Reading and Interpreting Graphs p. 35	ACTIVITY TWO A: Spreadsheet Games–Analysis of Long-Range Trends pgs. 33-34
ACTIVITY THREE: Who Wins the Race? p. S13	One-Meter Sprints: Compling the Results and Making Predictions p.45	ACTIVITY THREE A: Who Wins the Race? (Using a Photogate) p. 44
ACTIVITY FOUR: Understanding the Sprint p. S18	Carl Lewis's World Record 100-m Dash Average Speed, 10-m Intervals p. 52	
ACTIVITY FIVE: Acceleration p. S22	Constructing a Cork Accelerometer p. 60	ACTIVITY FIVE A: Investigating Different Types of Accelerometers pgs. 61-62 ACTIVITY FIVE B: Constant Acceleration p. 63
ACTIVITY SIX: Running a Smart Race p. S26	Final Lap: Kicker, You and Rabbit. Distance vs. Time p. 69	
ACTIVITY SEVEN: Increasing Top Speed p. S29	Alternative Solutions to Physics To Go p. 77	
ACTIVITY EIGHT: Projectile Motion p. S34	Assessment Rubrics for Two Falling Coins and the Rolling Chair pgs. 84-85	
ACTIVITY NINE: The Shot Put p. S38	Measuring the Acceleration Due to Gravity p. 96 Four-meter Tack Strip for Trajectory Model p. 98 Stick and String Model of Projectile Tratectory p. 99	
ACTIVITY TEN: Energy in the Pole Vault p. S44	Olympic Pole Vault: Actual Heights and Heights Predicted from Average Speed in Men's 100-m Dash, 1896 to 1992 p. 110	

1

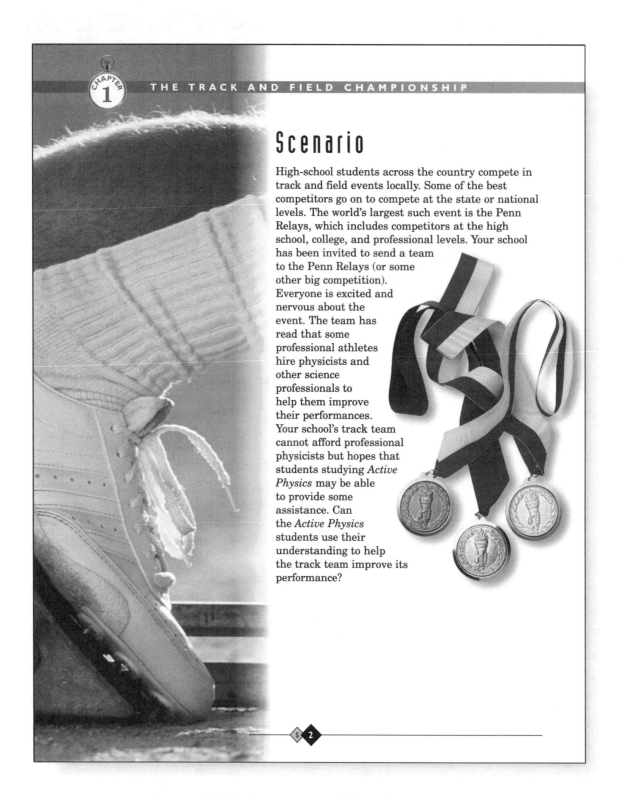

Scenario

High-school students across the country compete in track and field events locally. Some of the best competitors go on to compete at the state or national levels. The world's largest such event is the Penn Relays, which includes competitors at the high school, college, and professional levels. Your school has been invited to send a team to the Penn Relays (or some other big competition). Everyone is excited and nervous about the event. The team has read that some professional athletes hire physicists and other science professionals to help them improve their performances. Your school's track team cannot afford professional physicists but hopes that students studying *Active Physics* may be able to provide some assistance. Can the *Active Physics* students use their understanding to help the track team improve its performance?

Chapter and Challenge Overview

In this chapter students investigate simple kinematics using an invitation for the school's track team to participate at the Penn Relays as the driving scenario. We chose track and field, since almost every student can run, jump, or throw, while not all engage in team sports. We chose the Penn Relays, as opposed to the Olympics or a local school league championship, for two reasons. We have available the winning times and distances for male and female high school students at the Relays for many years. Second, it is unlikely that the students in your school can compete at an Olympic level, while the Relays are very possible.

If the Penn Relays are not well known, your own state or district track and field championships might be substituted, but then you will have to find some of the performance records and substitute them for the Relays' records.

It would be very useful to make contact with the school track coach before this chapter is started, possibly giving him/her a copy of the text before students start asking the coach questions about the material in it. The coach might even drop in on an early class and say a few words about the Penn Relays, the track team, and how he/she is looking forward to working with the class if they have any problems or ideas. You will need some data from the coach about your own school records and present team's records for certain events. For example, it would be useful to obtain data for the 100 m sprint, 1500 m and 3000 m run, 4 x 100 m relay, high jump, long jump, and shot put.

There are a number of places in this and subsequent chapters where it might be useful and interesting to have the entire class perform some athletic event and have it timed or measured and then have the results analyzed. We did not include this type of activity in the text since some students may feel embarrassed about their athletic ability. You are in the best position to judge if this type of activity is suitable for your students. There are usually a few students who are very willing to run, jump, and throw, and they can be your volunteers.

As you review the Challenge assignments, reassure the students that while they may feel incompetent now, by the end of the chapter they will have the necessary skills and vocabulary to respond adequately.

On the following pages of the Teacher's Edition there are suggestions on how to evaluate students on this material. It is very important at this time that the students be made aware of the method you are going to use and how you will evaluate their work. Have the students actively participate in deciding the criteria for evaluation.

The Physics To Go at the end of each section often contains more questions than should ever be assigned for homework. This section has been written in such a way as to give you a choice as to how much work, and the nature of the work the students will be expected to do each day out of class.

As you begin *Active Physics*, be aware that the same physics concepts appear repeatedly in different contexts. It is not necessary for the students to achieve total understanding the first time that they study speed, speed vs. time graphs, acceleration, trajectory motion, or potential and kinetic energy.

SPORTS

Challenge

Your challenge is to write a physics manual about track and field training for your high school team to help improve its performance. The manual should:

- **help students compare themselves with the competition;**
- **include a description of physics principles as they relate to track events;**
- **provide specific techniques to improve performance.**

Criteria

How will the manual be graded? What qualities should a good manual have? Discuss these issues in small groups and with your class. You may decide that some or all of the following qualities should be graded in your track and field manual:

- **physics principles**
- **inclusion of charts**
- **past records**
- **relevant equations**
- **definitions**
- **specific techniques**

Any advice you give should be understandable to athletes who have not studied physics. You can describe any activities you have done to explain how you know that the technique works, but you should not tell so much about each activity that the reader becomes bored.

Once you list the criteria for judging the track and field manuals, you should also decide how many points should be given for each criterion. If the group is going to hand in one manual, you must also agree on the way in which individuals will receive their grade. One method may be that the individual contributions toward the project receive 75% of the individual grade, and the group grade provides the remaining 25% of the individual grade. You should discuss different strategies and choose the one that is best suited to your school.

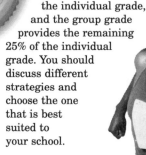

Assessment Rubric: Training Manual

Meets the standard of excellence. 5	• Significant information is presented in a consistently appropriate manner. • Details are specific and consistently effective. • Concepts within the chapter are integrated in appropriate places. • Additional research, beyond basic concepts presented in the chapter, is evident. • The writing holds the readers interest.
Approaches the standard of excellence. 4	• Significant information is most often presented in an appropriate manner. • Details are specific and effective. • Concepts within the chapter are integrated in appropriate places. • The writing holds the readers interest.
Meets an acceptable standard. 3	• Sufficient information is presented in an appropriate manner. • Details are general but effective. • Some of the concepts, presented within the chapter, are integrated in appropriate places. • The writing generally holds the readers interest.
Below acceptable standard and requires remedial help. 2	• Limited information presented in an inappropriate manner. • Details are limited and do not always pertain to the topic. • Few of the concepts, presented within the chapter, are integrated in appropriate places. • The writing generally does not hold the readers interest.
Basic level that requires remedial help or demonstrates a lack of effort. I	• Essential information is missing. • Details are limited and do not always pertain to the topic. • Few of the concepts, presented within the chapter, are integrated within the manual. • The writing is very difficult to follow. • Much of the manual remains incomplete.

For use with *Sports*, Chapter I

©1999 American Association of Physics Teachers

Assessment Rubric: Scientific Language

Meets the standard of excellence. **5**	• Scientific vocabulary is used consistently and precisely. • Sentence structure is consistently controlled. • Spelling, punctuation, and grammar are consistently used in an effective manner. • Scientific symbols for units of measurement are used appropriately in all cases. • Where appropriate, data is organized into tables or presented by graphs.
Approaches the standard of excellence. **4**	• Scientific vocabulary is used appropriately in most situations. • Sentence structure is usually consistently controlled. • Spelling, punctuation, and grammar are generally used in an effective manner. • Scientific symbols for units of measurement are used appropriately in most cases. • Where appropriate, most of the data is organized into tables or presented by graphs.
Meets an acceptable standard. **3**	• Some evidence that the student has used scientific vocabulary although usage is not consistent or precise. • Sentence structure is generally controlled. • Spelling, punctuation, and grammar do not impede the meaning. • Some scientific symbols for units of measurement are used. Generally, the usage is appropriate. • Limited presentation of data by tables or graphs.
Below acceptable standard and requires remedial help. **2**	• Limited evidence that the student has used scientific vocabulary. Generally, the usage is not consistent or precise. • Sentence structure is poorly controlled. • Spelling, punctuation, and grammar impedes the meaning. • Some scientific symbols for units of measurement are used, but most often, the usage is inappropriate. • No presentation of data by tables or graphs.
Basic level that requires remedial help or demonstrates a lack of effort. **1**	• Limited evidence that the student has used scientific vocabulary and usage is not consistent or precise. • Sentence structure is poorly controlled. • Spelling, punctuation, and grammar impedes the meaning. • No attention to using scientific symbols for units of measurement. • No presentation of data by tables or graphs.

Maximum = 10 Points

For use with *Sports*, Chapter 1

©1999 American Association of Physics Teachers

What is in the Physics InfoMall for Chapter 1?

If you have had much experience with the Physics InfoMall CD-ROM, you have probably done a few searches, and no doubt some of the searches have resulted in "Too many hits." Surprisingly, searching the entire CD-ROM with the keyword "sport*" does not give "too many" hits, but provides some interesting hits. Note that the asterisk is a wild character; this searches for any word beginning with "sport."

If you do the search just mentioned, the first hit is a resource letter ("Resource letter PS-1: Physics of sports," *American Journal of Physics,* vol. 54, issue 7) that discusses the published discussions on the physics of sports. According to this letter, "there is surprisingly little published information about the basic physics underlying most sports, even though the relevant physics is all classical." Included is a list of places you might find such information, including journals and books. The letter contains a list of specific references grouped by sport, such as these two for Track and Field: "Behavior of the discus in flight," J. A. Taylor, Athletic J. 12(4), 9, 45 (1932), and "Bad physics in athletic measurement," P. Kirkpatrick, Am. J. Phys. 12, 7 (1944). Also in this list of hits are several articles on the physics of specific sports, such as basketball (Physics of basketball, *American Journal of Physics*, vol. 49, issue 4). Another interesting article is "Students do not think physics is 'relevant.' What can we do about it?," in the *American Journal of Physics*, vol. 36, issue 12.

Given that the physics in sports is classical, you might search for student difficulties learning classical physics in general. One article you might find is "Factors influencing the learning of classical mechanics," *American Journal of Physics,* vol. 48, issue 12. Knowledge of such factors affecting learning can be a valuable tool. Perform other searches that meet your needs, and the InfoMall is very likely to provide good information. And we have not even opened the Textbook Trove yet!

As mentioned above, there is a resource letter that contains references to published information about the physics of various sports. Track and Field, the topic of Chapter 1, is one of the sports mentioned. Of course, each sport uses several different concepts in physics, and these are analyzed in the following activities. Each of these concepts can always be found in the textbooks in the Textbook Trove, although we might not make such specific references here.

ACTIVITY ONE
Running the Race

Background Information

The definition of average speed is introduced in Activity One:

Average Speed=Distance x Time

where the terms on the right-hand side of the equation mean the total distance travelled by an object and the total time required to travel that distance.

The standard units of measurement for distance and time are, respectively, the meter and the second. Therefore, the standard unit of average speed is meters per second. In some cases, subdivisions and multiples of the meter are used for measuring distance (e.g., centimeters and kilometers). Other units of measurement are used in cases where common practice involves measuring distance in feet, yards, or miles and measuring time in hours, resulting in units for average speed such as cm/s, km/hr, ft/sec, mi./hr.

It is important to note that the above definition of average speed does not take into account any variations in speed which an object may have as it travels along its path; the average speed is, simply, the total distance travelled divided by the total time taken to travel the distance. Therefore, an object's average speed may be more or less representative, or descriptive, of what really happens during particular cases of motion.

For example:

Suppose that a driver operates a vehicle on a super-highway for an hour with only minor variations in speed due to traffic. If the odometer shows that 65 miles were travelled during the hour, the average speed, 65 miles per hour, probably is quite representative of the vehicle's speed at most instants during the hour of travel.

Suppose that the same driver stopped for ½ hour to fix a flat tire, causing the 65 mile trip to require 1½ hours. According to the definition, the average speed for this trip is (65 mi.) / (1½ hr) = 43 mi./hr. True, the driver "averaged" 43 mi./hr, but the speedometer probably showed 43 mi./hr at only a few instants during the trip. This average speed is less representative of how the car moved (or, at some times, did not move) during the trip.

The second of the above two examples raises the caution that "average speed" should not be confused with "instantaneous speed;" the latter is an object's speed at a particular instant, which will be introduced later in this chapter.

This activity, and many others in this chapter, involves track events in which runners start from rest (zero speed) and "build up" (accelerate) to a "top speed" which is maintained at a reasonably constant level for the remainder of the race. The "rest start" usually causes the average speed for the entire race to be less than the runner's "top speed." The discrepancy between a runner's average speed and top speed becomes less for races of greater distance because the period of acceleration occupies a less significant part of a long race than a short race.

Active-ating the Physics InfoMall

Speed is one of the first concepts introduced in introductory physics. It is also one that causes students problems. If you search the InfoMall for "student difficult*" OR "student understand*" you will find several articles that deal with research into how students learn fundamental concepts in physics. Some of these are "Investigation of student understanding of the concept of velocity in one dimension," *American Journal of Physics* vol.48, issue 12 (see "Diagnosis and remediation of an alternative conception of velocity using a micro-computer program," *American Journal of Physics*, vol. 53, issue 7 for a discussion of this); "Research and computer-based instruction: Opportunity for interaction," *American Journal of Physics*, vol. 58, issue 5 (if you look at other articles in this volume, you will find "Learning motion concepts using real-time microcomputer-based laboratory tools," *American Journal of Physics*, vol. 58, issue 9); and more. These last two articles make specific mention of the use of computers in teaching physics. This is a trend that is gaining strength and shows great promise. Look for more such articles in journals that are newer than the CD-ROM. As a starting point, you can always look in the annual indices of the physics journals and look under the names of the authors of the articles mentioned above.

One of the strengths of the InfoMall is the ability to search the entire database quickly. Don't think that searching is the only, or even the best, method for finding what you want. For example, suppose you want a demonstration of lab activity that uses the concept of speed. Go to the Demo & Lab Shop and choose almost any of the selections. If you choose

Physics Lab Experiments and Computer Aids, you may notice that choice number 8 is Straight-Line Motion at Constant Speed (the next choice involves acceleration, which may also be useful). This shows (with a nice graphic) how to use a toy bulldozer in the lab. Choice 7 shows how you can use strobe photographs in a lab as well.

If you want to find additional problems for your students, you may browse through Teachers Treasures and find that there is a section on Speed Problems. Or you can look in the Problems Place, where you will find *Schaum's 3000 Solved Problems in Physics* (among others). You can find plenty of speed problems (with their solutions!).

Planning for the Activity

Time Requirements

- One class period.

Materials Needed

For the class:

- metric tape measure or meter stick
- chalk stick or tape for marking 5-m intervals along running surface
- stopwatches (4 minimum, 8 preferred)

For each group:

- calculator

Advance Preparation and Setup

An area will be needed to serve as a straight running "track." The area needs to be a minimum of 30 m (about 100 feet) in length, allowing 20 m for running and a stopping distance of at least 10 m. Tape (or some other marking material) must be able to be placed on the running surface at 5-m intervals. The track must be wide enough for one person at a time to run while other persons serving as timers stand alongside the track and the entire class observes. Noise can be expected. An outdoor area such as a sidewalk or parking lot may be preferred to accommodate noise and to avert scheduling problems. Indoors, a gymnasium or hallway could be used.

If you do not have an adequate supply of stop-watches (see Materials Needed), you may wish to check in advance to see if class members or others have wrist watches having a stopwatch function which could be used instead of stopwatches.

Teaching Notes

Active Physics uses a modified constructivist model. By confronting students' misconceptions and by having them do hands-on exploration of ideas, we seek to replace their misconceptions with correct perceptions of reality. In order to do this, a consistent scheme is integrated into the course activities to elicit the students' misconceptions early in any activity.

Students' current mental models are sampled by one or more What Do You Think? questions. Students are not expected to know a "right" answer. Answers are provided for your use only. These questions are supposed to elicit from students their beliefs regarding a very specific prediction or outcome, and students should commit to a written specific answer in their logs.

Some students in the class may be aware that the average speed of a runner is determined by dividing the total distance travelled by the total elapsed time. Most runners change speed while running. In a short dash starting from rest, a runner does not reach top speed instantly; therefore, more time usually is required to run the first half of the total distance than the last half. When running a long distance, a runner may tire and run a lower speed during the final part of the distance, taking more time to travel the second half of the total distance than the first half.

Stage one or more practice runs wherein all students serving as timers simultaneously measure the time required for one runner to travel 10 m. Use the variation in times measured by different individuals to establish consensus on the accuracy to which times should be reported - probably to the nearest 1/10 s. Establish that distance can be measured more accurately than time in this activity, and that the accuracy of the time measurements, in turn, places a limitation on the accuracy of speeds calculated using the time measurements; see the sample data below for how this works. Be aware that students tend to believe that if a stopwatch used in this activity shows, for example, a time of 3.37 s for a student to run 15 m, the time is known to the nearest 1/100 of a second when, in reality, student reaction times probably limit the accuracy to less than 1/10 s.

This is not an appropriate time to get bogged down in discussion of precision, accuracy and significant figures, but it also would be unwise to allow students to believe that they know the value of average speed to greater accuracy than they really do. A good way to put this is to say that the accuracy to which the speed is known can be no better than the least accurate measurement involved in calculating the speed; in this case, the least accurate measurement is time. During each 5-m interval we know the time to only one significant figure; therefore, the speed during each interval should be reported to only one significant figure.

You may wish to point out that the unit of measurement for average speed, m/s, is a "derived" unit, meaning that it is made up from a combination of "base," or fundamentally defined, units, the meter and the second.

You may also wish to show students the mathematical operations involved in transforming the equation for average speed to solve for any of the three variables, pointing out that this offers the advantage of needing to remember only one form of the equation.

The average speed of record holders at the Penn Relays varies from 9.6 m/s for the men's 100-m dash to 4.9 m/s for the women's 5,000-meter race. Speeds generally decrease as the distance of race events increases.

NOTES

1

Activity Overview

In this activity students measure the time it takes a runner to reach each 5-m mark along a 20-m run. They then use their data to calculate the average speed for each interval. Students then compare the average speed for different segments as well as different runners.

Student Objectives

Students will:

- measure distance in meters along a straight line.
- use a stopwatch to measure the amount of time for a running person to travel a measured distance.
- use measurements of distance and time to calculate the average speed of a running person.
- compare the average speeds of a running person during segments of a run.
- compare the average speeds of different persons running along the same path.
- infer factors which affect the average speed of a running person.

ANSWERS FOR THE TEACHER ONLY

What Do You Think?

The average speed of a runner is determined by dividing the total distance travelled by the total elapsed time.

Most runners change speed while running. In a short dash starting from rest, a runner does not reach top speed instantly; therefore, more time usually is required to run the first half of the total distance than the last half. When running a long distance, a runner may tire and run at a lower speed during the final part of the distance, taking more time to travel the second half of the total distance than the first half.

THE TRACK AND FIELD CHAMPIONSHIP

Activity One
Running the Race

WHAT DO YOU THINK?

Very few people in the world can run 100 meters in less than 10 seconds.

- **How can you measure a runner's speed?**
- **Does running twice as far take twice as much time?**

Record your ideas about these questions in your *Active Physics log*. Be prepared to discuss your responses with your small group and the class.

FOR YOU TO DO

1. Your teacher will indicate the location of the "track." Place a mark on the track at the 0-, 5-, 10-, 15-, and 20-m positions.

2. Place students with stopwatches at the 5-, 10-, 15-, and 20-m marks to serve as timers. Each timer should measure the amount of time for each runner to reach the timer's assigned mark.

3. Have someone serve as starter, saying "Ready, Set, Go!" for each runner. All watches should be started on the "Go" signal, and each watch should be stopped as the runner passes the timer's assigned mark. Measure the time for three runners.

   a) In your log, record the time it takes each runner to reach the 5-, 10-, 15-, and 20-m positions.

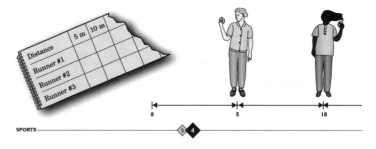

SPORTS

S 4

ANSWERS

For You To Do

1.-2. Student activity.

3. a) Sample data for one runner:

Distance	5 m	10 m	15 m	20 m
Runner #1 Time	1.6 s	2.5 s	3.3 s	4.0 s

4. Calculate the amount of time taken to run each 5-m interval. To calculate the time taken to run from the 5- to 10-m mark, you will need to subtract the time at the 5-m mark from the time at the 10-m mark.

✎ a) Use a table similar to the following to write the results in your log.

Distance	0-5 m	5-10 m	10-15 m	15-20 m
Runner #1 Time				
Speed				
Runner #2 Time				
Speed				
Runner #3 Time				
Speed				

5. Does running twice as far take twice as much time?

✎ a) Use data from the 20-m dashes to explain your answer in your log.

6. Calculate the average speed during each 5-m interval. Use the equation:

$$\text{Average speed} = \frac{\text{Distance}}{\text{Time}}$$

✎ a) Record the average speed during each 5-m interval in the table in your log.

SPORTS

ANSWERS

For You To Do (continued)

4. a) Sample data for one runner:

Distance	5 m	10 m	15 m	20 m
Runner #1 Time	1.6 s	0.9 s	0.8 s	0.7 s
Speed	3 m/s	6 m/s	6 m/s	7 m/s

5. a) Running twice as far does not necessarily take twice as much time. Students may note that in the case of a short-distance run, such as 20 m, it takes less than twice the time to run 20 m than to run 10 m. Students may also suggest that a great deal will depend on the distance run, that in longer races the runner may fatigue.

6. a) Using the sample data: Average speed = 20 m/4.0 s = 5.0 m/s

THE TRACK AND FIELD CHAMPIONSHIP

7. Use your data to answer the following questions:

a) In which distance interval did each runner have the greatest average speed? Circle the interval in your log for each runner.

b) Was the interval of greatest speed the same or different for different runners?

c) Which runner holds the record for the fastest 5-m interval?

d) Describe each runner's total dash in terms of speed during distance intervals.

e) Estimate the amount of time taken for each runner to reach maximum speed.

8. Calculate the average speed of each runner for the entire dash.

a) Record the average speed in your log.

b) If the three dashes had been an Olympic time trial, which runner would have won? What was the winning speed?

9. Use data from each runner's dash.

a) Write suggestions in your log for the runners to improve their performances.

PHYSICS TALK

Average Speed

The relationship between average speed, distance, and time can be written as:

$$\text{Average speed} = \frac{\text{Distance}}{\text{Time}}.$$

This relationship can be rearranged as follows:

$$\text{Distance} = \text{Average Speed} \times \text{Time}$$

$$\text{Time} = \frac{\text{Distance}}{\text{Average Speed}}$$

SPORTS

ANSWERS

For You To Do (continued)

7. a-e) Students will probably find that the greatest average speed was during the final distance intervals. The initial time interval will be greater because the student is accelerating from rest. Student average speeds will likely vary.

8. a-b) Answers will vary depending on experimental results.

9. a) Student could suggest that the runner improve his/her start. Also, students could observe if a runner slows down for the approach of the finish line.

ACTIVITY ONE: Running the Race

FOR YOU TO READ

The Scenario for this chapter mentioned that the Penn Relays is the world's largest track and field event. It includes athletes at the high school, college, and professional levels. Examine the record times of male and female athletes in running events at the Penn Relays in the table to the right.

Many students in your class may be able to run 100 m in less than 15 s. Some members of your school's track and field team might be able to do the same in under 12 s. Do a calculation to show yourself that even a slow person who could run 100 m in 15 s could, theoretically, break the Penn Relays record for the 1,500-m race if it is true that "running 15 times farther takes 15 times as much time." It will be necessary to convert the Penn Relays record given in minutes and seconds into seconds. (Example: A time of 4:08 for the mile equals 248 s.)

Obviously, not only speed but also stamina is involved in winning a race. An athlete needs to be able to run at high speed and have the stamina, or strength, to keep the speed as high as possible for the entire race. Both speed and stamina can be improved through training and knowledge. Athletes who hold records are those who run both fast and smart.

Penn Relay Record Times		
Distance (meters)	Time–Men (minutes:seconds)	Time–Women (minutes:seconds)
100	0:10.47	0:11.44
200	0:21.07	0:23.66
400	0:45.49	0:52.33
800	1:48.8	2:05.4
1500	3:49.67	4:24.0
mile	4:08.7	4:49.2
3000	8:05.8	9:15.3
5000	15:09.36	16:59.5

S 7   **SPORTS**

ANSWERS

For You To Read

A person who can run 100 m/15.0 s would have an average speed of 6.67 m/s. If that speed were maintained for a 1,500 m race, the time would be Time = Distance/Speed = 1,500 m/6.67 m/s = 225 s = 3:45 minutes. This would break the record time for men (3:49.67) in the Penn Relays 1,500 m race, and it would break the record for women in that race (4:24.0).

ANSWERS

Physics To Go

1. a) 1,500 m/229.7 s = 6.53 m/s

   b) There might be students who can reach the same speed.

   c) A student probably could not maintain such a speed for a distance of 1,500 m.

2.   100 m : 8.74 m/s
     200 m : 8.45 m/s
     400 m : 7.64 m/s
     1,500 m : 5.68 m/s

   The average speeds of runners decrease as the distance of races increases.

3. Compare record times for school track team to record times for Penn Relays; also compare average speeds by event. If you are unable to obtain records from your school's track team, sample data is provided in Additional Materials.

4. Expect students to indicate that it probably would not be fair or meaningful to compare data for high school running events to world records due to differences in athletes' ages, levels of training, and other factors.

5. More time would be required during the first half of the run because the runner is starting from rest and building up to top speed during the first half, therefore having a lower average speed than during the last half.

6. Split times provide ability to calculate speeds during parts of a race, allowing runners to have detailed information which could be used to improve performance.

THE TRACK AND FIELD CHAMPIONSHIP

REFLECTING ON THE ACTIVITY AND THE CHALLENGE

You know from measurements of your classmates' running that the person who travels the entire distance of a race in the least time wins. The winner also has the highest average speed for the overall race. You also know that the speed of each runner in your class changed during the run.

Search for patterns in the distances and times of men and women who hold records in the Penn Relays. Do the speeds vary with the distance of the events? What information could you give your school's track and field team about the Penn Relays right now? What further information do you need?

PHYSICS TO GO

1. a) Calculate the average speed of the male who holds the Penn Relays record for the 1,500-m run.
   b) From the data you gathered, are there students in your class who can reach the same speed as the male 1,500-m record holder?
   c) Do you think the fastest student in your class could run the 1,500 m in record time?

2. Calculate the average speeds of women who hold Penn Relays records in the 100-, 200-, 400-, and 1,500-m runs. What is the pattern of speeds?

3. Find some record times for your school's track team in running events and compare them with the Penn Relays' record times. Also compare the average speeds for races of equal distance for your school with those for the Penn Relays.

4. Is it fair to compare data for high-school running events with data for world records? Why or why not?

5. How do the amounts of time taken by a champion 100-m runner to travel the first and last 50 m of a sprint compare? What is the basis for your answer?

6. Track coaches often measure "split times" for runners during races. For example, a runner's time might be measured every 100 m during a 400-m race. How could "splits" be useful for helping runners improve both speed and stamina?

 S 8

NOTES

Sample Data of Records from High School Track Meets

Girls' Records for Track Events

Track Event	Name	Time	Year
100 Meter Dash	Carol C.	13.0	1989
200 Meter Dash	Antoinette C.	27.1	1986
400 Meter Dash	Meg K.	62.2	1984
800 Meter Run	Amy C.	2:19.3	1983
1500 Meter Run	Julie S.	4:47.2	1983
3000 Meter Run	Dawn E.	10:24.4	1985
100 Meter Hurdles	Holly F.	16.2	1987

Boys' Records for Track Events

Track Event	Name	Time	Year
100 Meter Dash	Bill C.	10.5	1985
200 Meter Dash	Chris B.	21.2	1984
400 Meter Dash	Jerome W.	46.9	1988
800 Meter Run	Mike S.	1:50.5	1983
1500 Meter Run	Mike S.	3:46.9	1982
3000 Meter Run	Mike S.	9:01.7	1983
100 Meter Hurdles	Art M.	13.5	1976
400 Meter Hurdles	Sherwin S.	52.7	1986

Girls' Records for Field Events

Field Event	Name	Distance	Year
Shot Put	Judy B.	33'1/2"	1998
Discus	Naomi J.	108'10"	1991
Javelin	Naomi J.	126'0"	1992
Long Jump	Tonya J.	17'1/2"	1974
Triple Jump	Roxanne R.	31'4 3/4"	1991

Boys' Records for Field Events

Field Event	Name	Distance	Year
Shot Put	C.J. H.	64'5 1/4"	1986
Discus	Phil C.	183'3"	1997
Javelin	Andy S.	225'0"	1986
Long Jump	Matt S.	24'9 1/4"	1990
Triple Jump	Sanya O.	15'8"	1996

Sample Data of Records from High School Track Meets

Girls' Records for Relays

Event	Times	Total Time	Year
400 Meter Relay	14.2	51.6	1987
	13.3		
	11.0		
	13.1		
800 Meter Relay	27.9	1:49.7	1997
	27.2		
	27.0		
	27.7		
1600 Meter Relay	63.3	4:14.9	1985
	63.7		
	62.7		
	65.3		
3200 Meter Relay	2:21.8	9:22.55	1985
	2:20		
	2:21.2		
	2:19.0		

Boys' Records for Relays

Event	Times	Total Time	Year
400 Meter Relay	10.7	42.4	1994
	10.6		
	10.6		
	10.5		
800 Meter Relay	21.8	1:25.4	1987
	21.5		
	21.3		
	20.8		
1600 Meter Relay	49.8	3:12.7	1966
	48.0		
	47.4		
	47.5		
3200 Meter Relay	1:56.2	7:45.8	1995
	1:56.8		
	1:57.1		
	1:55.7		

For use with *Sports*, Chapter 1, ACTIVITY ONE: Running the Race
©1999 American Association of Physics Teachers

ACTIVITY TWO
Analysis of Trends

Background Information

The concept of average speed is extended from Activity One. Students are asked to use graphing techniques (curve fitting and extrapolation) to explore how the performance of athletes has improved over a century of time, and how the average speed of runners varies with the distance of a race.

It is of value for students to estimate the best fit of a curve to data points on a graph and to extrapolate the curve of a graph "by hand," as directed in the activity. However, it also is of great value for students to experience the power of technology-assisted graphing techniques. Routines are available for both personal computers and graphing calculators which essentially automate graphing procedures; only the ordered pairs of data and the range of values for the axes of the graph are required to be entered. Once entered, the "best fit" curve is displayed, and both interpolation and extrapolation can be accomplished by simple commands.

If students are to perform extensive graphical analysis of data from your school's track team, doing so manually would be discouragingly tedious. This would be an excellent example of using a personal computer or graphing calculator as a tool.

If you are not familiar with technological tools for graphing, it is possible that a math, science, or computer science teacher at your school already has the knowledge and tools needed to accomplish this. Seek their help. If you don't have the time to learn how to use the devices, a highly able student may be very willing to learn to use the tools and to teach you and the class how they work.

If the technological tools are not available at your school, they can be found in most science catalogs, in advertisements in professional journals, and on display at science education conferences. These technological tools are the "state of the art," exist in great variety, and are becoming quite inexpensive. If you and your students are not using them, perhaps it's time you did. (See also *Active*-ating the Physics InfoMall.)

Active-ating the Physics InfoMall

For You to Do Step 3 suggests using computers or graphing calculators to examine trends. In addition to the articles mentioned in Activity One on the use of computers, you can find "Student difficulties with graphical representations of negative values of velocity," in *The Physics Teacher*, vol. 27, issue 4 on the InfoMall. This was one of the articles found in the search using "student difficult*" OR "student understand*". Also found in that search is "Student difficulties in connecting graphs and physics: Example from kinematics," in the *American Journal of Physics*, vol. 55, issue 6. Don't let these titles make you think that graphs are a bad thing! It is good to be aware that students do not always understand graphs, a valuable tool in the study of physics. This step also suggests extrapolation. For discussions on extrapolation, you can perform a search using the keyword "extrapolation". One of the hits is Chapter 11 of the textbook *Physics for the Inquiring Mind*. This warns that "both interpolation and extrapolation are easier if the graph is a straight line. Even then they are not equally safe as sources of information, and asks "Which of the two, interpolation and extrapolation, would you as a scientist value more?"

Planning for the Activity

Time Requirements

• One class period.

Materials Needed

For each group:
• calculator

For each student:
• graph with plotted data (curve not drawn), "Speed versus Year, Men's Olympic 400-meter Dash" (provided)
• graph with plotted data (curve not drawn), "Speed versus Distance, Men and Women, Penn Relays" (provided)
• local track team records (running events)

Advance Preparation and Setup

Allow time for reproduction of the blackline masters "Speed versus Year, Men's Olympic 400-meter Dash" and "Speed versus Distance, Men and Women, Penn Relays" found on the pages following this activity.

Obtain and reproduce records of the performances of members of your school's track team. Information about running events is needed now, and later it would be useful to have information about all track and field events. Your school's track coach should be able to provide this information. If possible, the coach could be invited to present the information to your class; this would require additional time beyond the one class period needed to complete this activity. If your school does not have a track team, perhaps you can obtain data from another school. Sample data are also available at the end of the previous activity.

Teaching Notes

There is no "right" trend line or curve to sketch through the points on either graph; some estimates of the trends shown by the data would be better than others, depending on the methods used to establish "best fits" to the data. Students should be encouraged to sketch a straight line or a curve through the array of data points on each graph which makes the most sense to them; students should be discouraged, however, from simply connecting consecutive data points with straight line segments to produce a "saw tooth" curve because doing so involves no interpretation of the data. It is desired that students take the "risk" to sketch lines or curves which show their impressions of the trends of the data.

When students have completed their work with the graphs, you may wish to share with them the examples shown of best fit lines generated for the same data using a computer program. The computer-generated graphs use sophisticated routines to estimate "best fits" to data and to accomplish extrapolations, but should not be taken as "absolute truths."

It is indicated in Reflecting on the Activity and the Challenge that you, the teacher, will help students obtain data on the performances of members of your school's track team in various events; students will need such data in order to proceed. If your school does not have a track team, try to get data from a neighboring school. Sample data is also provided in this Teacher's Guide.

If you have ready access to a computer and a spreadsheet program such as MS Excel, you may wish to do the alternative Activity Two provided on the pages following this activity.

Assessment

See the Performance-based Assessment Rubric: Reading and Interpreting Graphs following this activity.

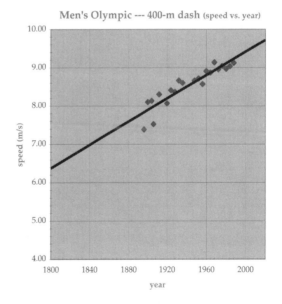

NOTES

Activity Two
Analysis of Trends

WHAT DO YOU THINK?

Current trends indicate that women will start outrunning men in 65 years.

- **Is it useful to compare track records over many years of time?**
- **Can future track records be predicted based on past performances?**

Record your ideas about these questions in your *Active Physics log*. Be prepared to discuss your responses with your small group and the class.

FOR YOU TO DO

1. Look at the graph "Speed Versus Year: Men's Olympic 400-m Dash." The average speed of runners is shown on the vertical axis, and the year in which the race was run is shown on the horizontal axis. Take a moment to be sure you understand that the plotted points show a 100-year history of the speeds of male athletes in the Olympic 400-m dash.

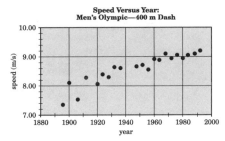

Speed Versus Year: Men's Olympic—400 m Dash

Activity Overview

Students in this activity use graphical analysis of authentic track records to explore the trend of average speed in the 400-m dash over the past 100 years. Students also analyze the speed vs. distance graph to determine what happens to average speed as the distance of the race increases.

Student Objectives

Students will:

- sketch a best fit curve or trend line on a graph on which points have been plotted.
- extrapolate graphs to predict trends beyond available data.
- create graphs of data presented in tabular form.
- calculate the average speed of a runner given distance and time.
- calculate the time required for a runner to travel a specified distance given the runner's average speed.
- interpret information presented in graphical form.

ANSWERS FOR THE TEACHER ONLY

What Do You Think?

Track records in particular events show clear trends over time.
Within reason, extrapolation allows prediction of future performances in track events.

a) Copy the graph into your log. Sketch either a straight line or a smooth, curved line through the data points to show what you think is the shape of the graph. Do this for the points plotted from 1896 to 1996.

b) Comment in your log about what you see as the trend of average speed in the 400-m dash over the past 100 years.

c) Make the best guess you can to sketch how the graph would continue to the year 2020. This process of going beyond the data is called *extrapolation*. Try it.

d) According to your extrapolation, what will be the speed of the winning runner in the men's Olympic 400-m dash in the year 2020?

2. Look at the graph "Speed Versus Distance: Men and Women, Penn Relays." Notice that the average speed of runners is shown on the vertical axis and the distance of the race is shown on the horizontal axis. Be sure you understand that there are two sets of plotted points, one for men and one for women. Also be sure you understand that the plotted points show how average speed varies with distance of the race for both men and women.

Speed Versus Distance:
Men and Women, Penn Relays Records

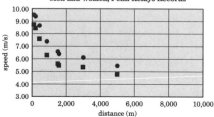

a) Copy the graph into your log. Sketch the shapes of the graphs for men and women by connecting the plotted points with either a straight line or a curved line, according to your choice.

ANSWERS

For You To Do

1.a) There is no "right" trend line or curve to sketch through the points; some estimates of the trends shown by the data would be better than others, depending on the methods used to establish "best fits" to the data.

b) The average speed has increased over the past 100 years.

c) Students will use their trend line or curve to extrapolate data.

d) Computer-generated extrapolations (as shown in this Teacher's Guide) indicate a winning speed of about 9.6 m/s for the Olympic 400-m dash in the year 2020. Accept any reasonable answers from the students.

2.a) There is no "right" trend line or curve to sketch through the points; some estimates of the trends shown by the data would be better than others, depending on the methods used to establish "best fits" to the data.

ACTIVITY TWO: Analysis of Trends

b) Comment on the trends of average speed for both men and women as the distance of races gets longer and longer.

c) Extrapolate from the graphs to predict what the record speed at the Penn Relays would be for 10,000-m races for men and women.

d) Try to use extrapolation to find a race distance for which men and women would run at the same average speed. Comment on your attempt.

3. The kind of data analysis you performed in this activity can be made much easier with a computer or graphing calculator. Basic distance and time data from track records can be entered into a computer spreadsheet software program or a graphing calculator. The computer or calculator can then be instructed to calculate speed from distance versus time data and to display graphs. The graphs can be analyzed to show trend lines, extrapolations, and other information. Your teacher or someone else familiar with computers and calculators may be able to help you use such devices as tools for data analysis.

REFLECTING ON THE ACTIVITY AND THE CHALLENGE

Your plan to help your school's team at the Penn Relays will only be as good as your knowledge of current performances in track and field. Also, knowledge of trends in how performances are changing with time in various events is also important.

Research shows that women are improving their track performances about twice as fast as men. This is because more and more women are participating in running as a sport than ever before. Certainly, you will want to consider the possibility of either a male or female athlete bringing home a medal from the Penn Relays for your school. Data can help you decide which athletes at your school will have the best chance.

Perhaps it's time to get some data about the performance of athletes at your school so that you can begin forming a strategy to help your team. Your teacher will help you with this.

ANSWERS

For You To Do (continued)

b) As the distance of the races increase, the average speed decreases.

c) Students will use their trend line or curve to extrapolate data.

d) Computer-generated extrapolations (as shown in this Teacher's Guide) indicate winning speeds of about 5.0 m/s for men and 4.3 m/s for women if a 10,000 meter race for each gender were held at the Penn Relays. Accept any reasonable answers from the students.

3. Students use a computer spreadsheet or a graphing calculator.

ANSWERS

Physics To Go

1. From data on the school's track team, students will calculate the average speed for each event and superimpose the data on the graphs of speed vs. distance for the Penn Relays used in the activity. Students will critically compare school data to Penn Relays data.

2. Using a speed of 9.6 m/s obtained from computer-assisted extrapolation and the equation Time = Distance/Speed, Time = 400 m/9.6 m/s = 41 s. Student results may vary from this.

3. Using a speed for men of 5.0 m/s and a speed for women of 4.3 m/s , both obtained from computer-assisted extrapolation, and the equation Time = Distance/Speed, the times are:
   - for men, 10,000 m/5.0 m/s = 2,000 s.
   - for women, 10,000 m/4.3 m/s = 2,300 s.

   Student results may vary from the above.

4. Distance = Speed x Time = 10.14 m/s x 9.86 s = 100 m.

5. a) Use the equation
   Speed = Distance/Time
   1964: 400 m/52.00 s = 7.7 m/s
   1968: 400 m/52.00 s = 7.7 m/s
   1972: 400 m/51.08 s = 7.8 m/s
   1976: 400 m/49.29 s = 8.1 m/s
   1980: 400 m/48.88 s = 8.2 m/s
   1984: 400 m/48.83 s = 8.2 m/s
   1988: 400 m/48.65 s = 8.2 m/s
   1992: 400 m/48.83 s = 8.2 m/s

b) Students provide graph.

c) Until 1980 the average speed has been increasing. During the past few years the average speed has showed no significant change.

d) Students extrapolate speed from their graph.

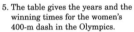
THE TRACK AND FIELD CHAMPIONSHIP

PHYSICS TO GO

1. Get distance and time data for the best performances of your school's boys' and girls' track teams for races of various distances.

   a) Calculate the average speed for each event.
   b) Plot the average speeds as data points on the Penn Relays graph used in "For You to Do." Be sure to keep male and female data separate.
   c) Connect the points to show the shapes of the graphs for males and females.
   d) For which race events does the graph for males at your school come closest to touching the graph for men at the Penn Relays? What does this mean? Do the same analysis for females at your school and the Penn Relays.

2. In "For You to Do" you extrapolated from a graph to predict the speed of the winner of the men's 400-m dash in the Olympics in the year 2020. Predict the winning time for that race. (Hint: See "Physics Talk" in Activity 1.)

3. In "For You to Do" you extrapolated from two graphs to predict the average speeds of men and women who would win if a 10,000-m run were held at the Penn Relays. Predict the time of the winning man and woman.

4. A runner had an average speed of 10.14 m/s for 9.86 s. Calculate the distance the runner traveled.

5. The table gives the years and the winning times for the women's 400-m dash in the Olympics.

   a) Calculate the winning speeds.
   b) Plot a speed vs year graph for the data.
   c) What is the trend of the average speeds over the past years?
   d) From your graph extrapolate what the winning speed will be in 2020.

Year	Time (seconds)
1964	52.00
1968	52.00
1972	51.08
1976	49.29
1980	48.88
1984	48.83
1988	48.65
1992	48.83

S 12

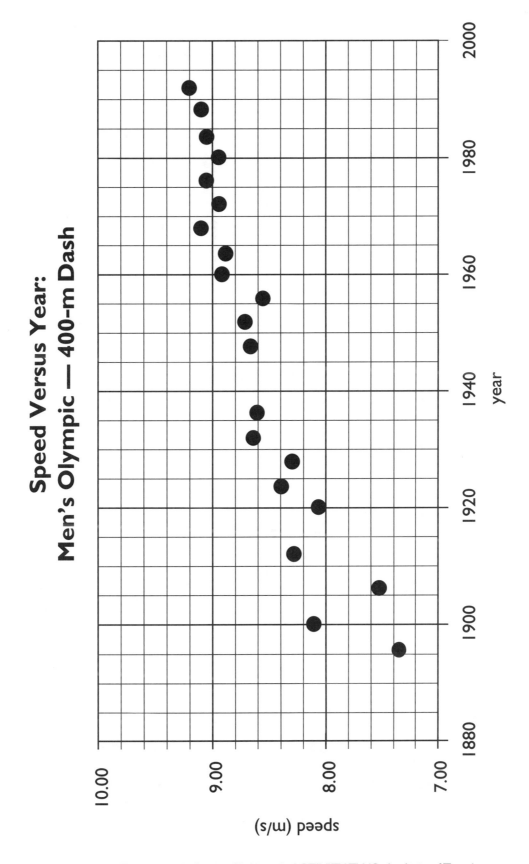

Speed Versus Year:
Men's Olympic — 400-m Dash

For use with *Sports*, Chapter 1, ACTIVITY TWO: Analysis of Trends
©1999 American Association of Physics Teachers

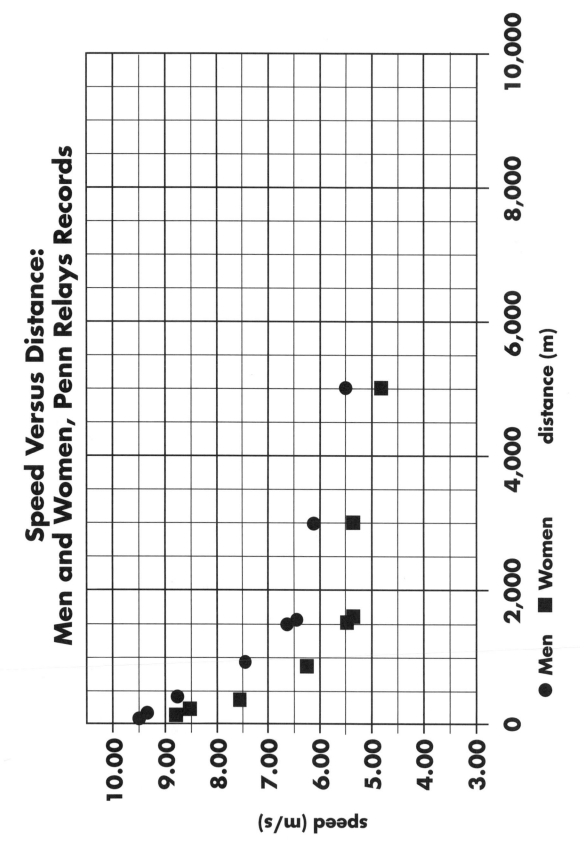

Speed Versus Distance:
Men and Women, Penn Relays Records

Activity Two A

Spreadsheet Games—Analysis of Long-Range Trends

FOR YOU TO DO

In this activity you will need a computer (Mac or MS DOS) and the spreadsheet program MS Excel.

1. Scientist often take a set of data and try to explain what happened in the past or predict what will happen in the future. A convenient method for doing this is to use a computer spreadsheet. Open the file track.xls and click the tab at the bottom of the screen that says "400-m dash." The winning time for each Olympic year is listed as well as the average speed of the runner during the race. To the right of the screen is a graph of Speed versus Year. As you can see from the graph, the speed of the runners has been generally increasing over the years.

a) Can the speed of future races be predicted from the graph?

NOTE: When using the spreadsheet, to enter a number click in the box, type the number (do not include units or commas) and press <Enter>. If you get the message, "Locked cells cannot be changed," it means that you did not click in the proper box. Try again.

2. The first thing that has to be done is to try to determine how speed and time are related mathematically. In this exercise let's guess that the relationship is a simple straight line. With the aid of a computer, you can create the best straight line to represent this data.

   Click the mouse on the screen button named "Trend line." When you do this, the computer calculates the best-fit line. You will notice that the line is extended into the future and back into the past. This is called extrapolation.

a) If this trend continues, what will be the speed of the winning runner in the year 2020?

3. Using the equation below, calculate the winning time in the year 2020?
   distance (400 m) = $\frac{time}{speed}$

a) Record your calculations and answer in your log.

b) Do you think this trend will continue? Explain your answer.

c) Do you still think that the speed versus year graph is straight, or should it be curved?

d) If you think the graph should be curved, sketch in your log what your guess is for the shape of the graph.

4. Open the file track.xls and click the tab at the bottom of the screen that says "Penn Relays." The winning time for each event is listed for both men and women. In addition, the average speed for each race has been calculated. To the right of the screen is a graph of Speed versus Distance for the races. It is easy to see from the graph that the speed of the runners generally decreased with increasing race distance. Record your predictions for the following questions:

a) Can this data be used to predict unknown information? For instance, what will be the winning speed for a much longer race?

b) If the distance is long enough, will men and women run at the same speed?

5. The first thing that has to be done to answer such questions is to try to determine how speed and race distance are related mathematically. In the 400-m dash activity, the data looked like a straight line. However, this data does not look at all straight. One of the advantages of spreadsheets is that they have the capability to try to calculate a "best fit" and extend the line past the data.

Click the mouse on the screen button named "Trend line." When you do this, the computer calculates the best-fit line. You will notice that the line is also extended to a race distance of 25,000 m; this is called extrapolation.

a) If this trend continues, is there a distance at which men and women will run at the same speed? Explain.

b) Based on the graph, what would be the record speeds (men and women) for a 10,000 meter race? Explain.

c) Using the equation below calculate the time for this race.

$$\text{distance (400 m)} = \frac{\text{time}}{\text{speed}}$$

d) The equation calculates time in seconds. Convert this to minutes and seconds.

6. Compare the records at the Penn Relays to the world records.

a) Add the data points for the world records to the Penn Relays graph.

7. Click the mouse on the screen button named "World Records."

a) Based on the graph, do the Penn Relay runners get closer to the world records, further, or stay about the same?

b) Based on the graph, what should be the world record for the 25,000-m run? Determine speed from the graph and then calculate time using the above equation.

c) Convert the time to hours, minutes, and seconds.

Stretching Exercises

Use the spreadsheet to try to compare the best runner in the history of your school with the Olympic runners of the past.

• Look up your school record for the 400-m dash.

• To enter the value on the spreadsheet for the 400-m dash, click in the box pointed to by the arrow.

• Type the school record.

• Click the mouse on the screen button named "Plot Your Record." You should now see a dot on the graph. The position of this dot was determined by calculating the average speed for the runner and finding where that speed appeared along the trend line.

• Using the plotted point for your school record, determine what year your high school runner would have been competitive in the Olympic 400-m dash.

Performance-Based Assessment Rubric:
Reading and Interpreting Graphs

Criteria	Excelling	Good	Basic
Makes comparisons of changes in speed over time.	Working independently, is capable of identifying trends and describe how relay speeds are changing over time.	Requires some assistance to identify trends and describe how relay speeds are changing over time.	Can read graphing coordinates but is unable to make generalizations about how speeds in relays have changed over time.
Makes comparisons of two independent data sets.	Working independently, is able to compare race speeds for men and women at any given year.	Is able to group coordinates for two sets of data, but requires some assistance to compare race speeds for men and women at any given year.	Is unable to group coordinates for independent data sets and is unable to compare race speeds for men and women at any given year.
Compares rates of change for two independent data sets.	Working independently, is capable of comparing the rate at which race speeds are changing for men and women from graphical data.	Is able to explain rate of change, but requires some assistance in reading graphical data to compare the rate at which race speeds are changing for men and women.	Is unable to explain rate of change or make comparisons about how race speeds are changing for men and women from graphical data.
Uses extrapolation to make predictions from a single data set.	Working independently, is capable of making predictions by extrapolating graphical information to predict relay speeds for 10, 000 m.	Requires some assistance to make predictions by extrapolating graphical information to predict relay speeds for 10,000 m.	Can read graphing coordinates but is unable to extrapolate information to predict relay speeds for 10,000 m.
Using extrapolation to make predictions from two data sets.	Working independently, is capable of making predictions by extrapolating graphical information from two sets of data to find a race distance for which men and women would have the same average race speed.	Requires some assistance to make predictions by extrapolating graphical information from two sets of data to find a race distance for which men and women would have the same average race speed.	Is unable to make predictions by extrapolating graphical information from two sets of data to find a race distance for which men and women would have the same average race speed.

For use with *Sports*, Chapter 1, ACTIVITY TWO: Analysis of Trends
©1999 American Association of Physics Teachers

ACTIVITY THREE
Who Wins the Race?

Background Information

This activity continues to apply the concept of average speed, but the particular cases of motion chosen for analysis intentionally involve acceleration. Two major points about average speed should be made through the experiences provided by the activity:

In any race, the runner having the highest average speed wins the race; all runners travel the same distance, but the winner finishes in the least amount of time.

A particular runner's average speed for an entire race may or may not be representative of the runner's speed at various instants during the race.

It is important to maintain a distinction between an instant in time and a time interval. Be clear in your own mind, and encourage your students also to be clear, that the calculation of average speed involves a time interval, the elapsed time between two instants in time.

When this activity has students measure the average speed during a very small time interval which involves only a small part of an object's total "trip," the average speed begins to approximate the instantaneous speed which happened sometime during the small interval. For your information (not suggested to be shared with students at this time), an object's average speed during an infinitesimally small time interval is the object's instantaneous speed:

Instantaneous speed = limiting value, as Δt approaches zero, of the ratio $\Delta d/\Delta t$

where Δd is the distance travelled during the time interval Δt

For your information as the teacher (not recommended to be brought up to students at this time), the final speed of each car at the end of each 1-m run in For You To Do, step 4 should depend on the overall change in the height from the beginning to the end of each track.

Speed is proportional to $\sqrt{\Delta h}$, where Δh is the change in height.

Active-ating the Physics InfoMall

In addition to the references from the previous activities on speed and graphing, we need information on acceleration for this activity. Again, we can look at the list of hits for the search using "student difficult*" or "student understand*". On that list is "Investigation of student understanding of the concept of acceleration in one dimension," *American Journal of Physics*, vol. 49, issue 3.

This activity uses a ticker-tape timer extensively. You can search the InfoMall using "ticker" and "tape" and "timer" and get a short list of hits, many of which are not very useful. But notice that the book *Teaching Physics: A guide for the Non-specialist* (from the Book Basement) appears a number of times. This book describes the ticker-tape timer, but also points out some problems associated with the timer. Another hit on this list is "Resource Kit for the New Physics Teacher", found in the Pamphlet Parlor. If you look at the table of contents for this pamphlet, you will find that there are sections that you may find useful for other activities as well. One of the parts of this is a worksheet on the "Analysis of a falling body tape." Contained in this is a nice method for converting the tape to a graph of position versus time. Also in this search, you can find that there is an alternative to ticker-tapes.

Physics Lab Experiments and Computer Aids suggests that strobe photographs can be used to replace ticker-tapes. Depending on what equipment you have, you may or may not want an alternative. A little more searching (you may want to broaden your search by using only the words "ticker" and "timer") finds the article "Can pupils learn through their own movement? A study of the use of a motion sensor interface," in *Physics Education*, vol. 26, issue 6. This mentions that the motion sensor is compared favorably with the ticker-tape timers by students.

Planning for the Activity

Time Requirements

• One class period.

Materials Needed

For the class:

- tabletop or counter top surfaces for 1-m horizontal runs of toy cars (one station per group preferred)
- set of four tracks assembled arranged as shown on *Sports* text, page S15 (1 set minimum; 2 sets preferred)

For each group:

- Hot Wheels ™ car (or equivalent)
- ticker-tape timer and supplies
- tape for attaching paper strip of timer to toy car
- meter stick
- calculator

Advance Preparation and Setup

The apparatus intended for this activity is toy Hot Wheels ™ cars and tracks. You may substitute other carts and tracks, but Hot Wheels ™ are preferred because the cars have excellent bearings, and the interlocking, flexible track sections easily may be made into sloped tracks having the shapes specified for this activity.

If you do not have this particular toy, there is high probability that your students, colleagues, or friends would have a collection which you may be able to borrow, or even inherit. Clamping devices for placement of the track are available for the toy sets and are convenient, so accept those, too.

Teaching Notes

It would be worth class discussion of the questions presented in What Do You Think? to detect if any students do not strongly agree that the runner having the highest average speed is the winner of a race. If some students do not recognize this, it is clear they have not internalized the meaning of average speed and need clarification. Hopefully this activity will help them accomplish that.

The blackline cartoon shown on the following page of the Teacher's Edition may be reproduced to help emphasize the concept of average speed. Encourage students who enjoy drawing cartoons to make up their own cartoons to convey the misconceptions of average speed. Also, suggest that the students examine closely the humorous illustrations presented in the *Active Physics* books. Have the students identify which physics concept is being

illustrated. We hope that you and your students will find many of the drawings helpful and informative, as well as amusing.

This activity should convey to students the value of split times for analyzing the performances of runners in races. Split times for runners would be a good basis for discussion among members of the class, the track coach, and members of the track team. Perhaps some students could volunteer to serve as timers for gathering split time data for the school's athletes in practice sessions or at track meets — volunteers for such purposes often are needed and welcomed by coaches.

Unless you are sure that all of the ticker-tape timers operate at the same frequency, you will want each group of students to carry a single timer (and steady power supply) with them from station-to-station to assure that the timer's frequency does not enter as a variable in the activity.

Each group first will make runs of the car, pulling a timer tape, on a horizontal surface. A sufficient number of timers and surfaces to accommodate all groups simultaneously would be ideal (the floor could be used as a surface for this part of the activity).

Groups should rotate among sloped track stations. To prevent need to wait at stations, duplicate sets of tracks could be used.

For the runs down the tracks, you should be aware that need may exist to use a guide for the timer strip at the bottom end of the first sections of Tracks 2 and 4. A smooth, cylindrical object such as a pen could be held in horizontal orientation, crossing above the track with room for the car to pass beneath it. The timer strip then would "feed" under the smooth cylinder without lifting as the car continues on the final section of the track.

Refer to *Active*-ating the Physics Infomall for further suggestions on the use of the ticker-tape timer. An alternate Activity Three, using a photogate, is presented in the pages following this activity in the Teacher's Edition.

"Seventy-five miles per *hour*? Impossible — I only left home five minutes ago."

ACTIVITY THREE: Who Wins the Race?

Activity Three
Who Wins the Race?

WHAT DO YOU THINK?

The fastest human can't go as fast as a car traveling at 25 miles per hour.

Who wins the race?

• **The runner with the highest finishing speed?**

• **The runner with the highest average speed?**

• **The runner with the greatest top speed?**

Take a few minutes to write answers to these questions in your *Active Physics log*. Discuss your answers with your small group to see if you agree or disagree with others. Be prepared to discuss your group's ideas with the entire class.

◆ 13   SPORTS

Activity Overview

In this activity students produce and use a distance versus time record of a toy car moving on a variety of sloped tracks to explore changing speed.

Student Objectives

Students will:

• measure short time intervals in arbitrary units.

• measure distances to the nearest millimeter.

• use a record of an object's position vs. time to calculate the object's average speed during designated position and time intervals.

• measure and describe changes in an object's speed.

• relate changes in the speed of an object traveling on a complex sloped track to the shape of the track.

ANSWERS FOR THE TEACHER ONLY

What Do You Think?

The runner with the highest average speed wins the race.

For You To Do

1. Student activity.

2. a-c) As a result of the horizontal run of the car, students should recognize that equal spacing of the dots on the tape indicates that the car is moving with constant speed. If the students have pulled the cars at a relatively constant speed, the spacing between the dots will be the same all along the tape.

d) If the spacing stays the same, the car is moving at a constant speed. If the spacing varies, the car must be speeding up or slowing down.

e) Student answers will vary.

THE TRACK AND FIELD CHAMPIONSHIP

⚠ Use only the tape provided. Do not substitute other paper.

Start ⟵――――― Motion of tape

⟵1 tick⟶
⟵――2 ticks――⟶
⟵―――3 ticks―――⟶
⟵――――4 ticks――――⟶

FOR YOU TO DO

1. Your teacher will show you how to use a ticker-tape timer to record a toy car's position and the time it takes to move. Thread a piece of paper tape about 1 m long in the timer, and attach one end of the tape to the car. Turn on the timer, and pull the car at a nearly constant speed so that the tape is dragged completely through the timer.

2. Examine the pattern of dots that the timer makes on the tape. Assume that the timer makes dots that are separated by equal amounts of time. The amount of time from one dot to the next will be called a "tick." Obviously, a tick in this case is some small fraction of a second. You will use the tick as a unit of time in this activity and not worry about converting it to seconds.

🖎 a) Do you agree that the distance from one dot to the next on the tape is the distance that the car travelled during one tick of time? Check with your teacher if you have difficulty with this. When you understand this concept, record it in your log.

🖎 b) Do you agree that if you measured the distance from one dot to the next in a unit such as centimeters, you would know the car's speed during that part of the motion in "centimeters per tick"? (Remember, average speed equals distance divided by, or per unit of, time.) When you understand this concept, record it in your log.

🖎 c) Is the spacing between the dots about the same all along the tape, or does the spacing vary?

🖎 d) What would it mean if the spacing stays about the same? If the spacing varies?

🖎 e) Find the part of the tape that shows the beginning of the car's motion where the dots are far enough apart to be seen clearly, and mark one dot as the first dot for analysis. Also locate the last dot made before the tape left the timer at the end of the motion. Measure the distance from the first clear dot to the last dot in centimeters (to the nearest $\frac{1}{10}$ cm, or 1 mm). Also count the number of tick time intervals (the total number of spaces between dots) from the first to the last dot. Record these measurements in your log.

ACTIVITY THREE: Who Wins the Race?

f) Use the data from part (e) to calculate the average speed of the car in "centimeters per tick." Record your work in your log.

3. Make another "run" with the car, using another paper tape, but this time try to pull the car to make it go faster and faster along its path.

a) Compare the pattern of dots on the tape for this run to when you tried to pull the car at constant speed. In your log, explain differences between the two records of motion in terms of position, speed, and time.

b) Choose a section near the middle of the "faster and faster" tape that is 5 tick intervals long (count 5 spaces between dots, not 5 dots) and mark the beginning and end of the 5-tick time interval. Measure the distance between the first and last dots of the interval. Calculate the average speed during the interval in "centimeters per tick." Record your work.

c) How would the average speed compare if you were to measure it for a similar interval earlier in the run? Later? How do you know? Write your answer in your log.

4. Now let's race! Your "runner" is going to race on each of four tracks set up as shown. You are going to use the techniques learned above to record and analyze each race.

Track 1: This is a simple incline 1.00 m long, supported by a meter stick along its length to keep the track from sagging. One end is raised so that h = 0.10 m.

Track 2: This time the incline is 0.40 m long, with a 0.60-m level section for the remaining part. As for track 1, h = 0.10 m.

Track 3: The setup is the same as for track 1, except that h = 0.15 m.

Track 4: Set the first height at 0.20 m. At about 0.50 m along the track, slope the track up so that the second height is 0.10 m.

SPORTS

For You To Do
(continued)

f) Students will use the relationship Speed = distance/time to calculate the average speed along the horizontal track. The unit for speed will be cm/tick.

3.a) Students should recognize that increasing or decreasing spacing between the dots indicates, respectively, that the car is "speeding up" or "slowing down." Some students may volunteer the terms "acceleration" and "deceleration" to the latter cases, and that is fine. You may wish to point out that the formal terms will be introduced in Activity Five. Students should relate the ideas that when the dots are further apart, the car has travelled a greater distance, since the time between the dots is the same, the speed must have increased.

b-c) The average speed will vary along the track. It would be slower at the beginning and faster towards the end.

For You To Do

(continued)

4. a) Accept any reasonable student answer. Have the students record their prediction in their logs before they begin the race.

5. a) For the final speed students will find the speed during the last few ticks at the end of the tape. For top speed, students will calculate the speed in the length of the tape where the ticks are closest together. On Track 1 and 3 the top speed and the final will be the same. Students will find the overall average speed by dividing the total distance travelled along the track by the total amount of time taken (the total number of ticks).

b) Check that students have correctly answered part c) in their logs before they proceed to complete this chart.

c) Neglecting friction (an unrealistic ideal in this case):

Tracks 1 and 2 should produce equal final speeds. Friction along the final horizontal section of Track 2 probably will cause a lower final speed for that track.

Tracks 3 and 4 should produce equal final speeds, higher than the final speeds for Tracks 1 and 2.

The highest top speed should be reached on Track 4 at the bottom of the first half of the Track, followed by Track 3 at the end of the run, where the top speed should equal the final speed. Tracks 1 and 2 should produce equal top speed, equal to the final speeds.

The highest average speed should be found for Track 4, followed by, in decreasing order of average speed, Tracks 3, 2 and 1.

THE TRACK AND FIELD CHAMPIONSHIP

a) Predict which track will produce the winning result if your car is released and allowed to run 1 m along each track, starting from the top. Record your prediction in your log.

5. Run your toy car on each track. In each case:

Allow the force of gravity to do the pulling by simply releasing the car at the start of the run.

Have the car pull at least 1 m of tape through the timer. This may require adding a "leader" to the 1-m tape length to allow for the distance between the car and the timer at the start of the race.

At the beginning of each run, the timer should be started and then the car should be released.

After each run, mark the track number on the tape, mark the first clear dot made at the beginning of the run, and place a mark on the tape 1 m (100 cm) beyond the first clear dot.

a) For each race, explain in your log how you will analyze the tapes to measure the final speed, the top speed, and the overall average speed in centimeters per tick.

b) Make the necessary measurements, and do the calculations to fill in the speed values in a table similar to the one below.

Track Number	Final Speed	Top Speed	Average Speed
1			
2			
3			
4			

c) Which track produced the winning run in the big race? How can you tell? Explain your answer in your log.

REFLECTING ON THE ACTIVITY AND THE CHALLENGE

Now you can see that it is the details of what happens during a race that determines who wins. The distance of a race and the time taken to run it do not reveal what a champion does along the way to win races consistently.

ACTIVITY THREE: Who Wins the Race?

Speed within most races varies. The sprinters who get up to top speed quickly and maintain their speed throughout a race often win. Those who start quickly and "fade" at the end of the race often lose.

Helping athletes at your school analyze their performances in terms of speed during parts of a race will be needed if they are to compete with the best runners.

PHYSICS TO GO

1. Describe a procedure that you could use to convert one "tick" of the timer used in this activity into seconds of time. How could you find out how many "ticks" equal one second?

2. What would the spacing of dots look like for a ticker-tape timer record of an object that is slowing down in its motion?

3. From what you observed and measured during this activity, describe how the speed of a toy car behaves as it travels:
   a) on a straight ramp that slopes downward.
   b) on a level surface when the car already has some speed at the beginning.
   c) on a straight ramp that slopes upward.

4. Aisha and Bert are running at constant speeds, Aisha at 9.0 m/s and Bert at 8.5 m/s. They both cross a "starting line" at the same time. The "finish line" is 100 m away.
   a) How long does it take Aisha to finish the race?
   b) How long does it take Bert to finish the race?
   c) Where is Bert when Aisha crosses the finish line?
   d) By how many meters does Aisha finish ahead of Bert?

5. The Penn Relays women's high-school record for the 1,500-m run is 4 min, 24.0 s. The women's high-school record for the mile (1,609 m) run at the Penn Relays is 4 min, 49.2 s. In which race did the record holder have the greatest average speed in meters/second?

6. Salina runs the 200-m race for the school's track team. She runs the first 100 m at 9.0 m/s. Then she hears her classmates cheer, "GO, Salina, GO!" and runs the final 100 m at 10.0 m/s.
   a) Calculate the time for Salina to run the first 100 m.
   b) Calculate the time for Salina to run the final 100 m.
   c) Calculate Salina's average speed for the entire race.

S 17

Physics To Go

1. A wall clock or wristwatch could be used to measure seconds of time as a tape is pulled through a ticker-tape timer. The number of dot intervals created during, for example, 10 s could be used to calculate how many dots are made by the timer per second. This process is called calibration.

2. An object slowing down would produce a pattern of dots which would have the distance from one dot to the next decreasing along the tape.

3. a) The speed increases steadily.

   b) The speed is constant, perhaps slowing down gradually due to friction.

   c) The speed decreases steadily.

4. a) 100 m / 9.0 m/s = 11.1 s

   b) 100 m / 8.5 m/s = 11.8 s
   c) (Bert's speed) x (Aisha's time)
      = 8.5 m/s x 11.1 s = 94.3 m
      from start line.

   d) 100 m - 94.3 m = 5.7 m

5. 1,500 m / 264.0 s = 5.681 m/s
   1.609 m / 289.2 s = 5.564 m/s
   The 1,500 m runner had the highest average speed.

6. a) 100 m / 9.0 m/s = 11.1 s

   b) 100 m / 10.0 m/s = 10.0 s

   c) 200 m / (11.1 s + 10.0 s)
      = 200 m / 21.1 s = 9.48 m/s

1

Activity Three A

Who Wins the Race? [Using a Photogate]

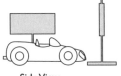

Side View   Front View

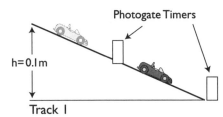

Photogate Timers

h=0.1m

Track 1

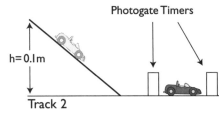

Photogate Timers

h=0.1m

Track 2

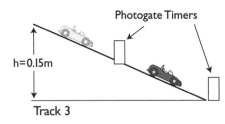

Photogate Timers

h=0.15m

Track 3

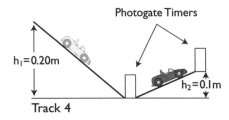

Photogate Timers

h₁=0.20m

h₂=0.1m

Track 4

FOR YOU TO DO

1. The speed of a car at a particular point on its track can be measured using a photogate timer. A photogate timer consists of a small light source and a light detector. When the light beam is interrupted, the detector can signal a timing device to start. When the light beam reaches the detector again, it can turn the timer off.

   A piece of cardboard 5.0 cm long attached to the top of a car will be used to interrupt the light beam as the car goes through the photogate.

   Set up a piece of track and the photogate, and try out the operation of the timing system. Be sure only the cardboard interrupts the photogate beam. Give the car a gentle push to give it a small speed.

   a) Record the time for the 5.0 cm card to pass through the photogate system.

   b) Divide the length of the piece of cardboard (0.050 m) by the time interval reading from the photogate to calculate a value for the speed of the car as it passed through the photogate. Record this value.

2. Reset the system. Give the car a higher speed.

   a) Record the time and distance values, and calculate the speed.

   b) Does the calculation also show that the speed was higher this time? If not, review your procedures and find out what went wrong.

3. The speeds of cars at various points along four different one-meter tracks will be measured. Set up four tracks as shown at left.

   a) Predict which track will produce the winning result if your car is released and allowed to run 1 m along each track, starting from the top. Will it be the track that produces the highest final speed? Will it be the track that produces the highest top speed? Record your prediction in your log.

4. Run the race.

   a) Record your results in the tables on the next page.

   b) Based on the results of your race, write a statement informing the track team about who wins the race and why. Include any hints you have for improving their performances for running.

For use with *Sports*, Chapter 1, ACTIVITY THREE: Who Wins the Race?

©1998 American Association of Physics Teachers

One-Meter Sprints

TRACK	MID GATE TIME (SECONDS)	END GATE TIME (SECONDS)	DISTANCE (METERS)
1			0.050
2			0.050
3			0.050
4			0.050

TRACK	MID GATE SPEED (M/S)	END GATE SPEED (M/S)
1		
2		
3		
4		

Compiling the Results and Making Predictions

TRACK	TOP SPEED (M/S)	FINAL SPEED (M/S)	PLACE PREDICTIONS
1			
2			
3			
4			

TRACK	BEST TOP?	BEST FINAL?	BEST AVERAGE?	WINNER
1				
2				
3				
4				

For use with *Sports*, Chapter 1, ACTIVITY THREE: Who Wins the Race?

ACTIVITY FOUR
Understanding the Sprint

Background Information

Instantaneous speed is introduced in this activity as the slope of a distance versus time graph at a particular instant; this is approached only semi-quantitatively.

Prior to introduction of instantaneous speed, a histogram is used to allow comparison of the average speed of a sprinter during each 10-m interval of a 100-m dash. Be careful how you interpret the histogram; it is tempting to think of the average speed during each 10-m interval as the speed at a particular instant, such as the beginning of the interval — this is not true. The data allows calculation of only the average speed during each 10-m interval, and the particular instant(s) at which the average speed value may have occurred as the instantaneous speed during the interval cannot be determined from the histogram.

To satisfy yourself that you understand that the slope of the distance versus time graph at a particular instant gives the instantaneous speed at that instant, you may wish to measure the slope of the graph at one or more points. To do so at a particular time, such as 3 s:

- Identify the point on the curve corresponding to $t = 3$ s.
- Use a straightedge to estimate the orientation of a line tangent to (having the same direction as) the curve at $t = 3$ s, and draw the tangent.
- Mark two points on the tangent line (recommended: points corresponding to $t = 2$ s, $t = 4$ s, which establishes the horizontal "run" of the slope as 4 s - 2 s = 2 s.
- From each of the two marked points on the tangent line, draw a horizontal line to intersect the distance axis. The difference in the two intersections establishes the vertical "rise" of the slope (approximately 30 m - 10 m = 20 m in this example).
- Calculate the instantaneous speed at $t = 3$ s as the "rise divided by the run," approximately 20 m / 2 s = 10 m/s.

You may wish to use the same method to measure the instantaneous speed at another, earlier time when the slope obviously is different.

Taking the slope of a curve is another task that can be accomplished much easier, and more accurately, using a computer or graphing calculator.

Active-ating the Physics InfoMall

Do a search with keyword "sprint*". You will get a great list of references, including "Effect of wind and altitude on record performance in foot races, pole vault, and long jump," *American Journal of Physics,* vol. 53, issue 8, and "How Olympic records depend on location," *American Journal of Physics,* vol. 54, issue 6. Also in this list of hits is Chapter 2 Describing Motion in *The Fascination of Physics* from the Textbook Trove. This has an informative discussion of speed, velocity, and sprinting, including acceleration. A glance through this chapter finds more about strobe photographs. (This technique may be deserving of its own search on the InfoMall.) If you look at the questions at the end of this chapter in the textbook, you will find that question C9 compares distance runners with sprinters. If you click on the "answer" link, you will get the answer to the question, which can be found in the *Instructor's Guide for The Fascination of Physics* in the Study Guide Store. This search also finds some problems from the Problem Place that you may wish to check out and use for your class. You are urged to do this search, and others, yourself to find how much information is available on the InfoMall.

Planning for the Activity

Time Requirements

- One class period.

Materials Needed

For each student:
- sheet of graph paper, any grid size

For each lab group:
- calculator

Advance Preparation and Setup

No significant advance preparation is required for this activity. If you need to construct accelerometers for the next activity, you may wish to get started now. (See the Materials Needed for Activity Five.)

Teaching Notes

You may wish to have students work in their groups, but each student should produce and analyze the histogram and graph as directed in For You To Do.

You may wish to provide the students with a copy of the graph from S19, found on page 49, to complete.

Encourage students to get involved with the school's track coach and members of the track team. Suggest specific proposals that students could make which would be of mutual benefit to students for the chapter challenge and to the team for improved performances.

Activity Overview

In this activity students create and then compare a distance versus time graph and a bar graph of speed versus time, to analyze in detail the speed of a runner during a race.

Student Objectives

Students will:

- calculate the average speed of a runner given distance and time.
- produce a histogram showing the average speeds of a runner during segments of a race; analyze changes in the runner's speed.
- produce a graph of distance versus time from split time data for a runner.
- estimate the slope of a distance versus time graph at specified times.
- recognize that the slope of a distance versus time graph at a particular time represents the speed at that time.

ANSWERS FOR THE TEACHER ONLY

What Do You Think?

In his world record 100-m dash, Carl Lewis reached his top speed about halfway into the race in terms of both time and distance. World class sprinters run fastest during the last part of a race; the first part of the race involves acceleration from rest and, therefore, lower average speed.

THE TRACK AND FIELD CHAMPIONSHIP

Activity Four
Understanding the Sprint

WHAT DO YOU THINK?

It was not believed to be humanly possible to run a mile in less than four minutes until Roger Bannister of England did it in 1954.

- **How much time does it take to get "up to speed"?**

Take a few minutes to write an answer to this question in your *Active Physics log*. Discuss your answer with your small group to see if you agree or disagree. Be prepared to discuss your group's ideas with the class.

FOR YOU TO DO

1. Carl Lewis established a world record for the 100-m dash at the World Track and Field Championships held in Tokyo, Japan, in 1991. The times at which he reached various distances in the race (his "split times," or "splits") are shown in the table below.

Distance (meters)	0.0	10.0	20.0	30.0	40.0	50.0	60.0	70.0	80.0	90.0	100.0
Time (seconds)	0.00	1.88	2.96	3.88	4.77	5.61	6.45	7.29	8.13	9.00	9.86

SPORTS　　　 18

ANSWERS

For You To Do

1. a - b) (see chart)

Distance (m)	Time Interval (s)	Average Speed (m/s)
0.0 - 10.0	1.88	5.3
10.0 - 20.0	1.08	9.3
20.0 - 30.0	0.92	10.9
30.0 - 40.0	0.89	11.2
40.0 - 50.0	0.84	11.9
50.0 - 60.0	0.84	11.9
60.0 - 70.0	0.84	11.9
70.0 - 80.0	0.84	11.9
80.0 - 90.0	0.87	11.5
90.0 - 100.0	0.86	11.6

ACTIVITY FOUR: Understanding The Sprint

❧ a) In your log, copy and complete the table to the right. Use subtraction to calculate the time taken by Carl Lewis to run each 10 m of distance during the race.

❧ b) Calculate Lewis's average speed during each 10 m of the race. The values of the time interval and the average speed have been entered in the table for the first 10 m of the dash.

Distance (meters)	Time Interval (seconds)	Average Speed (meters/second)
0.0–10.0	1.88	5.32
10.0–20.0		
80.0–90.0		
90.0–100.0		

2. Use the data you created for the above table to make a bar graph to give you a visual display of Carl Lewis's average speed during each 10 m of his world-record 100-m dash. Use a piece of graph paper set up as shown to the right.

❧ a) Tape or copy the bar graph in your log.

3. Analyze the bar graph to answer these questions:

❧ a) At what position in the dash did Lewis reach top speed? How close can you state that position to the nearest meter? to the nearest 10 m? Explain your answer.

❧ b) How well did Carl Lewis keep his top speed once he reached it? Did he seem to be getting tired at the end of the race? Give evidence for your answers.

❧ c) Can you tell how fast Lewis was going at an exact position in the race, such as at 15.0 m or 20.0 m? Why or why not?

❧ d) It took 9.86 s for Lewis to run the entire 100 m. Calculate his overall average speed. Draw a horizontal line across the bar graph at an appropriate height to represent the average speed for the entire race. Compare the height of each bar on the graph with the height of the line. Explain what the comparisons mean.

4. Use the "splits" given at the start of this activity to plot a graph of Carl Lewis's position versus time.

❧ a) On a piece of graph paper, make a vertical distance scale from 0 to 100 m and a horizontal time scale from 0 to 10 s. Plot each position at the appropriate time and connect the points to show what you think is the shape of the graph. Tape or copy the graph in your log.

5. Compare the distance versus time graph with the bar graph of speed versus distance.

Carl Lewis's World Record 100-m Dash Average Speed, 10-m Intervals

(bar graph with vertical axis Average Speed (m/s) 0–12, horizontal axis Distance (m) 0-10 10-20 20-30 30-40 40-50 50-60 60-70 70-80 80-90 90-100)

◆ 19   SPORTS

For You To Do
(continued)

2. a)

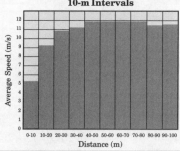

Carl Lewis's World Record 100-m Dash Average Speed, 10-m Intervals

3. a) Lewis reached his top average speed per interval, 11.9 m/s, at the interval 40-50 m. The refinement of the histogram allows "resolution" only to the 10-m interval.

b) Lewis's speed "faded" only slightly during the final two 10-m intervals.

c) No. The histogram is an assembly of average speeds during 10-m intervals; it does not give information about the speed at exact positions along the 100 path.

d) Lewis's overall average speed for the dash was 100-m / 9.86 or s = 10.1 m/s. His average speed during the first interval was only about half of the overall average speed. At the second interval he had almost reached the overall average speed, and thereafter he exceeded the overall average speed.

4. a)

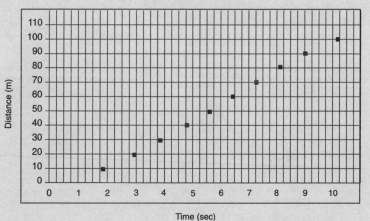

Carl Lewis's World Record for the 100-m Distance vs. Time

THE TRACK AND FIELD CHAMPIONSHIP

a) When the distance versus time graph is curving early in the run, do the bars on the bar graph change in height or are they fairly steady in height? What does this comparison mean? When the graph is climbing in a straight line, what is happening to the heights of the bars on the bar graph? What does this comparison mean?

b) Someone said, "The slope, or steepness, of a distance versus time graph at any instant is the speed at that instant." Do you believe this statement? Why or why not?

c) Describe the slope, or steepness, of the distance versus time graph each second (1.00 s, 2.00 s, 3.00 s, and so on) during Lewis's record run. You may use the word *low* to describe the slope at the first second into the race.

REFLECTING ON THE ACTIVITY AND THE CHALLENGE

In this activity you saw two different ways to analyze in detail the speed of a runner during a race. If you had split-time information for runners in sprint events for your school's track team, you could help the runners find out, for example, if they are "letting up" at the end of a race or how rapidly they are reaching top speed.

With more knowledge about details of their performances, your school's runners may find that they can improve parts of their races.

PHYSICS TALK

$$\text{Speed} = \frac{\text{Distance traveled}}{\text{Time elapsed}}$$

On a distance versus time graph, the speed is equal to the slope of the graph.

ANSWERS

For You To Do *(continued)*

5.a) When the graph is curving early in the run (increasing in slope), the bars of the histogram at corresponding positions and times are increasing in height; both representations indicate that Lewis was accelerating.

When the graph is climbing on about a straight line, the bars of the histogram at corresponding positions and times are of fairly steady height; both representations indicate that Lewis had a fairly constant speed.

b) Formally (not expected of students here), the slope of the distance versus time graph at a particular instant gives the "instantaneous speed."

c) With some latitude for individual differences in visual impression, a sequence would be: 1 s : low, 2 s : higher, 3 s : even higher, 4 to 10 s : highest (about the same at all times 4 through 10 seconds).

ACTIVITY FOUR: Understanding The Sprint

PHYSICS TO GO

1. If you were to design a track for a toy car to run as in Activity 3 to simulate Carl Lewis's 100-m dash, what shape would you design for the track?

2. For long distances, humans can run at a constant speed of about 6 m/s, pigs at about 4 m/s, and horses at 20 m/s. Sketch a distance-versus-time graph with three lines showing a person, a pig, and a horse running for 100 s.

3. Sketch a distance versus time graph for a person who is not moving at all.

4. In a 2 × 100 m relay race, Joan ran the first 100 m at a speed of 5 m/s and then Rami ran the next 100 m at a speed of 10 m/s. What was the average speed for the entire relay race? (Hint: It is not 7.5 m/s.)

5. Do you think it is possible for a runner to keep increasing speed for an entire race as a strategy to win?

6. Examine what excellent runners do in long-distance races. On the right is a chart of Eamonn Coghlan's split time and total time every 200 m in a mile (1,609 m) race.

Distance (m)	Split Time (s)	Total Time (s)
0	0.00	0.00
200	29.23	29.23
400	29.87	59.10
600	30.09	89.19
800	30.25	119.44
1,000	29.88	149.32
1,200	29.90	179.22
1,400	29.38	208.60
1,600	29.38	238.15

a) Use the total times listed to calculate Coghlan's average speed for distances of 200, 400, 800, and 1,000 m distances during his run.

b) Compare Coghlan's average speeds for distances of 200, 400, 800, and 1,000 m with the average speeds of world-record holders at the same distances. (World-record average speeds for various distances are listed in Activity 1.) What patterns do you see in the comparison? At what distances was Coghlan's speed getting closest to world-record speed?

c) Use Coghlan's split times to identify the distance interval (0–200 m, 200–400 m, 400–600 m, and so on) when he had the greatest average speed. Also identify the interval of lowest average speed.

d) You found out in this activity that Carl Lewis slowed down slightly near the end of his record 100-m dash. Did Eamonn Coghlan do the same thing near the end of his mile run? Use data in your answer.

e) How could you use data about Coghlan's performance to give advice to members of your school's track team who enter long-distance events?

S 21

ANSWERS

Physics To Go

1. The track would have a downward slope for about the first half of the distance and a flat section for about the last half - the last section could have a slight upward slope near the very end of the track.

2.

3. For zero speed, the distance versus time graph would correspond to the time axis.

4. Joan: 100 m / (5 m/s) = 20 s

   Rami: 100 m / (10 m/s) = 10 s

   Average speed = 200 m / 30 s
   = 6.7 m/s

   Be alert to a misconception: taking the "average" of average speeds, in this case 15 m/s, is not valid because the lower speed in this case was maintained for more time than the higher speed.

Some students may find this tempting, but it is not a valid method.

5. Maintaining a significant acceleration for a significant amount of time is hardly possible.

6.a) 200 m : 6.84 m/s
     400 m : 6.76 m/s
     800 m : 6.70 m/s
     1,000 m : 6.70 m/s

   b) All of Coghlan's average speeds are less than world record speeds for corresponding distances; however, a clear pattern exists which is that Coghlan's speeds get closer to world record average speeds as the distance increases (his speed up to 1,000 m is 6.7 m/s; the world record average speed for that distance is 7.55 m/s).

   c) Coghlan had the greatest average speed when his split time was lowest, during the interval 0-200 m; he had the least speed when his split time was greatest, during the interval 600-800 m.

   d) No, Coghlan did not slow down near the end of the race; indeed, he accelerated, as shown by the decreased splits for the last two 200-m intervals.

   e) One piece of useful advice which could be derived from Coghlan's race would be to save some energy for a "kick" of acceleration near the end of the race.

Carl Lewis's World Record 100-m Dash Average Speed, 10-m Intervals

Distance (m)	Time Interval (s)	Average Speed (m/s)
0.0 - 10.0	1.88	5.3
10.0 - 20.0		
20.0 - 30.0		
30.0 - 40.0		
40.0 - 50.0		
50.0 - 60.0		
60.0 - 70.0		
70.0 - 80.0		
80.0 - 90.0		
90.0 - 100.0		

NOTES

ACTIVITY FIVE
Acceleration

Background Information

$a = \Delta v / \Delta t$ where Δv is the change in speed accomplished in the time interval Δt.

The standard unit of acceleration is m/s^2. If this unit is mysterious to you, see discussion of the unit in the For You To Read section of the student text for this activity.

It is possible that you will encounter acceleration expressed in nonstandard units such as "feet per second squared" (the acceleration due to gravity in the British system of units is 32 ft/s^2; some students may have heard about this number and may bring it up in discussion). Another nonstandard unit of acceleration used in this country for automobiles, with which some students may be familiar, is "miles per hour per second."

Regardless of the unit used to express acceleration, the meaning is the change in speed per unit of time.

Strictly speaking, the above definition applies to "average acceleration." The acceleration of runners in track events changes with time. For example, a runner in a 100-m dash typically begins a race with high acceleration, and then the acceleration decreases, falling to zero when the runner reaches and maintains a fairly constant "top speed." If the overall change in speed, Δv, during the early part of the race is the difference between zero speed (at the start) and the top speed, the average acceleration is Δv divided by the amount of time, Δt, taken to reach top speed. The runner's "instantaneous acceleration" (the acceleration at a particular instant) would at some times be less than the average acceleration and at other times more than the average acceleration. Just as average speed may or may not be representative of the detailed behavior of speed during a journey, average acceleration may or may not be representative of detailed variations in speed.

Do not assume that all cases of acceleration involve constant acceleration (a steady, constant rate of change of speed). Constant acceleration, also referred to as uniform acceleration, happens only under special conditions, including, for example, free fall and circular motion at constant speed.

Notice that the symbol "v" is used for speed in the definition of acceleration. This is because the term "velocity" is used in place of the term "speed" in more advanced treatments of motion where a directional property is included to treat velocity as a vector quantity. To avoid confusion for students who may study physics at a more advanced level in the future, it would be best for you to use the word "speed" and the symbol "v" and to avoid using the word "velocity."

Active-ating the Physics InfoMall

You may wish to search the InfoMall with keywords "acceleration" and "misconception*". This generates a slightly different list than the search done earlier (you may find more information here that you can use in Activities One-Four). One interesting hit is "Physics that textbook writers usually get wrong," in *The Physics Teacher*, vol. 30, issue 7. A passage from this reference says that "Deceleration means a decrease in speed, so if the velocity is positive, a negative acceleration is a decrease in velocity, and therefore a deceleration. A ball thrown upward ($v > 0$) has an acceleration downward ($a < 0$) and therefore decelerates. However, the same ball, as it falls ($v < 0$), is subject to the same negative acceleration (a is downward), and therefore gains speed; it is not decelerated." This is often a cause of confusion with students. (See the Physics To Go questions, especially 4 and 5.)

You may also find "Common sense concepts about motion," *American Journal of Physics,* vol. 53, issue 11, to be interesting. Perform a search using the keyword "accelerometer" for some information regarding these devices. Note that Chapter 6 of *Teaching High School Physics* (in the Book Basement) has a nice graphic of a simple cork accelerometer you can easily build yourself. Don't forget that we found some good references for acceleration earlier, including "Investigation of student understanding of the concept of acceleration in one dimension," *American Journal of Physics,* vol. 49, issue 3.

Planning for the Activity

Time Requirements

• One class period.

Materials Needed

For each group:

- cork accelerometer (1 per group minimum)
- cart rigged to be pulled by falling weights (see diagram of cart, string, pulley, weight in student text)
- device for attaching accelerometer to cart
- weights to provide reasonable accelerations of cart (2 different amounts of weight to provide two reasonable, contrasting accelerations)

Advance Preparation and Setup

If you do not have cork accelerometers, you may wish to construct one per group using the directions given at the end of this activity.

Other kinds of accelerometers may be substituted, but doing so will require that you alter the directions to students in For You To Do to comply with the behavior of the kind of accelerometer used.

Teaching Notes

Expect that some students will have difficulty grasping the meaning of acceleration. It is a "rate of a rate"— a compound abstraction that can be expected to present intellectual difficulty for young minds (and some old minds). Be patient, and hope that the variety of experiences which students will have involving acceleration will allow each to come to understand it.

Similar to the definition of acceleration, some students can be expected to have difficulty understanding the unit of acceleration. It is strongly suggested that you check their understanding of the mathematical steps involved in arriving at meters per second squared as the "short" version of the unit.

Activity Overview

In this activity students are introduced to acceleration using a simple accelerometer.

Student Objectives

Students will:

- understand the definition of acceleration.
- understand meters per second per second as the unit of acceleration.
- use an accelerometer to detect acceleration.
- use an accelerometer to make semi-quantitative comparisons of accelerations
- distinguish between acceleration and deceleration.

ANSWERS FOR THE TEACHER ONLY

What Do You Think?

Many races are won during the period of acceleration at the beginning of the race.

The runner that is out of the blocks first, that is, the one who has the quickest response time after the starting pistol is fired, has an advantage. All other things being equal, that runner will win the race.
This is particularly important in short distance races.

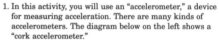

Activity Five
Acceleration

WHAT DO YOU THINK?

Accelerating out of the starting blocks is important if a runner is going to win a race.

- **If all runners in a dash have equal top speeds and none "fades" at the end of the race, what determines who wins?**
- **How is response time a factor in determining who wins the race?**

Write your answers to these questions in your *Active Physics log*. Be prepared to discuss your ideas with your small group and other members of your class.

FOR YOU TO DO

1. In this activity, you will use an "accelerometer," a device for measuring acceleration. There are many kinds of accelerometers. The diagram below on the left shows a "cork accelerometer."

Explore how the accelerometer works by holding it in your hands in front of you so that you can look down to see the top of the cork through the bottle and liquid. Hold the accelerometer upright so that the cork is centered in the bottle.

2. Take the accelerometer for a walk. Observe any movement of the cork, especially as you start from a resting position and speed up (accelerate). Walk at a fairly constant speed, and then slow down to a stop (decelerate). Try it a few times—starting, walking, and stopping at normal rates.

   a) Record your observations of the cork's movement at each stage in your log.

3. Repeat the above walk and observe what happens if you start faster, if you walk faster at a constant speed, and if you stop faster.

   a) Record your observations in your log.

⚠ **Use only plastic bottles to construct accelerometers.**

A cork attached to a string floats in a liquid-filled bottle.

SPORTS ———————————————— ◆ S 22 ◆

ANSWERS

For You To Do

1. Student activity.

2. a) While exploring the accelerometer, students should discover that the cork "leans" in the direction of acceleration, and, the greater the acceleration, the greater the amount of lean of the cork.

3. a) The greater the acceleration, the greater the amount of lean of the cork.

4. Repeat the walk in step 2, but walk backward.

   a) Record your observations in your log.

5. Use your observations to answer the following questions in your log.

   Describe the *amount* and the *direction* the cork leans in each of the following situations:

   a) standing at rest
   b) low acceleration while walking forward
   c) high acceleration while walking forward
   d) low constant speed while walking forward
   e) high constant speed while walking forward
   f) high deceleration (slowing down) while walking forward
   g) low deceleration while walking forward

6. Someone said, "Deceleration while walking forward is the same as acceleration while walking backward."

   a) Do you agree or disagree? Use your observations of the accelerometer for your answer.

7. Set up a system to take the accelerometer for a ride on a cart that is being pulled by a falling weight, as shown in the diagram to the right. Hang a weight on the string and allow it to pull the cart as you observe the accelerometer. Record your answers to the questions below in your log.

   a) Does the cart appear to accelerate? Does the accelerometer tell you that it's accelerating? How can you tell?

   b) Does the accelerometer show that the acceleration is constant or changing? How can you tell?

    Keep area under the falling mass clear.

8. Repeat step 7 using a larger weight to pull the cart. Answer the following questions in your log.

   a) How does the acceleration for the large weight compare with the acceleration for the small weight? How can you tell?

   b) Which produced a more steady, constant acceleration: using the falling weights or walking with the accelerometer? What evidence do you have for your answer?

$\diamond$ 23   **SPORTS**

ANSWERS

For You To Do

(continued)

4. a) The cork leans in the direction of the acceleration. In this case, rearward.

5. a) Standing at rest; Zero lean.

   b) Low acceleration while walking forward; Low lean, forward direction.

   c) High acceleration while walking forward; High lean, forward direction.

   d) Low constant speed while walking forward; Zero lean.

   e) High constant speed while walking forward; Zero lean.

   f) High deceleration (slowing down) while walking forward; High lean, rearward direction.

   g) Low deceleration while walking forward; Low lean, rearward direction.

6. a) Yes. In both cases the cork leaned rearward.

7. a) The accelerometer indicates that the cart is accelerating because there is a lean.

   b) The accelerometer shows a lean of a fixed amount during each run.

8. a) The amount of lean should be greater for the greater acceleration provided by the larger weight.

   b) Students probably discovered earlier that it is very difficult to walk with constant acceleration; gravity does a much better job of providing constant acceleration.

PHYSICS TALK

Acceleration

The relationship between acceleration, speed, and time can be written as:

$$\text{Acceleration} = \frac{\text{Change in speed}}{\text{Time interval}}$$

Using symbols, the same relationship can be written as:

$$a = \frac{\Delta v}{\Delta t}$$

where

a = acceleration,

Δv = change in speed, and

Δt = the time interval, or change in time, for the change in speed to happen.

Example:

A sprinter at the start of a race increases speed from 0 m/s to 5.0 m/s as the clock runs from 0 s to 2.0 s. Find the acceleration.

$$\text{Acceleration} = \frac{\text{Change in speed}}{\text{Time interval}}$$

$$= \frac{5.0 \text{ m/s} - 0 \text{ m/s}}{2.0 \text{ s} - 0 \text{ s}}$$

$$= \frac{5.0 \text{ m/s}}{2.0 \text{ s}}$$

$$= 2.5 \text{ (m/s)/s}$$

Mathematically, meters per second per second is equal to meters per second squared. Therefore, when you use "m" and "s" as abbreviations for meters and seconds, you can shorten the unit of acceleration to m/s². All of the following ways of stating the unit of acceleration are the same:

(m/s)/s   m/s/s   m/s²

When speaking about acceleration, you can describe the unit as "meters per second squared" or "meters per second every second."

REFLECTING ON THE ACTIVITY AND THE CHALLENGE

You now know a lot more about acceleration. Since a race always begins with a speed of zero and ends with runners in motion, acceleration is part of every race. Depending on the distance of a race, acceleration may go up, go down, or disappear several times during the race. One runner may have more acceleration than another runner at the start of a race, or one runner's acceleration may change during a race. Do the athletes on your school's track team know about acceleration? If not, maybe they should, and perhaps you can help them.

PHYSICS TO GO

1. Is there anything in nature that has constant acceleration?

2. If Carl Lewis were to carry a cork accelerometer during the start of a sprint, describe what the accelerometer would do.

3. Did Carl Lewis accelerate for the entire 100 m of his world-record dash? Explain his pattern of accelerations.

4. If you are running, getting tired, and slowing down, are you accelerating? Explain your answer.

5. When you throw a ball straight up in the air, what is the direction of its acceleration while it is going up? While it is coming down?

6. What additional tips could you give your school's track team as a result of this activity?

INQUIRY INVESTIGATION

Repeat the activity in which the cart is pulled by falling weights so that you can get detailed information about the cart's position and/or speed at several points during its motion. Try to get numerical data to find out whether or not the cart has constant acceleration. If the acceleration is constant, measure the cart's acceleration.

 S 25

Physics to Go

1. Objects in a state of free fall have constant acceleration due to gravity.

2. In principle, the cork would lean forward, but, in reality, it would waver wildly if carried by Lewis.

3. Carl Lewis accelerated during approximately the first half of his record dash. He also had minor deceleration and acceleration near the end of the dash.

4. Slowing down, also called deceleration, can be thought of as negative acceleration; an amount of speed is being subtracted per unit of time.

5. At all times during which a ball is in the air, the direction of its acceleration is downward.

6. Advice to sprinters: (a) accelerate to top speed as soon as possible, (b) maintain top speed to the end of the race.

Constructing a Cork Accelerometer

A cork accelerometer can be constructed from a jar having a tight-fitting lid, a cork on a string, and liquid. Some provision such as glue must be made for attaching the end of the string to the inside of the jar lid.

A plastic jar and lid is essential. Do not use a glass jar due to near certainty of breakage. A pint-size jar works well, and you should select one which has a relatively flat bottom (not depressed in the molding process) so that the cork can be seen through the bottom of the jar when the system is inverted (lid side down) for use.

A thicker-than-water liquid such as vegetable oil is preferred to slow down the response of the cork to acceleration.

Any cork which will float in the liquid will work; brightly colored fishing bobs work very well and have provision for attaching the string.

Modern multipurpose glue of the "Goop" variety works well to attach the end of the string to the inside of the jar lid.

The lid can be glued permanently to the jar to deter the student who can't resist the temptation of opening the jar.

It adds convenience to glue the jar lid to a piece of wood slightly larger than the lid so that the accelerometer can be clamped to a cart to prevent "flying accelerometers." Alternatively, the accelerometers may be placed in a heavy base such as a coffee can filled with sand, to prevent the accelerometer from flying off.

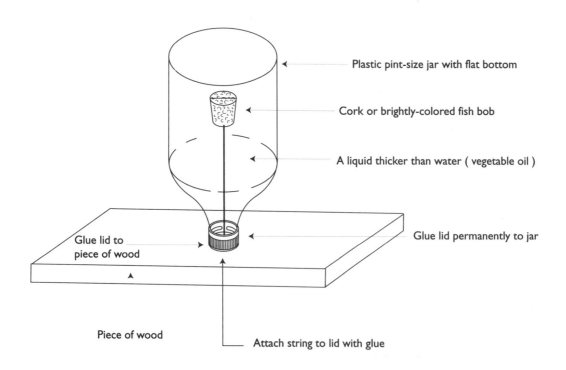

Plastic pint-size jar with flat bottom

Cork or brightly-colored fish bob

A liquid thicker than water (vegetable oil)

Glue lid to piece of wood

Glue lid permanently to jar

Piece of wood

Attach string to lid with glue

For use with *Sports*, Chapter 1, ACTIVITY FIVE: Acceleration
©1999 American Association of Physics Teachers

ACTIVITY FIVE A
Investigating Different Types of Accelerometers

Bubble level at rest on a
horizontal surface with the
bubble in the center.

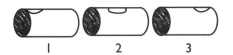

FOR YOU TO DO

1. A device which detects acceleration is an accelerometer. It can respond to changes in speed. One kind of accelerometer is made by enclosing a liquid in a container. A tubular bubble level fits that description, so it can be used as an accelerometer. When the bubble level is at rest on a horizontal surface, the bubble is in the middle. Try it.

2. Place the bubble level on a free-wheeling cart that is at rest.

3. Stick the level to the cart with clear tape or plastic clay. If necessary, adjust the level so that the bubble is in the middle when the cart is at rest.

4. Now pull the cart along the table so that the bubble stays in the middle after you get the cart into motion.

   a) Describe the kind of motion that keeps the bubble in the middle. Does it make any difference whether the motion is slow or fast?

   b) Which of the diagrams best describes the position of the bubble in the following situations:

   • As the cart is just put into motion to the right?
   • As the cart is brought to an abrupt halt when it is moving to the right?
   • As the cart is brought to an abrupt halt when it is moving to the left?

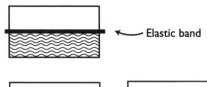

← Elastic band

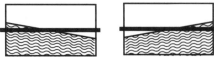

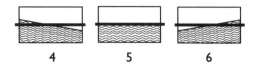

4   5   6

5. Another kind of accelerometer is made from two sheets of clear plastic with a thin space between them filled with colored water. An elastic band can be used as a reference to mark the position of the water surface when the accelerometer is a rest on a level surface.

6. When this kind of accelerometer indicates acceleration, the water forms half an arrowhead. Try it on a cart.

a) Does this "arrow" point in the direction of the acceleration or in the opposite direction?

b) Compare the behavior of this accelerometer with the behavior of the bubble level accelerometer.

c) Which bubble level accelerometer diagram (1, 2 or 3) is equivalent to each of these accelerometers (4, 5 or 6)?

7. An extremely simple pocket accelerometer can be constructed from a piece of string and a steel washer (one among many possibilities).

a) Match the behavior of this kind of accelerometer (7, 8, 9) with the other two kinds of accelerometers (1, 2, 3) and (4, 5, 6). Which string and washer is equivalent to which bubble level diagram?

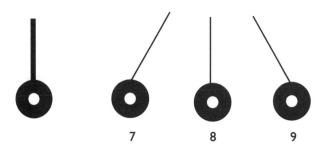

7   8   9

ACTIVITY FIVE B
Constant Acceleration

FOR YOU TO DO

1. Now that you can detect acceleration, you can try to achieve constant acceleration to see what affect this kind of motion might have on a race. Before you personally try for constant acceleration, it may be useful to observe some controlled accelerations. Set up a system so that a cart can be pulled along by a falling weight.

2. Put an accelerometer on the cart. Also set up a "Smart Pulley" for one of the string guides as shown.

✎ a) Does the falling weight accelerate the cart at a constant rate?

✎ b) Does changing the amount of weight used to pull the cart affect the acceleration?

3. Nature runs the race with a constant acceleration. A falling object pulls a cart so that the speed of the cart increases at a constant rate. Although you may find it extremely difficult to run at constant acceleration, the falling mass does it well.

Select a weight to pull the cart so that it accelerates at a comfortable pace. Now try to keep pace with the cart as it goes along.

✎ a) Can you keep up with the cart when the maximum available weight is used to pull it?

4. Without using a pace cart, try moving with constant acceleration over a prescribed course of several meters. To see how well you do at holding a constant acceleration, attach a string around your waist and guide it over a "Smart Pulley" system.

5. Run your race, and then look at the computer display of the acceleration.

✎ a) What should you look for on the graph of acceleration vs. time to show that you achieved constant acceleration?

✎ b) Is it possible to run a race at constant acceleration? Why or why not?

✎ c) While you are slowing down, are you accelerating?

Smart Pulley

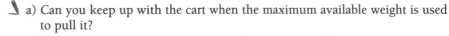

For use with *Sports*, Chapter 1, ACTIVITY FIVE : Acceleration
©1999 American Association of Physics Teachers

ACTIVITY SIX
Running a Smart Race

Background Information

No new physics principles are introduced in this activity, but the activity presents opportunity to reinforce speed as the slope of a distance versus time graph introduced in Activity Four.

If you have not examined the graph "Final Lap: Kicker, You, and Rabbit. Distance versus Time" (provided for you as a transparency master in this Teacher's Edition) from the standpoint of the slopes of plots for the three runners, it is suggested that you do so. Clearly, the slopes of the three lines exist in the same relationship as the speeds of the runners. It would be good to ask your students how they could tell "at a glance" which runner has the highest speed. It is the runner whose distance vs. time graph has the greatest (steepest) slope.

Active-ating the Physics InfoMall

Most references that help with this activity have already been mentioned, including Chapter 6 of *Teaching High School Physics* (in the Book Basement), which is on Curved Motion.

Planning for the Activity

Time Requirements

• One class period.

Materials Needed

For each student:
• graph paper (1 piece)
For each group:
• calculator
For the class:
• overhead projection transparency, "Final Lap: Kicker, You and Rabbit. Distance versus Time"

Advance Preparation and Setup

Produce an overhead projection transparency of the graph "Final Lap: Kicker, You and Rabbit. Distance versus Time." A master copy is included at the end of this activity.

Check that you have an overhead projector available for use.

Teaching Notes

Students may suggest some of the following strategies for a distance race. For example, they may suggest that the runner start off as fast as they can and then, as they run out of steam, they will slow down. The danger here is that the runner may become so tired that they are barely able to complete the race. An experienced runner who likes this fast start strategy will still not run at top speed in order to leave some energy available for the last lap. Some runners attempt to run the entire race at a constant speed. From experience, they have learned how to pace themselves to be able to do this and not slow down in the last lap. Other runners set their own pace a little slower and when they reach the last lap they kick into high gear and finish the race in a blaze. The danger here is that another runner may have a stronger kick at the end. Therefore, it is not only important to analyze your team's strategy but also advisable to obtain as much information about the opponent's strategy as possible.

You may wish to have students work in their groups, but each student should produce and analyze the graph as directed in For You To Do.

Following initial discussion of What Do You Think? and introduction of the problem to be solved in For You To Do, it is suggested that you circulate among students as they proceed through the stepwise instructions presented in For You To Do. Finally, the overhead projection transparency can be used to provide closure for the activity.

The completed distance versus time graph and answers to most questions associated with the graph are shown on the overhead transparency. Answers also are given below for your convenience or in case you decide not to use the transparency.

One way to use the transparency to interpret the graph to yield answers to the questions in the student text (the answers also can be found algebraically, without using the graph) is to place a straightedge on top of the transparency and "shift" the plots of the rabbit and kicker so that the endpoints of their lines correspond to the endpoints of the "you" line on the graph. Of course, the new lines for the rabbit and kicker each must be parallel to the original line for each runner (changing the slope would indicate a speed change). Draw the lines on the transparency. For analysis:

Notice that the redrawn line for the rabbit intercepts the distance axis at about 18 m, the maximum allowable "lead" the rabbit may have at the beginning of the final lap for "you" to win.

Notice that the redrawn line for the kicker, if the same length as the original line is used, does not intercept the distance axis. Use a pen and straightedge to extend the distance axis downward, below the origin. Next, extend the beginning end of the redrawn line for the kicker to intercept the extension of the distance axis. The intercept will be at minus 13, the minimum necessary lead "you" must have over the kicker at the beginning of the final lap to win.

This activity has provided an example of how a long-distance runner can develop a strategy to win a race based on knowledge of his or her capabilities and knowledge of patterns in past performances of opponents. Specific data similar to the kind used in this activity is needed to assist runners of the school's track team.

Activity Overview

In this activity students create a strategy to win a 1,500-m race against two opponents: one who starts fast, and one who finishes fast.

Student Objectives

Students will:

- calculate average speed given distance and time.
- produce a graph of distance versus time.
- calculate distance given average speed and time.
- compare distances travelled by runners traveling at different speeds in equal amounts of time.

What Do You Think?

In a sprint there are two distinct parts to the race. A drive phase (where the runner accelerates) and a cycle phase (where the runner tries to maintain a constant speed). A sprinter doesn't have enough time to apply strategy in the midst of a dash; turning the head to see where an opponent is could break concentration enough to cause a sprinter to lose the race.

A long-distance runner, by contrast, can apply strategy during a race. World-class long-distance runners are known to adjust speed during races to carry out plans based on knowledge of their own capabilities, the past performances of opponents, and what opponents are doing during a race.

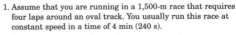

THE TRACK AND FIELD CHAMPIONSHIP

Activity Six

Running a Smart Race

WHAT DO YOU THINK?

Good sprinters are usually not good distance runners.

- **How do strategies for winning sprints and distance runs differ?**

Record your ideas about this question in your *Active Physics log*. Be prepared to discuss your response with your small group and the class.

FOR YOU TO DO

1. Assume that you are running in a 1,500-m race that requires four laps around an oval track. You usually run this race at constant speed in a time of 4 min (240 s).

   Assume you are running against two opponents, one who starts fast (a "rabbit") and another who finishes fast (a "kicker"). You have studied their split times from earlier track meets. From their splits, you have discovered that both runners, like you, usually run the 1,500 m in 4 min. But there are differences:

   - **the kicker runs the final lap in 58 s.**
   - **the rabbit runs the final lap in 63 s.**

ACTIVITY SIX: Running a Smart Race

a) Calculate and record the distance of each lap of the race.

b) Calculate and record the time it takes you to run one lap.

c) In your log, record how you think you can develop a strategy to win the race.

2. Make a graph of distance versus time for the three runners (rabbit, kicker, and you) for the final lap of the race:

a) Scale distance vertically from 0 to 375 m and scale time horizontally from 0 to 63 s.

b) Assume that all runners are at $d = 0$ when $t = 0$ and that all runners keep a constant speed during the final lap.

c) Use the individual times to plot a line on the graph for each runner. Label the lines as rabbit, kicker, and you.

3. Look at the distance versus time graph. The graph shows that all three runners are "even" (375 m remaining to run) at the beginning of the final lap.

a) In what order will the runners finish the race?

b) By how much time will the winner finish ahead of the second-place runner? The third-place runner?

c) By how much time will the second-place runner beat the third-place runner?

4. Calculate the speed of each runner during the final lap of the race in m/s.

a) Record each speed in your log: speed of rabbit, speed of kicker, your speed.

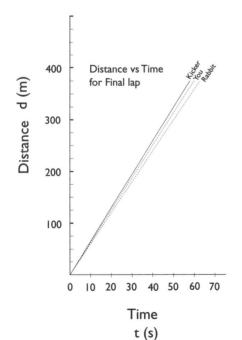

S 27 — SPORTS

For You To Do

1.a) Distance of each lap:

1,500 m/4 = 375 m

b) Time to run one lap:
240 s/4 = 60 s

c) Student answers will vary. See Teaching Notes for some suggested strategies.

2.a-b) Students scale and label graph correctly.

c) Students provide graph similar to the one shown.

3.a) For an "even" start of the final lap, the order of finish would be: kicker, you, rabbit.

b) The kicker would finish 13 m ahead of you, and 31 m ahead of the rabbit.

c) The second place runner (you) would finish 3.0 s ahead of the third place (rabbit) runner.

4.a) Rabbit's speed: 5.95 m/s
Your speed: 6.25 m/s
Kicker's speed: 6.47 m/s

Distance vs Time for Final lap

Kicker
You
Rabbit

Distance d (m)

400

300

200

100

0 10 20 30 40 50 60 70

Time
t (s)

For You To Do

(continued)

5. a) Rabbit's distance in 60 s:
5.95 m/s x 60 s = 357 m
Kicker's distance in 60 s:
6.47 m/s x 60 s = 388 m

6. a) You must begin the final lap less than 18 m behind the Rabbit to win.

b) You must begin the final lap more than 13 m ahead of the Kicker to win.

THE TRACK AND FIELD CHAMPIONSHIP

5. You run the final lap in 60 s. During those 60 s, how far will the rabbit and the kicker travel at the speeds calculated above?

   a) Calculate and record each distance in your log: distance traveled by the rabbit in 60 s, distance traveled by the kicker in 60 s.

6. Review the data. Determine and record each of the following.

   a) How far behind the rabbit can you be at the beginning of the final lap to win the race?

   b) How far ahead of the kicker must you be at the beginning of the final lap to win the race?

REFLECTING ON THE ACTIVITY AND THE CHALLENGE

It is a good idea to "scout" your opponents and know what they can do. It is wise to develop strategies to compete against runners based on information about their past performances. Sometimes, however, opponents may change their strategies based on their knowledge of *your* past performance. Then it is necessary to change your strategy while running the race.

PHYSICS TO GO

1. Explain what a runner must do to win a race against a "rabbit."

2. Explain what a runner must do to win a race against a "kicker."

3. Harry usually runs the final 200 m of a 1,600-m race in 30.0 s. How far ahead of Eamonn Coghlan should Harry plan to be at the beginning of the final 200 m if Coghlan has the same split times listed in question 6, "Physics to Go," in Activity 4?

S 28

Physics to Go

1. To win against a rabbit, you can allow the rabbit to be ahead of you during a race, but not too far.

2. To win against a kicker, you must be sufficiently ahead of the kicker when the kicker accelerates near the end of the race.

3. Coghlan's split for final 200m: 29.38 s

Coghlan's speed for final 200m: 200.0 m / 29.38 s = 6.807 m/s

Harry must lead Coghlan at beginning of final lap by a distance equal to

[(Coghlan's speed) x (Harry's time)] - 200.0 m
= [(6.807 m/s x 30.00 s) - 200.0 m] = (204.2 m - 200.0 m) = 4.2 m

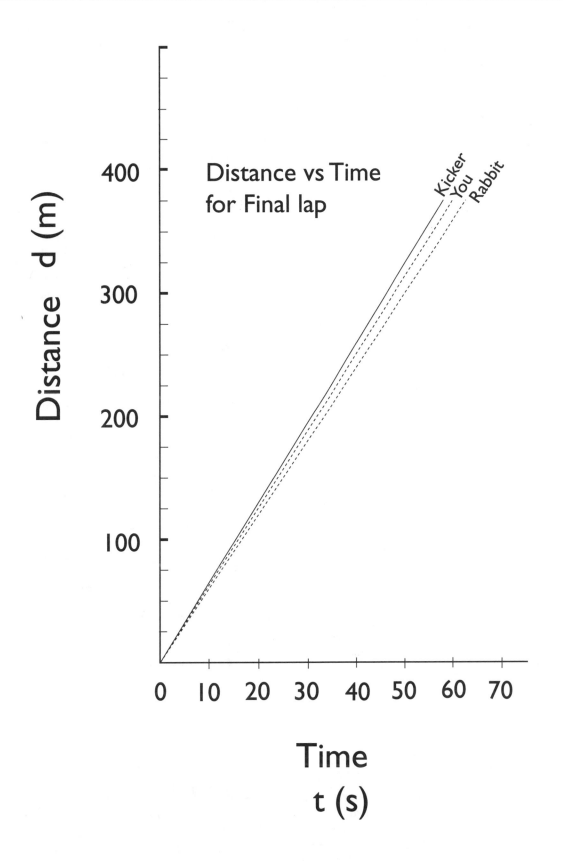

ACTIVITY SEVEN
Increasing Top Speed

Background Information

An alternate way of calculating a runner's speed is introduced in this activity.

Speed = Stride frequency x Stride length

Dimensional analysis presented in the Physics Talk section of the student text for this activity shows that the above equation is equivalent to:

Speed = Distance/Time

A caution is offered regarding the unit "strides per second" to express stride frequency and the unit "meters per stride" to express stride length in this activity. The caution is based on a subtle, sometimes misunderstood aspect of the definition of frequency.

The formal unit of frequency is the "hertz" (abbreviated Hz) which is defined as:

1 hertz = s^{-1} = $1/s$

A common misconception about the hertz as a unit of measurement arises when a frequency of, for example, 60 Hz is expressed as 60 "cycles per second," or 60 "vibrations per second," or, in the case of this activity, some number of "strides per second." While it is true that terms such as "cycles," "vibrations," "strides" or other descriptive nouns may enhance communication, it is essential to recognize that, to formally comply with the definition of frequency, a runner making, for example, 2 strides per second should have his frequency expressed as "2/s." Similarly, if the same runner makes strides which are of average length 3 m, the proper expression of stride length would be, simply, "3 m." Clearly, dimensional analysis using proper, formal units shows that stride frequency multiplied by stride length yields speed:

2/s x 3 m = 6 m/s

Carrying descriptors such as "strides" or "cycles" in expressions of frequency will do no harm here, but will if, in the future, the student uses such descriptors when performing dimensional analysis of complex equations involving frequency in areas such as electricity or quantum mechanics. For example, suppose a student wants to calculate the energy of a photon of light corresponding to a frequency of 4.0×10^{14} Hz using the below equation and, through bad habit, substitutes 4×10^{14} cycles/second as the frequency:

Energy of photon = Planck's constant x frequency

= $(6.6 \times 10^{-34}$ joule-second) x $(4.0 \times 10^{14}$ cycles/second)

= 2.6×10^{-19} joule-cycles

The student has a problem, because the answer should be in joules, not joule-cycles; this problem would not exist if the student had expressed the frequency properly, substituting 4×10^{14}/second for 4.0×10^{14} Hz as the frequency.

This may seem "nit picky," but the author actually has observed students struggle while trying to get "cycles" to cancel during dimensional analysis of calculations involving frequency. It seems a disservice to plant misconceptions which are avoidable; therefore, it is suggested that you call students' attention to "per second" and "meters" as the proper units for, respectively, stride frequency and stride length, in this activity.

Active-ating the Physics InfoMall

Does the information regarding the cheetah sound interesting? Search the InfoMall with the word "cheetah" and you get two hits. Both are interesting, but you should look at the second one, *Many Magnitudes: A Collection of Useful and Useless Numbers*, in the Utility Closet. This has a LONG list of interesting numbers that students may enjoy reading.

The relationship that velocity is the product of frequency and wavelength occurs often in physics. You can find this in most textbooks on the InfoMall. However, "frequency" is a term that occurs so often that you will need to limit your search (either to one or two stores, or by adding more keywords to your search) to avoid getting "too many hits."

Planning for the Activity

Time Requirements

• One class period.

Materials Needed

For the class:

• VCR and monitor
• *Active Physics Sports* content video (segment: runner on football field)

- metric tape measure or meter sticks
- chalk stick or tape for marking intervals on surface of walking lanes
- stopwatch (1 minimum, 2 preferred)

For each group:

- calculator

Advance Preparation and Setup

An area will be needed to serve as a walking "track." The area needs to be a minimum of 12 m in length, which may be larger than your classroom. If needed, arrange to use a larger room, hallway, or outdoor area. Noise can be expected. Tape (or some other material) must be able to be placed on the walking surface to mark intervals on two walking "lanes," each 12 m in length. One lane must be marked at 0.50-m intervals and the other at 0.75-m intervals. The entire class should be able to observe as volunteers walk along the lanes with controlled stride lengths.

A VCR player and monitor should be reserved to be used for this activity.

Teaching Notes

Data collection from the video, calculations based on the data and discussion of Physics Talk to verify the equation Speed = Stride Frequency x Stride Length is suggested as a whole-class activity. This will provide that all students will complete steps 1-5 of For You To Do at the same time and be ready to proceed as a total class to the next part of the activity which involves the walking with controlled stride lengths.

It would provide the most efficient use of time to have a few volunteers walk the lanes with controlled stride lengths while other volunteers measure time with a stopwatch. The data collected for the volunteers could be used as data for all class members. However, all students should try walking both lanes to have direct experience with controlling one's stride.

After walking the lanes, students may complete answers to the questions while working in groups, alone, or as homework.

Having just been introduced to the relationship between speed, stride frequency, and stride length, most students probably haven't thought about the relationship of stride length and stride frequency to acceleration. Perhaps you should point out that

when a runner accelerates, either stride frequency, stride length, or both, must be changing. Encourage students to think of ways that they could actually observe, measure and do calculations involving these variables for track team members, and use the results to improve the performance of the school's athletes in races.

NOTES

Activity Seven

Increasing Top Speed

WHAT DO YOU THINK?

A cheetah can reach a top speed of 60 miles per hour (about 30 m/s).

• **What can a runner do to increase top speed?**

Record your ideas about this question in your *Active Physics log*. Be prepared to discuss your responses with your small group and the class.

FOR YOU TO DO

1. Watch the video of a runner.

2. Use information from the video to answer the following questions. Record your data and show your calculations in your log.

 a) Use the total distance traveled and the total time to calculate the runner's speed in yards per second.

$$\text{Speed (yards/second)} = \frac{\text{Distance (yards)}}{\text{Time (seconds)}}$$

(Yards/second are used since the running is being done on a football field.)

◆ S 29 ◆ ————— SPORTS

ANSWERS

For You to Do

1. Students watch video.

2. a) Distance = 50.0 yards
       Time = 14.3 seconds
       Speed = 50.0 yd/14.3 s
             = 3.50 yd/s

Activity Overview

In this activity students run their own race, exploring the relationship between speed, stride length, and frequency.

Student Objectives

Students will:

• calculate the average speed of a runner given distance and time.

• measure the frequency of strides of a running person.

• measure the length of strides of a running person.

• apply the equation (Speed = Stride Frequency x Stride Length) to calculate the speed of a running person.

• recognize that either the equation (Speed = Distance / Time) or the equation (Speed = Stride Frequency x Stride Length) may be used to calculate a runner's speed with equivalent results.

• infer ways in which stride length and stride frequency can be adjusted by a runner to increase speed.

ANSWERS FOR THE TEACHER ONLY

What Do You Think?

Students should understand that to win a race, they must have a faster average speed than their opponents. In a sprint, a runner may win against an opponent who has a faster top speed by exploding out of the starting blocks with tremendous acceleration. In a distance race, a runner may win against an opponent with a faster top speed by having greater stamina, running more efficiently, and being able to keep a reasonable speed for a longer time.

Even though a runner can compensate for not having the fastest top speed, the runner with the greatest top speed will clearly have an advantage. Students probably will not have given previous thought to stride length and stride frequency as factors which affect a runner's speed. They may look ahead to the title and early parts of For You To Do and give responses based on their first impressions of how these variable affect speed.

ANSWERS

For You to Do

(continued)

b) Number of strides = 37 strides
Time = 14.3 s
Stride Frequency = 37 strides / 14.3 s = 2.6 strides/s

c) Average stride length: 25 yd / 18 strides = 1.4 yd/stride

d) Speed = Stride Frequency x Stride Length
= 2.6 stride/s x 1.4 yd/stride
= 3.6 yd/s

3. a-b) The speeds determined by the two methods, 3.5 yd/s and 3.6 yd/s, differ by less than 3%; within error of measurement, they appear to be equal speeds.

4. Student activity.

5. a-d) The answers involving walking with controlled stride length will depend on the frequency with which the individuals walk. Observers will need to count the number of strides (determined by number of marked intervals; 16 strides for the 0.75 m stride length), and a volunteer must time the duration of each 12-m walk.

THE TRACK AND FIELD CHAMPIONSHIP

b) Count the number of strides taken by the runner during the entire run. Use the number of strides and the total time to calculate the runner's stride frequency in strides per second.

$$\text{Stride frequency (strides/second)} = \frac{\text{Number of strides}}{\text{Time (seconds)}}$$

c) Calculate the average length of one stride for the runner. To do so, measure the length of several single strides and then calculate the average length per stride. The unit for your answer will be yards/stride.

d) Calculate the runner's speed using the following "new" equation for speed. Your answer will be in yards/second.

Speed = Stride Frequency × Stride Length.

3. Compare the results of using the two equations for calculating speed.

a) Do the equations agree on the runner's speed? How good is the agreement?

b) How could you explain any difference in results?

4. You will test the "new" equation at the "track." Your teacher will show you where to set up the track. Place marks at intervals of 0.75 m along a 12-m track.

5. Starting from the "zero" mark, walk so that you step on each mark. Count the number of strides to complete the walk, and use a stopwatch to measure the total time.

a) Record your data in your log.

b) Calculate your speed using the distance/time equation.

c) Calculate your stride frequency.

$$\text{Stride frequency (strides/second)} = \frac{\text{Number of strides}}{\text{Time (seconds)}}$$

d) Calculate your speed using the stride frequency and stride length.

Speed = Stride Frequency × Stride Length.

ACTIVITY SEVEN: Increasing Top Speed

6. What happens to your speed if you change your stride length?

🖎 a) Record what you think will happen to the speed if you decrease the length of the stride.

7. Mark the track at 0.50 m intervals and walk again, stepping on each mark. Again measure the time and count the number of strides.

🖎 a) Record the data in your log.

🖎 b) Calculate your speed using both equations.

🖎 c) How well do your speeds calculated by both methods on this track compare?

8. Compare your performances on the two different tracks.

🖎 a) Was your speed on the track which had 0.75 m stride lengths about the same as, or different from, your speed on the track which had 0.50 m intervals? Why?

🖎 b) If you must step on each mark, what would you need to do to make your speed on the second track equal to your speed on the first track? Express your answer in numerical terms.

PHYSICS TALK

Calculating Speed Using Frequency and Length

If it is really true that speed = frequency × stride length, then it should also be true that the relationship produces an answer that has a unit of speed, such as meters/second. If frequency is measured in strides/second and if length is measured in meters/stride, then

$$\frac{strides}{second} \times \frac{meters}{stride} = \frac{meters}{second}.$$

The equation produces an answer that has a unit of speed.

For You to Do

(continued)

6.a) Have students enter their predictions before beginning the activity.

7.a-c) The answers involving walking with controlled stride length will depend on the frequency with which the individuals walk. Observers will need to count the number of strides (determined by number of marked intervals; 24 strides for the 0.50 m stride length), and a volunteer must time the duration of each 12-m walk.

8.a-b) Since a particular walker probably will not adjust frequency to offset the change in stride length from one track to the other, the speeds for an individual on the two tracks should not be expected to be equal. To make the speeds equal, the frequency on the track having the 0.50 m stride length would need to be 0.75/0.50 = 1.5 times higher than the frequency on the track having the 0.75 m stride length.

1

<small>ANSWERS</small>

Physics to Go

1. a) 1.8 strides/s x 2.0 m/stride
   = 3.6 m/s

   b) 200 m / 3.6 m/s = 56 m/s

2. (6.0 m/s) / 1.5 m/stride
   = 4.0 strides/s

3. 4.0 strides/s x 1.6 m/stride
   = 6.4 m/s

4. 2.0 strides/s x 0.65 m/stride
   = 1.3 m/s

5. The graphs generally show that speed decreases with age and that stride length also decreases with age, which is consistent with the equation:

   Speed = Stride Frequency x Stride Length.

   Trends of the effects of age on Stride Frequency could be explored by selecting pairs of data points for corresponding age and gender on the two graphs and then dividing the speed by the stride length to determine the frequency.

6. Hurdle events in track require that hurdlers establish, and sometimes adjust between hurdles, stride lengths so that a they arrive "on step" at a desired "launching point" in front of each hurdle. Missing the launching position may cause the hurdler to trip on the hurdle, which usually results in losing the race.

<small>THE TRACK AND FIELD CHAMPIONSHIP</small>

REFLECTING ON THE ACTIVITY AND THE CHALLENGE

Distance runners learn from their coaches how to make conscious changes in stride length and frequency to improve their performances. To increase speed, the "trick" is to increase one—either frequency or stride length—without decreasing the other. Good runners know how to increase either, or even both, when needed in a race. Think of experiments that you could do with members of your school's track team to help them to learn to use frequency and stride length to win races.

PHYSICS TO GO

1. A runner's stride length is 2.0 m and her frequency is 1.8 strides/s.
   a) Calculate her average speed.
   b) What would be her time for a 200-m race?

2. A runner maintains a constant speed of 6.0 m/s. If his stride length is 1.5 m, what is his stride frequency?

3. If the runner in Question 2 increases his stride length to 1.6 m without changing his stride frequency, what will be his new speed?

4. If a marching band has a frequency of 2.0 strides/s and if each stride length is 0.65 m, what is the band's marching speed?

5. The graphs on the next page are reproduced from an article which reported the characteristics of runners who are 30 or more years of age.

   Write a statement which explains the information on the two graphs. Include your own inferences about what happens to stride frequency with increasing age.

6. The running events called "hurdles" present special problems involving stride length and frequency. In both 100- and 400-m races, runners must jump over hurdles placed at regular intervals along the track. Discuss special techniques that hurdlers must develop to make sure that they are ready to jump when they reach each hurdle.

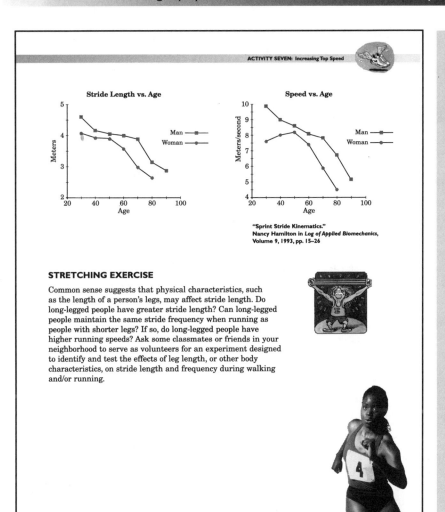

Stride Length vs. Age

Speed vs. Age

"Sprint Stride Kinematics."
Nancy Hamilton in *Log of Applied Biomechanics*,
Volume 9, 1993, pp. 15–26

STRETCHING EXERCISE

Common sense suggests that physical characteristics, such as the length of a person's legs, may affect stride length. Do long-legged people have greater stride length? Can long-legged people maintain the same stride frequency when running as people with shorter legs? If so, do long-legged people have higher running speeds? Ask some classmates or friends in your neighborhood to serve as volunteers for an experiment designed to identify and test the effects of leg length, or other body characteristics, on stride length and frequency during walking and/or running.

ANSWERS

Alternative Solutions to Physics To Go

Alternate forms of solutions to Physics To Go questions 1, 2, 3 and 4 which comply with correct formal expression of frequency and length are shown in brackets. See "Background Information for the Teacher" regarding a caution about including descriptive terms such as "strides" in calculations involving frequency.

1. a) 1.8 strides/s x 2.0 m/stride
= 3.6 m/s
[1.8/s x 2.0 m = 3.6 m/s]

   b) 200 m / 3.6 m/s = 56 s

2. (6.0 m/s) / 1.5 m/stride = 4.0 strides/s
[(6.0 m/s) / 1.5 m = 4.0/s]

3. 4.0 strides/s x 1.6 m/stride
= 6.4 m/s
[4.0/s x 1.6 m = 6.4 m/s]

4. 2.0 strides/s x 0.65 m/stride
= 1.3 m/s
[2.0/s x 0.65 m = 1.3 m/s]

ANSWERS

Stretching Exercise

Student Activity.

ACTIVITY EIGHT
Projectile Motion

Background Information

New phenomena introduced in this activity include:

- free fall
- projectile motion

It is suggested that you read the Physics Talk section of the student text for this activity to see the meaning of terms used in the below background information.

Free Fall: All objects in a state of free fall, regardless of mass, shape, or other characteristics, experience a uniform (constant) acceleration of approximately 10 m/s/s in the downward direction. (Notice that the condition made in Physics Talk in the student text would not include objects which experience a significant force due to air resistance as being in a state of free fall.)

Projectile Motion: Inertia (an object in motion remains in motion unless a force acts to cause a change in motion) causes a projectile to retain any horizontal motion which it has at the instant it is launched, and it retains that motion in the form of constant speed.

Gravity causes a projectile to exhibit vertical motion which matches free fall, whether or not the projectile has simultaneous horizontal motion.

This activity does not approach projectile motion quantitatively, but it is clearly demonstrated that a coin which simply falls strikes the floor in the same amount of time as a coin launched horizontally from the same height at the same instant. The only way this can happen is if both objects fall downward with identical motions. The fact that one coin simultaneously moves horizontally (at constant speed) as it falls affects where it lands, not when.

This activity further demonstrates that a ball thrown upward does not have its (accelerated) upward and downward motion affected by any horizontal motion which it may have at the time of launch. The horizontal motion (constant speed) is independent of the vertically accelerated motion. If a student sitting in a stationary chair throws a ball upward, it lands in her hand; if the chair is moving at constant speed throughout the vertical toss,

inertia demands that the ball also lands in the student's hand.

Quantitative treatment of projectile motion, with additional background for the teacher, is presented in the next activity.

Active-ating the Physics InfoMall

The obvious keywords to use for a search related to this activity are "projectile motion." This search is particularly rewarding. Look at the information from the textbooks *Modern College Physics* and *Foundations of Modern Physical Science*. Both have great discussions of projectiles, and they also have graphics that illustrate the independence of vertical and horizontal motion (Figure 3.4 in *Foundations*, and Figure 8A in *Modern College Physics*).

Our earlier search also produced the article "Aristotle is not dead: Student understanding of trajectory motion," in the *American Journal of Physics,* vol. 51, issue 4. Other nice references are *A Guide to Introductory Physics Teaching: Motion in Two Dimensions* in the Book Basement, and "Thoughts on projectile motion," in the *American Journal of Physics*, vol. 53, issue 2.

In For You To Do, step 1: Equipment for a similar experiment can be found in the Cenco catalog (in the Catalog Corner). This item is on the list generated from the "projectile motion" search. It is also item number 76525 in the catalog, in case you wish to browse for yourself.

For step 4, you should look at Figures 3.1 and 3.3 in Chapter 3 of *Foundations of Modern Physical Science* (in the Textbook Trove) for this activity.

Planning for the Activity

Time Requirements

- One class period.

Materials Needed

For the class:
- chair with wheels
- tennis ball (or similar)

For each group:

- coins (2)
- meter stick (to serve as "cue stick" if needed)

Advance Preparation and Setup

SAFETY PRECAUTION: A chair on wheels is used for a demonstration in this activity. You will want to locate a chair which is as stable and sturdy as possible to prevent a "crash." An office chair having a 4 or 5 wheel base and arm rests is suggested.

Plan to control the situation, preferably with pushing the chair as a student rides in the chair and throws a ball upward and then catches it when it falls. Do not let students play with the chair; it could be very dangerous.

Teaching Notes

Help students appreciate that both a shot put and a human body can be a projectile. Both will follow a path through the air called a trajectory.

Set an amount of time for students to complete the work with coins in the first part of For You To Do, allowing sufficient time for the demonstration during the last part of the class period.

If students have poor aim using a flick of a finger to send the moving coin in a direction so that it barely "ticks" the stationary coin, have them try using a meter stick as a cue stick to launch the coin.

Be sure to have each student commit to a prediction of where the ball will land relative to the student in the moving chair. Expect the misconception that some students will predict that the student will "move out from under the ball as the ball flies straight up and down." If this prediction is not offered (some may believe it, but won't volunteer it), plant it as a suggested possibility. The demonstration will present a discrepancy which may cause such students to discard their misconception.

Have the students practice throwing the ball vertically and catching it again with the chair at rest first. Have the student throw the ball to several different heights. Lay out a straight line course of about 3 to 5 m. When the chair is put into motion and gets up to speed, have the occupant throw the ball upward gently and catch it again. Post observers along the side of the course so that the horizontal distance (range) from release to catch can be marked and measured. Repeat the measure-

ments to get a sense of the variation that can result from the differences from trial to trial.

In Reflecting on the Activity and the Challenge ask the class to develop a list of track and field events which involve projectile motion (long jump, high jump, hurdles, shot put, discus, javelin, hammer throw). Do fewer athletes pursue these events than others? Might these events offer higher probability of winning if less competition exists? Might some of these events have greater dependence on knowledge and technique and less dependence on physical ability than other events? Events involving projectile motion may have high potential to be improved through physics-based help.

You may wish to point out that Carl Lewis is one of the world's greatest long jumpers. The Jesse Owen 1936 Olympics performance in the sprints and the long jump also makes a very good story.

Activity Overview

Students explore the independence of vertical and horizontal components of a projectile's motion in this activity.

Student Objectives

Students will:

- understand and correctly apply the term "free fall."
- understand and correctly apply the term "projectile."
- understand and correctly apply the term "trajectory."
- understand and correctly apply, within physics context, the term "range."
- observe that all projectiles launched horizontally from the same height strike a level surface in equal times, regardless of launch speed (including zero launch speed).
- understand that the vertical and horizontal components of a projectile's motion operate independent of one another.
- infer factors which affect the range of a projectile.
- infer the shape of a projectile's trajectory.
- infer ways in which stride length and stride frequency can be adjusted by a runner to increase speed.

ANSWERS FOR THE TEACHER ONLY

What Do You Think?

The initial velocity (speed and direction) determines the range of a projectile; a difference in the elevations of the launch and landing points also would affect the range.

A 100 mph pitch thrown horizontally by a major league player will hit the ground in the same amount of time as a 10 mph pitch thrown horizontally from the same height by a child. It is a near certainty that many students would never believe this type of fact. The work with coins should present a discrepancy for them which, hopefully, will cause them to confront their misconceptions about projectiles.

CHAPTER 1   THE TRACK AND FIELD CHAMPIONSHIP

Activity Eight
Projectile Motion

WHAT DO YOU THINK?

Some track and field events involve launching things into the air such as a shot put, a javelin, or even one's body in the case of the long jump.

- **What determines how far an object thrown into the air travels before landing?**

Record your ideas about this question in your *Active Physics log*. Be prepared to discuss your response with your small group and the class.

FOR YOU TO DO

1. Place one coin at the edge of a table with about half of the coin hanging over the edge. Place the other coin flat on the table and use a finger to shoot this coin across the tabletop to strike the first coin. Aim "off center" so that the coin at the edge of the table drops straight down and the projected coin leaves the edge of the table with some horizontal speed. Repeat the event as many times as needed to record your answers to the following question in your log.

a) Which coin hits the floor first? (Hearing is the key to observation here, although you may also wish to rely on sight.)

2. Vary the speed of the projected coin.

a) How does its speed affect the amount of time for either coin to fall to the floor?

b) How does its speed affect how far across the floor the projected coin lands?

3. Use a box, stack of books, or a different table or countertop to vary the height.

a) Which coin hits the floor first?

b) How does increasing the height affect how far the projected coin travels horizontally as it falls?

4. Your teacher will supervise an activity in which one student sits on a chair that is moving at constant speed. While the chair is moving, the student will throw a ball straight up into the air and try to catch it when it comes down. The class will stand in a line along the track to observe the event, prepared to mark the horizontal distance (range) the ball travels from release to catch.

⚠ Make sure the path is clear before launching the coins.

ANSWERS

For You To Do

1. a) Both coins will hit the ground at the same time.

2. a) The initial speed has no effect on the amount of time for either coin to fall to the floor.

   b) The coin with the greater initial velocity will have a greater range, it will travel further across the floor.

3. a) The coin with the initial higher elevation will hit the floor second, the lower elevated coin will hit first.

   b) Increasing the height has no effect on the distance that the coin travels.

1

For You To Do

(continued)

4. a) It is important that students record their predictions before the activity is completed. This permits them to confront their misconceptions, if there are any. If students are not encouraged to commit to a prediction, they will often predict the "right answer" after the fact. Although students may learn to respond with the correct answer, they may still not have confronted their misconceptions.

b) The ball should land in the student's hands as if the chair were not moving because, due to inertia, the ball retains its horizontal speed as, independently, it flies up and down with accelerated vertical motion due to gravity.

c) Two factors affect the range of the ball: the vertical speed at which the ball is launched; the horizontal speed of the chair (and ball) when the ball is launched.

d) The shape of the trajectory will be a parabola (some students may guess this as the shape — encourage them by asking what they know about parabolas, and why that may be the shape of the trajectory).

---

a) In your log write your prediction of what you think will happen.

b) Write in your log what you observed about the ball's trajectory (shape of the ball's path) and the ball's approximate range (horizontal distance) for trials in which you varied the speed of the chair and the launching speed of the ball.

c) According to your observations, what factors affect the range of the ball?

d) According to your observations, what is the shape of the trajectory?

PHYSICS TALK

Projectiles and Trajectories

Physicists often work with objects that have been launched into the air in a state of "free fall." In a free fall, the main force acting on the object while it is in the air is the downward pull of the earth's gravity.

An object launched into the air is called a *projectile*. Examples of projectiles are a javelin, a shot put, or a broad jumper. The path that the projectile follows when launched into the air is called the *trajectory*.

REFLECTING ON THE ACTIVITY AND THE CHALLENGE

The first part of this activity (two falling coins) demonstrated that the time required for a coin to fall is independent of the horizontal speed. If two long jumpers rise to the same height, then they will remain in the air for identical times.

The second part of this activity (the rolling chair) showed that the faster the chair is moving, the farther the ball will travel horizontally. If a long jumper is able to increase horizontal speed, then the jumper will travel farther.

To maximize the distance a trajectory travels, you should try to maximize the horizontal speed and maximize the height it can rise. How can you use these conclusions from the activity to improve the performance of your broad jumper?

ACTIVITY EIGHT: Projectile Motion

PHYSICS TO GO

1. Draw a sketch of your two coins leaving the table. Show where each coin is at the end of each tenth of a second. Remember to emphasize that they both hit the ground at the same time.

2. Repeat the sketch of the two coins leaving the table, but this time have one of the coins moving at a very high speed.

3. It is said that a bullet shot horizontally and a bullet dropped will both hit the ground at the same time. Draw sketches of this (the bullet is like a very, very fast moving coin).

4. a) Survey your friends and family members to find out which they think will hit the ground first, a bullet that is dropped, or a fast-moving bullet.
   b) Explain why you think people may believe that the two coins hit the ground at the same time, but that they have a more difficult time believing the same fact about bullets.

5. Use evidence from your observations of the two coins in this activity to prove that a 100 mile/hour pitch thrown horizontally by a major league player will hit the ground in the same amount of time as a 10 mile/hour pitch thrown horizontally from the same height by a child.

6. Use evidence from your observations of the ball and chair in this activity to show the truth of the statement, "A projectile's horizontal motion has no effect on its vertical motion, and vice versa."

7. Write a note to your school's track coach describing how the information you learned in this activity could help the team's long-jump athletes.

S 37

Physics To Go

1. Students provide sketch.

2. Students provide sketch.

3. Students provide sketch.

4. a - b) Answers will vary. Encourage students to confront their own misconceptions when answering this question.

5. Both objects strike the ground at the same instant.

6. The ball would not have returned to the hands of the student riding in the chair if the horizontal and vertical motions were not independent of one another.

7. The long jumper should run as fast as possible to have great horizontal speed at launch, and then, upon launch, should jump straight upward with the greatest possible speed.

1

Assessment Rubric: Two Falling Coins

Student understands that the time required for a coin to fall is independent of the horizontal speed.

Place a check mark (√) in the appropriate box.

Descriptor	Task accomplished	Task not accomplished
1. Student records experimental data demonstrating that the speed at which a coin travels along a horizontal plane does not affect the time it takes to fall.		
2. Student records experimental data demonstrating that the height at which a coin is dropped affects the time it takes to fall.		
3. Student is able to draw a sketch showing that the speed at which a coin travels along a horizontal plane does not affect the time it takes to fall.		
4. Student is able to draw a sketch showing that a bullet shot horizontally from a gun and a bullet dropped at the same time, from the same height, will hit the ground at the same time.		
5. Student successfully conducts a survey of friends and family to uncover misconceptions about falling bodies. "What will hit the ground first, a bullet that is shot from a gun or one that is dropped from the same height?" • Data is collected. • Data is compiled. • A conclusion is provided.		
6. Student proposes a hypothesis explaining why so many people hold the misconception.		
7. Student uses experimental evidence to explain why a pitch moving at 100 mile/hour will hit the ground at the same time as one thrown 10 miles/hour along the same trajectory.		

Assessment Rubric:
The Rolling Chair

Student understands that the faster the chair moved, the farther the ball will travel horizontally.

Place a check mark (√) in the appropriate box.

Descriptor	Task accomplished	Task not accomplished
1. Student makes a prediction about how the velocity of the chair affects catching the ball.		
2. Student records experimental data in log book that shows the trajectory of the ball released from a moving chair.		
3. Student correctly identifies factors that affect the range of the tossed ball: • Height of the toss. • Velocity of the chair.		
4. Student observes and correctly records the path of the trajectory for the ball.		
5. Student uses experimental evidence to explain the statement: "A projectile's horizontal motion has no affect on its vertical motion."		
6. Student writes a note to the track coach explaining that if a long jumper is able to increase horizontal speed, then the jumper will travel farther.		

1

For use with *Sports*, Chapter 1, ACTIVITY EIGHT: Projectile Motion

©1999 American Association of Physics Teachers

ACTIVITY NINE
The Shot Put

Background Information

New material introduced in this activity includes:

- measurement of the acceleration due to gravity.
- calculation of an object's position and speed at any time after it enters a state of free fall.
- quantitative treatment of the horizontal (constant speed) and vertical (free fall) components of the motion of a projectile.
- modeling the trajectory of a projectile.
- variables affecting the height and range of a projectile.

Acceleration Due to Gravity

All objects in a state of free fall, regardless of mass, shape, or other characteristics, experience a uniform (constant) acceleration, g, of approximately 10 m/s^2 in the downward direction. (Falling objects which experience significant air resistance are not considered to be in a state of free fall.)

For your information (not for students' at this time), Earth exerts an amount of downward gravitational pull on objects at or near its surface which is proportional to the mass of each object. The outcome of this is that all objects experience the same amount of force per unit of mass and, therefore, acceleration. For example, Earth's pull on a 2 kg rock is twice as much as the pull on a 1 kg rock; the result is that both accelerate at the same rate during free fall.

Active Physics "rounds off" g to 10 m/s^2, which is within about 2% of the average value, about 9.80 m/s^2, on our planet. The acceleration due to gravity depends on the radial distance from the center of Earth, and, therefore, the value of g varies with location by as much as 4 cm/s^2 due to terrain differences (mountains, valleys) and the fact that Earth is not a perfect sphere (the equatorial diameter is greater than the polar diameter). Some argue that local variations in g, if taken into account, would sometimes have a seemingly poor performance in a track and field event such as the shot put actually be better than a longer throw made at a location having a lower value of g. Physicists have determined that variations in g sometimes could make a difference in records in field events.

Distance, Speed, Acceleration, and Time in Free Fall

This activity follows arguments originally presented by Galileo regarding free fall. Galileo hypothesized that free fall involved constant, or uniform, acceleration.

From the definition of constant acceleration, $a = \Delta v/\Delta t$, he reasoned that the speed attained by an object falling from a "rest start" at any time during its fall would be $v = \Delta v = a(\Delta t)$. (The speed, v, at the end of a time of fall Δt would be equal to the change in speed, Δv, because the initial speed for a rest start is zero.)

Galileo further reasoned that, since the speed of a falling object increases uniformly (or at a constant rate) with time (his hypothesis), the average speed for a time of fall Δt would be half of the speed attained at the end of a time of fall Δt:

(Average speed at the end of time interval Δt):

$$v_{ave} = v/2 = a(\Delta t)/2$$

Finally, Galileo reasoned that the distance of fall and the end of a time of fall Δt would be, simply, the average speed multiplied by the time of fall:

(Fall distance at the end of time interval Δt):

$$d = [a(\Delta t)]/2 \times (\Delta t) = (1/2)a(\Delta t)^2$$

The above reasoning was used as the strategy for developing the table in step 2 of For You To Do for this activity.

As an extension of the above for your understanding as the teacher, Galileo was limited in his ability to test his assumptions and reasoning. He did not have adequate instruments to measure acceleration and speed directly, so he devised a test of his thinking which involved what he could measure. From the last line in the above derivation, $d = (1/2)a(\Delta t)^2$, he solved for acceleration: $a = 2d / (\Delta t)^2$.

Rolling a sphere down a ramp to "dilute" the effect of gravity (he also assumed that a sphere rolling down a ramp also had constant acceleration, but less of it than an object in free fall) he used a crude timing device to measure the amounts of time for the sphere to roll, starting from rest, several measured distances along the ramp. Upon substituting pairs of distance and time measurements into the relation $a = 2d / (\Delta t)^2$, he obtained a fixed value for the right-hand side of the relation, proving that the acceleration indeed is constant.

The Trajectory of a Projectile

The steps used to develop the model of a trajectory in this activity provide a fine quantitative example of the independence of the vertical and horizontal components of the motion of a projectile. While it is not expected that students will predict the range, maximum height, time of flight, and other parameters of a projectile in terms of launch speed and direction, the following equations will prepare you to do so, if needed:

For a projectile launched horizontally at speed v from height h:

time of flight is $t = \sqrt{(2h/g)}$

range (horizontal distance) is: $R = vt = v\sqrt{(2h/g)}$

For the general case of a projectile launched from ground level at speed v at an angle q above the horizontal, and traveling over flat ground: speed in the horizontal direction is $v_x = v\cos q$ (remains constant)

initial speed in the vertical direction is $v_{yo} = v\sin q$

speed in the vertical direction at time t is: $v_y = (v\sin q) - gt$

horizontal position at time t is $x = (v\cos q)t$

vertical position at time t is $y = (v\sin q)t - (1/2)gt^2$

time to reach maximum height is $t_{max} = (v\sin q)/g$

total time of flight is $t = 2t_{max}$ (see t_{max}, above)

range (horizontal distance) is $R = (v^2\sin 2q)/g$

Note in the final of the above equations that $\sin 2q$ (and, therefore, also the range) has a maximum value of 1 when $q = 45$ degrees ($2q = 90$ degrees).

Active-ating the Physics InfoMall

Again, the title of the activity suggests a wonderful set of keywords: "shot put." There are articles on maximizing the put with calculus, without calculus, and with geometry. You may be interested in Figure 8F in the Projectiles section of *Modern College Physics* (in the Textbook Trove)

For You To Do, step 10: Search the InfoMall for "trajectory model" to see another portable version of this. There is even a nice figure on this one! (This is Mb-17 in the *Demonstration Handbook for Physics,* found in the Demo & Lab Shop. It is in the Mechanics section, under Trajectory Model.)

Planning for the Activity

Time Requirements

- It is suggested that two class periods be allowed for this activity. Try to accomplish measurement of the acceleration due to gravity and completion of the table of free fall distances during the first class period. During the second class period, construct and manipulate the trajectory model.

Materials Needed

For the class:

- chalkboard
- apparatus for measuring acceleration of gravity, teacher's choice
- 4-m tacking strip on wall
- labels to be placed on tacking strip: 0.00 s (1 label), 0.10 s (2 labels), 0.30 s (2 labels), 0.40 s (2 labels), 0.50 s (2 labels)
- pins or tacks for hanging paper clip/string assemblies from tack strip (9)
- ladder or high stool to elevate student to throw a ball from the height of the 4-meter strip
- 2.4-meter stick and string model of a trajectory
- chalk

For each group:

- calculator
- string
- lead fishing weights
- Post-It-Notes

Teaching Notes

The method used to measure the acceleration due to gravity is left to your choice due to the variety of equipment available in schools. Whatever method you use, inform students that the value used in *Active Physics* will be the "rounded" value, $10\ \text{m/s}^2$.

It may be possible to begin the trajectory model during the end of the first class period; if so, go ahead and continue it during the second class period. Reserve as much time as possible for students to "match" the model trajectory by throwing a ball and to use the portable stick model to make inferences about how launch angle affects range.

Depending on the number of student groups you may have established earlier, assign each group to one or more rows of the table in step 2 for the production of clip & string assemblies as diagrammed in the student text; ten assemblies are needed, corresponding to two assemblies for each of the non-zero rows of the table. Duplicate assignments will do no harm and would have the advantage of checking one group's work against another.

As each group completes its assigned clip & string assembly invite the group to hang its assembly (or more than one, if assigned) from the pin(s) corresponding to the assigned time(s) of fall to the right hand side of the pin labelled zero. Also have the group place a mark (or Post-It) on the chalkboard (or other surface, as may apply) at the bottom end of the hanged assembly.

When all groups have placed their assemblies and have marked the chalkboard, remove (and save) the assemblies and then have a volunteer draw a smooth curve which connects the marks on the chalkboard.

Assemble the entire class and choose a volunteer to throw a ball horizontally to try to match the curve. The volunteer should stand on a ladder or stool and throw the ball horizontally (to the right hand side) from the zero time mark. Allow the volunteer to practice until the trajectory of the ball superimposes well against the drawn curve on the chalkboard.

Then ask, "When the curve is matched, what must be the horizontal launch speed? How many meters per second? How can we tell?" Call attention to the argument presented in the student text step 5 (a) of For You To Do that the 0.40 m (40 cm) spacing of the pins on the tack strip was designed for a horizontal launch speed of 0.40 m/0.10 s = 4 m/s.

Next, use the clip & string assemblies again to construct a "mirror image" of the first trajectory to the left of the zero time mark to model. This will model an "arch" trajectory. Have a volunteer attempt to launch a ball from the bottom end of the mark corresponding to 0.50 s to the left of the zero mark to match the arch trajectory.

When the volunteer is able to match the trajectory with consistency, invite the class, to give the volunteer instructions to test the effects of launch angle and launch speed on the ball's trajectory.

Turn next to the "portable" 2.4-m Stick and String Model of a the same trajectory, prepared by you in advance. Explain that the model is identical to the one used above, but that an additional 0.10 s time interval with a fall distance of 1.8 m has been added to the model. Ask students to verify that the added

fall distance is correct, and then follow the steps given in For You To Do, as a demonstration for the entire class. Be aware that the endpoint of the range of a projectile launched across level ground in this model is the point where the curve of the trajectory crosses the same vertical level as the "zero time"(launching) end of the stick; at inclinations of the stick exceeding about 45 degrees the longest (1.8 m) string no longer reaches down to the "launch level," so it will be necessary to extrapolate the curve of the trajectory.

Close the activity by discussing the content of For You To Read.

Knowledge of the physics of projectile motion presents a host of possibilities for helping the school's track team. Note the example given in the student text for Reflecting on the Activity and the Challenge which describes the improvement in an Olympic shot putter's performance based on advice from a physicist.

NOTES

Activity Overview

In this activity students create and compare mathematical and physical models of a projectile's motion to see if there is anything they can learn which they can use to improve performance in events where trajectories occur.

Student Objectives

Students will:

- measure the acceleration due to gravity.
- apply the equation Speed = $g\Delta t$ to calculate the speed attained by an object which has fallen freely from rest for a time interval Δt.
- understand that the average speed at a specified time of an object which has fallen freely from rest is equal to half of the speed attained by the object at the specified time.
- apply the equation (Distance = Average Speed x Δt) to calculate the distance travelled by an object which has fallen freely from rest for a time interval Δt.
- use mathematical models of free fall and uniform speed to construct a physical model of the trajectory of a projectile.
- use the motion of a real projectile to test a physical model of projectile motion.
- use a physical model of projectile motion to infer the effects of launch speed and launch angle on the range of a projectile.

ANSWERS FOR THE TEACHER ONLY

What Do You Think?

Generally for a fixed launch velocity (speed and angle), the range will be greater for a higher launch point (as when launching from a tower) and less for a lower launch point (as when launching toward a rising hill).

The optimum range across level ground is attained for a 45° launch angle.

The range of a projectile across level ground is governed by the equation:
$R = (v^2 \sin 2q) / g$ where R is the range, v is the initial speed, q is the launch angle measured from the horizontal, and g is the acceleration due to gravity. (It is not suggested that this equation be presented to students with the intent of mastery, if at all.)

THE TRACK AND FIELD CHAMPIONSHIP

Activity Nine
The Shot Put

WHAT DO YOU THINK?

A world record shot put of 23.12 m was set by Randy Barnes of the USA in 1990.

- Will a higher or lower launch point of a projectile increase range?
- Will a particular launch angle increase range?
- Will a greater launch speed of a projectile increase range?

Record your ideas about these questions in your *Active Physics log*. Be prepared to discuss your group's responses with your small group and the class.

FOR YOU TO DO

1. Your teacher will provide you with a method of measuring the acceleration caused by the Earth's gravity for objects in a condition of free fall. One simple recommended method uses a "picket fence" and a photogate timer attached to a computer. The "picket fence" is dropped and the computer measures the time between black slats of the fence. The computer then displays the acceleration due to gravity. A second method uses a ticker-tape timer and a mass. This second method requires more class time for the analysis of data.

SPORTS S 38

ACTIVITY NINE: The Shot Put

a) In your log, describe the procedure, data, calculations and the value of the acceleration of gravity obtained. The acceleration of gravity is used often, so it is given its own symbol, *"g."*

2. In your log, make a table similar to the following. Some data already have been entered in the table to help you get started.

Time of Fall (seconds)	Speed at End of Fall (m/s)	Average Speed of Fall (m/s)	Distance (m)
0.0	0	0.0	
0.1	1	0.5	
0.2	2		
0.3			
0.4			
0.5			

a) Calculate and record in the table the speed reached by a falling object at the end of each 0.10 s of its fall.

For example:

Assume $g = 10 \text{ m/s}^2$
Speed = acceleration × time
Speed at the end of 0.2 s = $10 \text{ m/s}^2 \times 0.2 \text{ s} = 2 \text{ m/s}$

b) Calculate and record the average of all of the speeds the object has had at the end of each 0.10 seconds of its fall. Since the object's speed has increased uniformly from zero to the speed calculated above at the end of each 0.10 second of the fall, the average speed will be one half of the speed reached at the end of each 0.10 seconds of falling.

ANSWERS

For You To Do

1.a) Students provide procedure, data, and calculations for finding acceleration due to gravity.

2.a-b) The completed table of calculated values of speed, average speed and distance at 0.10 s intervals for an object falling from rest is shown below. (See chart. Significant digits have not been used for clarity.)

Since "g" is limited to one significant figure, 10 m/s², the tabled values calculated using g also should be limited to one significant figure. Strict adherence would suggest that the fall distance 1.25 m should be rounded to 1 m, but it perhaps is not advisable to distract students with that detail at this time - use your judgment on whether or not to bring this up.

Time of Fall (seconds)	Speed at End of Fall (m/s)	Average Speed of Fall (m/s)	Distance (m)
0.0	0	0.0	0.00
0.1	1	0.5	0.05
0.2	2	1.0	0.20
0.3	3	1.5	0.45
0.4	4	2.0	0.80
0.5	5	2.5	1.25

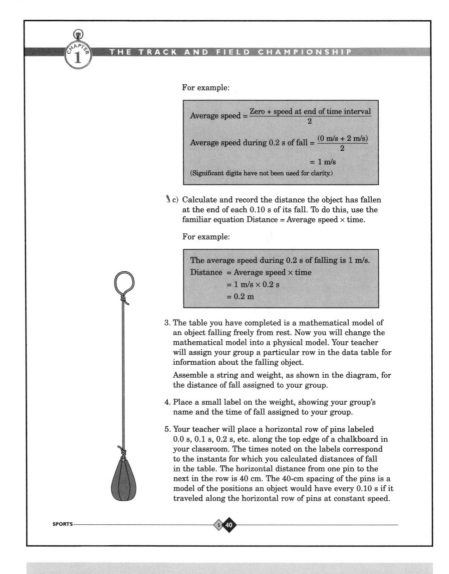

For example:

> Average speed = $\dfrac{\text{Zero + speed at end of time interval}}{2}$
>
> Average speed during 0.2 s of fall = $\dfrac{(0\ \text{m/s} + 2\ \text{m/s})}{2}$
>
> $= 1$ m/s
>
> (Significant digits have not been used for clarity.)

⟍ c) Calculate and record the distance the object has fallen
at the end of each 0.10 s of its fall. To do this, use the
familiar equation Distance = Average speed × time.

For example:

> The average speed during 0.2 s of falling is 1 m/s.
> Distance = Average speed × time
> $= 1$ m/s × 0.2 s
> $= 0.2$ m

3. The table you have completed is a mathematical model of
an object falling freely from rest. Now you will change the
mathematical model into a physical model. Your teacher
will assign your group a particular row in the data table for
information about the falling object.

Assemble a string and weight, as shown in the diagram, for
the distance of fall assigned to your group.

4. Place a small label on the weight, showing your group's
name and the time of fall assigned to your group.

5. Your teacher will place a horizontal row of pins labeled
0.0 s, 0.1 s, 0.2 s, etc. along the top edge of a chalkboard in
your classroom. The times noted on the labels correspond
to the instants for which you calculated distances of fall
in the table. The horizontal distance from one pin to the
next in the row is 40 cm. The 40-cm spacing of the pins is a
model of the positions an object would have every 0.10 s if it
traveled along the horizontal row of pins at constant speed.

ANSWERS

For You To Do (continued)

2. c) See data table on previous page.

3.-4. Student activities.

a) Calculate the horizontal speed by dividing the distance traveled, 40 cm, during each time interval, by 0.1 s. (Dividing a number by 0.1 is equivalent to multiplying the number by 10). Show your calculation and the result in your log.

b) Hang your string and weight assembly from the pin corresponding to the time assigned to your group. Place a small mark on the chalkboard at the bottom end of the string and weight assembly.

6. A volunteer from the class should draw a smooth curve which connects the marks on the chalkboard. A volunteer should try to match the path, a trajectory, by throwing a tennis ball horizontally from the point corresponding to time = 0.0 s. To match the trajectory, the ball will need to be thrown horizontally at the speed calculated in step 5 (a) above. This may require a few practice tries.

a) Write your observations in your log.

7. Create a "mirror image" of the trajectory by moving the 0.1–0.5 s pins to positions 40 cm to the left of the 0.0 s pin. Hang the string and weight assemblies, mark the chalkboard and connect the points to create the second half of an "arch-shaped" model of a trajectory.

8. A volunteer should try to throw a ball to match this trajectory. Have another person prepared to catch the ball.

a) What conditions seem to be necessary to match the trajectory? Write your observations in your log.

b) When a volunteer is able to match the trajectory, the class should agree upon and give the volunteer instructions to test, one at a time, the effects of launch speed and launch angle on the range of the projectile. Write your observations in your log.

9. Your teacher will show you a "portable" version of the row of pins used in step 5 above. It is different in two ways: one additional time interval of 0.1 s has been added (the row is 40 cm longer), and a string to show the distance of fall, 1.8 m, at a time of 0.6 s has been added.

ANSWERS

For You To Do (continued)

5. a) 400 cm/s or 4 m/s

b) Students mark end of string assembly on chalkboard.

6. a) Students record their observations about a volunteer throwing the ball.

7. Student activity.

8. a) Students throw the ball upward with the angle and speed to mirror the trajectory in step 6.

b) Students record their observations for a number of different angles and speeds, as suggested by the students.

9. Teacher activity. You will display portable model of stick and string model of the Trajectory of a Projectile.

10. Rest the end of the stick corresponding to 0.0 s on the tray at the bottom of the chalkboard while inclining the stick at an angle of 30°.

   a) Is the path indicated by the bottom ends of the string and weights assemblies a "true" trajectory? Have a volunteer try to match it. Record your observations.

   b) Repeat for angles of 45°, 60° and other angles of interest. Record your observations (it may be necessary to rest the lower end of the model on the floor to prevent the upper end from hitting the ceiling of the room).

   c) What angle of inclination predicts the greatest range for the projectile? Record your observation.

   d) Incline the stick to 90° (straight up)—do this outdoors if the ceiling is not high enough. What is being modeled in this case? Record your thoughts.

FOR YOU TO READ

The activities you have just completed demonstrate that a projectile has two motions which act at the same time and do not affect one another. One of the motions is constant speed along a straight line, corresponding to the amount of launch speed and its direction. The second motion is downward acceleration at 10 m/s² caused by Earth's gravity, which takes effect immediately upon launch. The trajectory of a projectile becomes simple to understand when these two simultaneous motions are kept in mind.

This activity also demonstrates the main thing that scientists do: create models to help you understand how things in nature work. In this activity you saw how two kinds of models, a mathematical model (the table of times, speeds, and distances during falling) and a physical model (the evenly spaced strings of calculated lengths) correspond to reality when a ball is thrown. For a scientific model to be accepted, the model must match reality in nature. By that requirement, the models used in this activity were good ones.

Technology as a Tool

Trajectories of projectiles can be modeled using a computer or graphing calculator. Your teacher, or someone else familiar with such devices, may be able to help you find computer software or enter equations into a calculator. These tools will allow you to manipulate variables such as launch angle, launch speed, launch height and range to enhance your ability to explore and understand projectile motion.

ANSWERS

For You To Do (continued)

10. a) Yes.

   b) All are "true trajectories".

   c) 45°.

   d) A ball thrown straight up.

ACTIVITY NINE: The Shot Put

REFLECTING ON THE ACTIVITY AND THE CHALLENGE

The information learned about projectile motion in this activity applies not only to the shot put but to any track and field event that involves throwing things into the air (including the self-launching of a human body as in the hurdles, long jump, or high jump). It has been reported that one Olympian who competed in the shot put increased his range in that event by nearly 4 meters, based on suggestions made by a physicist. You are now a physicist specializing in projectile motion. How will you help your school's track team?

PHYSICS TO GO

1. If the launching and landing heights for a projectile are equal, what angle produces the greatest range? Why?

2. Compared to a launch angle of 45°, what happens to the amount of time a projectile is in the air if the launch angle is:
   a) greater than 45°?
   b) less than 45°?

3. For a constant launch speed, what angle produces the same range as a launch angle of:
   a) 30°?
   b) 15°?

4. If you launch a projectile from a high building, the angle for greatest range is less than 45°. Explain why this is true.

5. If a shot putter releases the projectile 2 m above level ground, what angle produces the greatest range? 45°? More than 45°? Less than 45°? Why? How could you find the exact angle?

6. Analysis of performances of long jumpers has shown that the typical launch angle is about 18°, far less than the angle needed to produce maximum range. Why do you think this occurs?

7. You are familiar with Carl Lewis as a medal-winning sprinter. But he is also an Olympic gold medalist in the long jump. Why do you think he's successful in both events?

S 43

ANSWERS

Physics To Go

1. 45°.

2. a) More time.

   b) Less time.

3. a) The complement of the angle: 60°.

   b) 75°.

4. Mathematical proof of the statement is beyond the scope of this physics course, but it is true that the maximum range of a projectile launched from a high building — or a shot putter's hand above ground level — is produced by using a launch angle of less than 45°. The 2.4-m Stick and String Trajectory Model used in For You To Do can be used to demonstrate this phenomenon empirically. To do the demonstration, place the "launch" end of the stick on the tray at the bottom of a chalkboard with the stick elevated at 45°. Slide the inclined stick until the end of the 1.8-meter string hangs beyond the end of the tray and mark the position of the bottom end of the

1.8 m string on the wall; it should reach a level between 10 and 11 cm below the level of the chalk tray. Hold a meter stick in horizontal orientation against the wall at the level of the mark. Then depress the angle of the stick slowly and observe that the curve of the trajectory (as imaginary "filled in" between the bottom ends of the last two strings) crosses the meter stick at a greater range. It is a subtle, yet observable demonstration which answers the question.

5. Less than 45°, depending on the height of launch and the speed. As noted in problem #4 above, calculation of the exact angle is complex, but possible (in some college level physics textbooks this is referred to as the "extended projectile problem"). The optimum launch angle for most shot putters probably would be between 40° and 45°.

6. The horizontal running speed of a long jumper is much greater than the initial vertical speed that the jumper can attain; therefore, the angle is far less than 45 °.

7. Carl Lewis can run fast, which is half of the requirement for a good long jumper. Apparently he jumps vertically well, too.

Measuring the Acceleration Due to Gravity

You must select a method of measuring the acceleration due to gravity based on the equipment available at your school. Several possibilities exist:

- Sonic ranger connected to a computer or graphing calculator with software for analysis of motion data.
- Picket fence and timer.
- Mechanical release and foot pad attached to timer.
- Photocell gates attached to timer.
- Stroboscopic Photography.
- Calibrated Ticker-tape timer.

Manufacturers manuals and physics laboratory manuals abound which give detailed directions for measuring "g." The trick is to match the method to the equipment available at your school. It is possible that the mathematics teachers at your school have graphing calculators (and, possibly, motion sensors for the calculators) and that you haven't heard about their existence. If that is the case, by all means approach the teachers about using the calculators; if necessary, your budget may allow purchasing motion sensing attachments for the calculators.

NOTES

Four-meter Tack Strip for Trajectory Model

If you have a chalkboard a minimum of 4 m wide which has a cork strip across the top for tacking material to the strip, you are in business and need to do nothing more than prepare labels and pins to be placed at 40 cm intervals along the strip (see label specifications in equipment list). If you do not have such a setup, you need to create something which will serve the same purpose.

If you have a chalkboard without a tacking strip you could erect a wood strip across the top edge of your chalkboard (the strip can be in sections; it need not be one continuous strip). If the chalkboard is less than 4 m, it will probably serve if the expanse of wall on which it is mounted is 4 m wide.

If you do not have a chalkboard, or if yours is just too small, use a 4-m length of freezer paper taped to a wall, oriented about as a chalkboard would be on the wall and place a wood strip across the top edge of the paper (or tape labels and clip & string assemblies to the wall without using a wooden strip).

Whatever you use, it should appear as shown below:

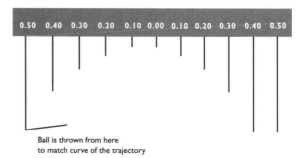

When the clip & string assemblies are in place for the first half of the model of the trajectory, the appearance of the model should be (lengths of clip & string assemblies not shown to exact scale) as shown:

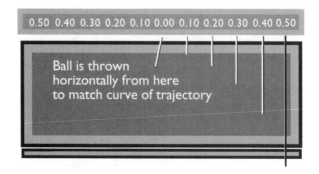

The appearance of the model for the second half of the trajectory should be as shown:

The portable trajectory model will have the same appearance as the right-hand half of the above model, except the strip on the wall is replaced by a longer (2.4 m instead of 2.0 m) stick, and a label for a time of 0.60 seconds is added with a string of length 1.8 meters at the 1.60 second mark to represent the distance of fall at that time.

2.4-meter Stick and String Model of the Trajectory of a Projectile Launched at a Speed of 4.0 m/s

• An 8-foot (2.44m) length of 2"x2" lumber serves as the stick   • Washers may be used to weight string ends

• Place time labels at 0.4 m (40cm) intervals along stick

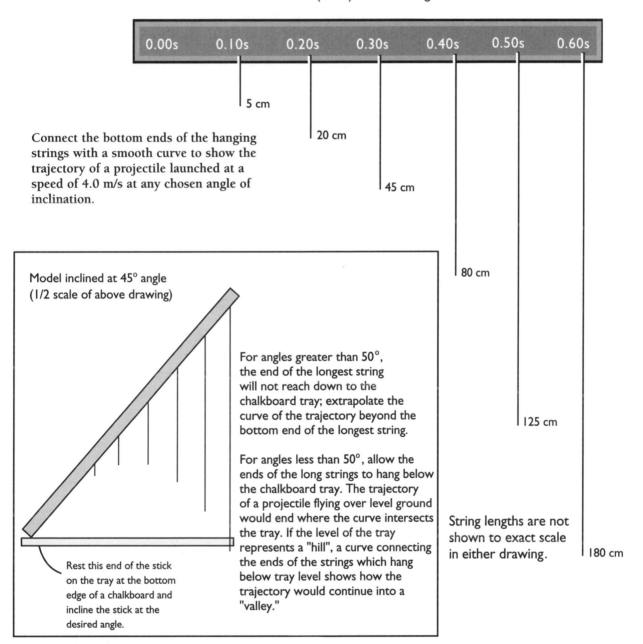

| 0.00s | 0.10s | 0.20s | 0.30s | 0.40s | 0.50s | 0.60s |

5 cm

Connect the bottom ends of the hanging strings with a smooth curve to show the trajectory of a projectile launched at a speed of 4.0 m/s at any chosen angle of inclination.

20 cm

45 cm

80 cm

125 cm

180 cm

Model inclined at 45° angle (1/2 scale of above drawing)

For angles greater than 50°, the end of the longest string will not reach down to the chalkboard tray; extrapolate the curve of the trajectory beyond the bottom end of the longest string.

For angles less than 50°, allow the ends of the long strings to hang below the chalkboard tray. The trajectory of a projectile flying over level ground would end where the curve intersects the tray. If the level of the tray represents a "hill", a curve connecting the ends of the strings which hang below tray level shows how the trajectory would continue into a "valley."

String lengths are not shown to exact scale in either drawing.

Rest this end of the stick on the tray at the bottom edge of a chalkboard and incline the stick at the desired angle.

For use with *Sports*, Chapter 1, ACTIVITY NINE: The Shot Put

©1999 American Association of Physics Teachers

ACTIVITY TEN
Energy in the Pole Vault

Background Information

Concepts involving energy are introduced in this activity, including:

- energy in three forms: kinetic energy, gravitational potential energy, and the energy stored in a spring.
- transformations among and conservation of the above listed forms of energy.

The treatment of the joule as the unit of energy in this activity is "soft" because groundwork for defining the joule in a mechanical context has not yet been established. If you have need for background about the joule, it is suggested that you look ahead to the Background Information for Chapter 2, Activity Three. There, you will find the definition of the newton as the unit of force which, in turn, is used to define the joule in terms of mechanical energy in its most elementary form, work.

At this stage without benefit of a definition of the joule, dimensional analysis can be used to satisfy yourself (and, only if it seems necessary, your students) that all of the equations given in the activity for forms of energy at least have the same unit:

Kinetic Energy = $1/2\ mv^2$

Unit: $(kg)(m/s)^2 = (kg)(m^2)/(s^2)$

Gravitational Potential Energy = mgh

Unit: $(kg)\ [m/(s)^2]\ m = (kg)(m^2)/(s^2)$

Potential Energy Stored in a Spring = $(1/2)kx^2$

(the spring constant, k, has a unit N/m which is equivalent to kg/s^2)

Unit: $(kg/s^2)\ m^2 = (kg)(m^2)/(s^2)$

As shown above, all of the forms of energy defined in the activity share the same unit, $(kg)(m^2)/(s^2)$. Since, for your information, 1 joule is the amount of work done when 1 newton of force is active through one meter of distance (1 joule = 1 newton x 1 meter) and since one newton is the force which will cause 1 kg to accelerate at 1 m/s^2, one joule also can be expressed as $(1\ kg\ m/s^2\ x\ 1m) = (kg)(m^2)/(s^2)$. Therefore, the equations are legitimate because the unit $(kg)(m^2)/(s^2)$ is equivalent to the standard energy unit, the joule.

The above information will prepare you to address questions about units, if they arise.

Active-ating the Physics InfoMall

In this final Chapter 1 activity, the title once again suggests the search keywords: "pole vault." And again you will not be disappointed. Try it and see.

In Physics Talk, we are introduced to Potential and Kinetic Energy. For more information, please consult any of the textbooks in the Textbook Trove. Also, you may wish to conduct searches of the InfoMall. Warning: you will want to limit your search to only a few stores or by using more keywords. If you search for "potential energy" you will get more than 1600 hits in the Textbook Trove alone!

Planning for the Activity

Time Requirements

- One class period.

Materials Needed

For each group:

- flexible plastic ruler or plastic strip
- starting ramp, directional
- centimeter ruler for making distance measurements
- penny coin
- sphere (marble or ball bearing)
- clamp(s) for holding plastic strip
- pen or pencil
- rubber band or tape
- rigid marking surface

Advance Preparation and Setup

You will need a supply (one per group) of 1-foot plastic rulers, or strips of plastic about the size of a ruler (the acetate strips used in many schools for demonstrating static electricity when the strips are rubbed with cloth would be ideal — if you do not have such strips, you may wish to see if other teachers have them). The rulers should be "floppy," having a thin, rectangular cross section which allows them to be bent with relative ease; safety caution: plastic rulers which are quite rigid may break and cause injury; do not use them for this activity.

You will also need marbles or ball bearings of mass such that, when rolled at reasonable speeds, they will produce reasonable, measurable amounts of bend in a clamped, flexible ruler upon striking the ruler near the tip.

You will need starting ramps which can be used to provide the range of speeds for the marbles; if you have no ramps, the groove between pages of an open book will serve as a ramp.

It is suggested that, in advance, you determine how the rulers will be clamped in the two orientations required for this activity (see diagrams in student text). The particular way of clamping the rulers will depend on the nature of the furniture in your classroom.

Teaching Notes

The activity is in two parts. The first part (steps 1 to 3 of For You To Do) simulates the vaulter transferring kinetic energy to the pole, causing the pole to bend; the second part simulates transfer of the potential energy stored in the bent pole to the vaulter as gravitational potential energy.

Before beginning the activity, conduct a brief discussion of restoring forces, explaining that it is believed that interatomic forces in solid objects (such as rulers) behave as miniature springs which connect atoms. When an object is deformed by application of an external force, as when bent, some of the springs are compressed and others are stretched. When the external force is removed, the springs relax to normal length, providing a "restoring force" which returns the object to its original shape. In some cases the external force exceeds the "elastic limit" of the material, causing the springs to break, and permanently deforming the object or breaking it into parts.

Comments on the transformation of kinetic energy into energy stored in a bent ruler (steps 1 to 3 of For You To Do):

It may not be clear to students that this part of the activity is meant to simulate the part of the pole vault event in which the vaulter "plants" the pole in the pit and "hangs on," stopping his running and causing the pole to bend as he and the pole rise due to rotation of the pole, the end of the pole in the pit

serving as the axis of rotation. The main transformation of energy active in this part of the pole vault is the transfer of the kinetic energy of the vaulter into potential energy stored in the bent pole. This is what is being simulated.

Keep in mind that the procedures ask students to find only a "semi-quantitative" relationship between the speed of the rolling ball and the amount of deflection of the tip of the ruler (e.g., "The faster the ball, the more the ruler bends.")

Some students may choose to be more quantitative when establishing the speed of the ball and choose, for example, ramps heights (or lengths along a straight inclined ramp) in a ratio such as : 1 : 2 : 3. Such thinking should be encouraged. However, the same students may hold the common misconception that ramp heights in a simple ratio such as 1 : 2 : 3 will produce speeds in the same simple ratio. If you detect that misconception, you should move to correct it. The information below would be helpful for that purpose:

Speed of ball leaving ramp:
$v = \sqrt{2gh}$

Therefore, the speed of the ball as it leaves the ramp is proportional to $\sqrt{h}$, and balls launched from heights in the ratio 1 : 2 : 3 would have speeds in the (same order) ratio of the square roots of 1, 2, and 3, approximately 1.0 : 1.4 : 1.7. The ramp heights which would produce speeds in the ratio 1 : 2 : 3 would be heights in the ratio of the squared values of 1, 2, and 3, the sequence 1 : 4 : 9.

Going on to consider the kinetic energy of the ball as it leaves the ramp, recall that the kinetic energy of the ball is proportional to the squared value of the ball's speed. Therefore, ramp heights in the ratio 1 : 2 : 3, producing speeds in the ratio 1 : 1.4 : 1.7 (as explained above), would result in kinetic energies in the ratio of the squared values of 1.0, 1.4 and 1.7, the ratio 1 : 2 : 3.

Therefore, selecting ramp heights (or distances along a straight ramp) in a 1 : 2 : 3 ratio results in kinetic energies, not speeds, in the ratio 1 : 2 : 3.

Finally, the kinetic energy of the ball is transformed into energy stored in the bent ruler (only for an instant, until the ruler "snaps back"):

$1/2 mv^2 = 1/2\ kx^2$

Solving for x, the deflection of the ruler:

$x = v\ \sqrt{m/k}$

1

Since m is constant for a chosen ball and since k is a constant for a chosen ruler, the deflection of the ruler, x, is proportional to the speed of the ball, v.

Therefore, in the likely scenario where students choose ramp heights in the ratio $1 : 2 : 3$ (resulting in ball speeds in the ratio $1.0 : 1.4 : 1.7$) the defection of the tip of the ruler should be in the same ratio as the speeds, $1 : 1.4 : 1.7$.

Maintain awareness as you interact with students during this activity (which is recommended) that some students will desire and perhaps even try to have the data "come out" to match their preconceptions (often misconceptions); keep them honest!

The actual data for this part of the activity will depend on the mass of the ball used, the speeds imparted to the ball, and the "stiffness"(k value) of the ruler used. Data for a sample case is not given due to the small likelihood that the case would resemble the conditions which you will have.

Comments on the transformation of the energy stored in a bent ruler into gravitational potential energy (step 4 of For You To Do):

Be sure that students understand that this part of the activity is intended to simulate that part of the pole vault in which the energy stored in the bent pole is transferred to increase the gravitational potential energy of the vaulter, "catapulting" the vaulter's body upward, hopefully high enough to clear the bar.

The relationship between the displacements of the end of the clamped ruler and the height to which a coin resting on the ruler should rise is:

$1/2\ k\ x^2 = mgh$

Solving for h, $\ h = k\ x^2 /\ 2mg$

Therefore, for a chosen ruler (constant k), chosen coin (constant m), and since g is a constant, the height to which the coin should fly should be proportional to the squared value of x. This suggests that displacements of the ruler (x values) of 1, 2 and 3 cm should result in heights of the coin which have relative values fitting a ratio of the squares of x,

$1 : 4 : 9$.

Depending on the stiffness of rulers available to you, you could adjust the x values specified in the student text to another set of values which would give reasonable amounts of ruler bend and coin height; for example you could specify x values of 0.5, 1.0 and 1.5 cm or you could use 2, 4, 6 cm. In any case, choose x's which are in the ratio $1 : 2 : 3$.

Again, sample data is not given because it would vary greatly with conditions.

Do not assume that students can absorb the energy equations presented in Physics Talk without your help. Help students link data gathered during For You To Do to the equations. The joule as the unit of energy is somewhat casually introduced in the sample calculation presented in Physics Talk because rigorous definition of the joule is reserved until force and work are formally defined in Chapter 2, Activity Three. The joule is not needed for any of the Physics To Go problems and should not be emphasized at this time.

NOTES

Activity Overview

In this activity students examine the conservation of energy as it occurs in a pole vault.

Student Objectives

Students will:

- understand and apply the equation Kinetic Energy = 1/2 mv^2.

- understand and apply the equation Gravitational Potential Energy = mgh.

- recognize that restoring forces are active when objects are deformed.

- understand and apply the equation Potential Energy Stored in a Spring = 1/2kx^2.

- understand and measure transformations among kinetic and potential forms of energy.

- conduct simulations of transformations of energy involved in the pole vault.

ANSWERS FOR THE TEACHER ONLY

What Do You Think?

Forms of energy involved in pole vault: kinetic energy, potential energy stored in a spring, gravitational potential energy (work, done by the vaulter using arm and upper body muscles is involved, but not mentioned in the activity).

Human limitations on the kinetic energy and the amount of work which the vaulter is able to provide as input limit the height.

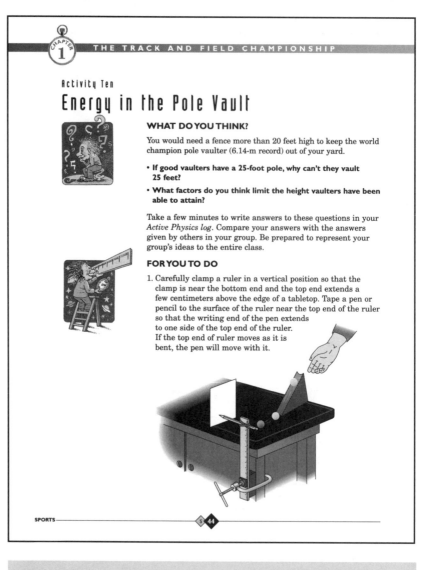

THE TRACK AND FIELD CHAMPIONSHIP

Activity Ten
Energy in the Pole Vault

WHAT DO YOU THINK?

You would need a fence more than 20 feet high to keep the world champion pole vaulter (6.14-m record) out of your yard.

- **If good vaulters have a 25-foot pole, why can't they vault 25 feet?**
- **What factors do you think limit the height vaulters have been able to attain?**

Take a few minutes to write answers to these questions in your *Active Physics log*. Compare your answers with the answers given by others in your group. Be prepared to represent your group's ideas to the entire class.

FOR YOU TO DO

1. Carefully clamp a ruler in a vertical position so that the clamp is near the bottom end and the top end extends a few centimeters above the edge of a tabletop. Tape a pen or pencil to the surface of the ruler near the top end of the ruler so that the writing end of the pen extends to one side of the top end of the ruler. If the top end of ruler moves as it is bent, the pen will move with it.

SPORTS   S 44

ANSWERS

For You To Do

1. Student activity.

ACTIVITY TEN: Energy in the Pole Vault

2. Set up a ramp as shown in the diagram. Three different starting points on a ramp will be used to roll a ball across the tabletop at three different speeds. Each time the ball rolls, it will strike the ruler near the top end, causing the ruler to bend. A marking surface held in contact with the tip of the pen or pencil will be used to measure the deflection.

3. Roll the ball at low, medium and high speeds to bend the ruler. In each case, measure the amount of deflection of the end of the ruler as indicated by the length of the pen mark.

   ⟍ a) If the rolling ball represents the running vaulter and the ruler represents the pole in the model, how does the amount of bend in the pole depend on the vaulter's running speed? Record your data and response in your log.

4. Carefully clamp a ruler flat-side down to a tabletop so that two thirds of the ruler's length extends over the edge of the table.

5. Place a penny on the top surface of the ruler at the outside end.

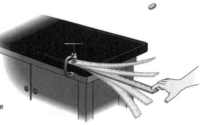

6. Use a second ruler to measure a 1 cm downward deflection of the outside end of the clamped ruler; that is, bend the ruler downward. Prepare to measure the maximum height to which the coin flies upward using the position of the coin when the ruler is relaxed at the "zero" vertical position of the coin. Release the ruler.

   ⟍ a) Record in your log the height to which the coin travels.
   ⟍ b) Repeat step 6 above for ruler deflections of 2 cm and 3 cm, and in each case, record the maximum height of the "vaulted" coin.

   ⚠ **Do not deflect the ruler excessively.**

   ⟍ c) How does the height to which the coin travels seem to be related to the amount of deflection of the ruler? (Remember, the coin is a projectile, in this case.)

S 45 —————————————————— SPORTS

Answers

For You To Do *(continued)*

2. Student activity.

3.a) The greater the speed of the vaulter, the greater the bend of the pole.

6.a-c) See Teaching Notes re: possible quantitative analysis. The greater the amount of deflection of the ruler, the greater the height of the coin.

FOR YOU TO READ

Conservation of Energy in the Pole Vault

The pole vault is a wonderful example of the Law of Conservation of Energy. The forms of energy are changed, or transformed, from one to another during a vault, but, in principle, the total amount of energy in the system of the vaulter and the pole remains constant. Food energy provides muscular energy for the vaulter to run, gaining an amount of kinetic energy equal to $\frac{1}{2}mv^2$, where m is the runner's mass and v is the running speed.

Some of the vaulter's kinetic energy is used to catapult the vaulter with an initial speed upward and the remaining kinetic energy is converted into work done on the pole to cause it to store an amount of potential energy $\frac{1}{2}kx^2$, where k is a constant associated with the "springiness" of the specific vaulting pole and x is the deflection of the end of the pole. As the bent pole straightens, its potential energy is delivered to the vaulter to increase the vaulter's gravitational potential energy by an amount, mgh, where m is the vaulter's mass, g is the acceleration due to gravity (10 m/s²) and h is the height to which the vaulter's body is lifted.

PHYSICS TALK

The terms and equations introduced in the above For You to Read are summarized for you here.

Potential energy of a spring:
$PE = \frac{1}{2}kx^2$ where k is the spring constant, and x is the amount of bending.

Gravitational Potential Energy:
$PE = mgh$ where m is the mass of the object,
g is the acceleration due to gravity,
h is the height through which the object is lifted.

Kinetic Energy:
$KE = \frac{1}{2}mv^2$ where m is the mass of the moving object,
and v is the speed of the object. Energy is expressed in a unit called a joule.

**REFLECTING ON THE ACTIVITY
AND THE CHALLENGE**

In this activity you learned that throughout the event of pole vaulting, energy changes from one form to another, but the total amount of energy in the system at all instants remains the same. (A small amount of energy may be lost from the system of the vaulter and pole by making a dent in the end of the pit, which stops the pole or by generating heat in the pole as it bends.)

This will be important for the athletes at your school to know. They will need to understand what they must do to gain the greatest height in a pole vault. By examining the equations for kinetic and potential energy, they will also be able to appreciate why there is a limit to the height that somebody can vault. The faster someone runs, the more kinetic energy they have. This kinetic energy makes the pole bend. The more kinetic energy there is, the more bend in the pole is expected. Potential energy stored in the bent pole will be transformed to increase the vaulter's gravitational potential energy. The more bend in the pole, the higher the vaulter goes. One key to success in the pole vault is to have the most kinetic energy. Another key to success is to bend the pole as much as possible. You may be able to use this knowledge to help a pole vaulter in your school improve performance.

PHYSICS TO GO

1. Describe the energy transformations in the shot put.

2. Describe the energy transformations in the high jump.

3. Assume that a vaulter is able to carry a vaulting pole while running as fast as Carl Lewis in his world record 100-m dash. Also assume that all of the vaulter's kinetic energy is transformed into gravitational potential energy. What vaulting height could that person attain? (Hint: Use the equation $\frac{1}{2}mv^2 = mgh$.)

4. Why doesn't the length of the pole singly determine the limit of vaulting height?

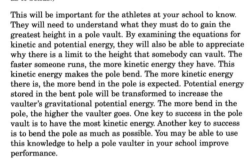

Physics To Go

1. The shot putter imparts a launching speed to the projectile, and therefore kinetic energy, by giving the shot two speeds which, since they are in the same direction, add together. One of the motions is provided by the spinning motion of the shot putter before release, and the other is provided by the thrusting action of the shot putter's arm. In both cases, the athlete does work, a form of energy (force times distance), which is transformed into the kinetic energy which the ball has upon release. The horizontal component of the shot's speed is maintained while the projectile is in the air; the vertical component of the speed can be used to calculate the part of the kinetic energy which will be transformed into gravitational potential energy, allowing prediction of the height to which the shot will rise at the peak of its flight.

2. The high jumper uses the leg muscles to jump upward with an initial kinetic energy which is transformed into potential energy, causing the jumper to rise to a height which hopefully is enough to clear the bar.

3. Lewis's average speed in the dash: 10.1 m/s.

   Solving for h in the equation given in the problem:

   $h = v^2/2g$ [m cancels]

   Since g is known to only two significant figures (10 m/s^2), round Lewis's speed to 10 m/s when substituting in the above equation:

   $h = (10\ \text{m/s})^2/ (2 \times 10\ \text{m/s}^2) = (100\ \text{m}^2/\text{s}^2) / (20\ \text{m/s}^2) = 5.0$ m

4. The vaulter's kinetic energy, determined by running speed, plus the amount of work which the vaulter does using the arms to lift the body determine the maximum height to which the vaulter can rise, regardless of the length of the vaulting pole.

ANSWERS

Physics To Go
(continued)

5. The amount of heat generated during the bending of the pole would "rob" part of the potential energy can be stored in the pole by causing it to bend less than it would if no heat were generated and by causing the pole to straighten with less "straightening" speed than if no heat were generated; overall, heating of the pole would reduce the vaulter's height.

6. $v = \sqrt{2gh} = \sqrt{2 \times 10 \text{ m/s}^2 \times 4.6 \text{ m}}$
$= \sqrt{92 \text{ m}^2/\text{s}^2} = 9.6 \text{ m/s}$

7. Notice that Emma's height was rounded to two significant figures in the above substitution.

$v = \sqrt{2gh} = \sqrt{2 \times 10 \text{ m/s}^2 \times 6.1 \text{ m}}$
$= \sqrt{120 \text{ m}^2/\text{s}^2} = 11 \text{ m/s}$

Notice that while Sergei's speed was only about 15% greater than Emma's speed, his vault height, 6.1 m compared to Emma's 4.6 m, was about 32% higher, this due to the squared effect of speed on kinetic energy; squaring the ratio of speeds verifies that the ratio of heights should be 1.32 : 1.00, as shown below

$[(6.1 \text{ m/s}) / (4.6 \text{ m/s})]^2 = (1.15)^2 = 1.32$

---

5. Some poles lose a significant amount of energy to heat as they flex. Use the Law of Conservation of Energy to explain how this would affect performance.

6. The women's pole vault world record as of spring 1997 was 4.55 m, set by Emma George. What do you estimate was Emma's speed prior to planting the pole? Use conservation of energy for your prediction.

7. Sergei Bubka held the world record for the pole vault as of spring 1997 at 6.14 m. How did Sergei's speed compare with Emma George's speed?

STRETCHING EXERCISE

You have used the transformation of kinetic energy into gravitational potential energy to predict pole-vaulting heights in some of the above problems. The same transformation was used to predict the men's pole-vault height for each Olympiad since 1896 using the speed in the men's Olympic 100-m dash during corresponding Olympiads. Something strange shows up in the analysis:

a) From 1896 to 1964, the actual height vaulted in each Olympiad was less than the height predicted from the speed in the 100-m dash in the same Olympiad, and

b) From 1968 to the present, the vault height in each Olympiad has been greater than the height predicted from the speed in the 100-m dash.

What changed? Better running shoes? Something else? Has the Law of Conservation of Energy been violated by pole vaulters? Try to find one or more reasons to explain this difference.

ANSWERS

Stretching Exercises

A table of detailed data showing actual and predicted pole vault heights based on the average speed for the men's 100-m dash for all Olympiads from 1896 to 1992 is provided on a Blackline Master in the Background Material for this activity. Reproduce it for students.

The "quantum leap" in pole vault performance was caused by the introduction of the flexible fiberglass pole between 1964 and 1968. The early poles made from wood and later aluminum were quite rigid; therefore, a significant amount of vaulter's kinetic energy was used inefficiently to do work which "tried" to bend the pole (and did bend the pole, but not by an amount which helped the vaulter store a significant amount of energy in the pole. Introduction of flexible poles allowed the vaulter's kinetic energy to be used more efficiently by transforming part of the kinetic energy into potential energy stored in the bent pole (later recovered when restoring forces caused the bent pole to straighten, carrying the vaulter along during the straightening process).

If the maximum height attainable by a vaulter depends only on the vaulter's kinetic energy at takeoff, it should seem discrepant to anyone who believes in conservation of energy that the height attained could be greater than the height predicted from the kinetic energy. If energy is conserved in the event, some "input" energy must be involved in addition to the kinetic energy input. Additional energy is involved: the vaulter uses arm muscles to do work to lift the body, using the pole in a way similar to the way a person lifts the body when using a "chinning bar" for exercise. So there is no mystery; the vaulter's muscles provide part of the energy to get over the bar. Clearly, vaulters have been doing so forever, but it is apparent in the data only since introduction of flexible poles.

SPORTS

PHYSICS AT WORK

Erv Hunt

COACH TO THE CHAMPIONS

This year when Erv Hunt begin his 25th season as the head coach of the University of California at Berkeley's track and field team, he will bring more than just a reputation as one of the foremost track and field coaches in the United States, he will also bring to his team the experience of having served as the US men's head track and field coach at the 1996 Olympic Games in Atlanta.

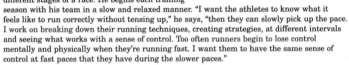

"We have a much better understanding of the science of running now, the physics of it, than we use to," states Erv, "and that is one of the reasons runners have improved over the years." Erv believes that it's important when training to understand pace work and intervals during the different stages of a race. He begins each training season with his team in a slow and relaxed manner. "I want the athletes to know what it feels like to run correctly without tensing up," he says, "then they can slowly pick up the pace. I work on breaking down their running techniques, creating strategies, at different intervals and seeing what works with a sense of control. Too often runners begin to lose control mentally and physically when they're running fast. I want them to have the same sense of control at fast paces that they have during the slower paces."

To be around and train some of the great athletes during his career has been an incredible experience for Erv. "For someone to reach that level of excellence makes them somewhat a different breed of person," and Erv has had the privilege to work with some of the best.

 49 — SPORTS

Olympic Pole Vault: Actual Heights and Heights Predicted From Average Speed in Men's 100-meter Dash, 1896 to 1992

Year	Time	Speed for 100 m race	Predicted Height (m)	Predicted Height (ft)	Actual Height (ft)	Actual Height (m)
1896	12	8.33	3.54	11.69	10.83	3.28
1900	11	9.09	4.22	13.91	10.83	3.28
1904	12	9.09	4.22	13.91	11.48	3.48
1908	10.8	9.26	4.37	14.43	12.17	3.69
1912	10.8	9.26	4.37	14.43	12.96	3.93
1920	10.8	9.26	4.37	14.43	13.42	4.07
1924	10.6	9.43	4.54	14.98	12.96	3.93
1928	10.8	9.26	4.37	14.43	13.77	4.17
1932	10.3	9.71	4.81	15.87	14.15	4.29
1936	10.3	9.71	4.81	15.87	14.27	4.32
1948	10.3	9.71	4.81	15.87	14.10	4.27
1952	10.4	9.62	4.72	15.57	14.92	4.52
1956	10.5	9.52	4.63	15.27	14.96	4.53
1960	10.2	9.80	4.90	16.18	15.42	4.67
1964	10	10.00	5.10	16.84	16.73	5.07
1968	9.95	10,05	5.15	17.01	17.71	5.37
1972	10.14	9.86	4.96	16.38	18.04	5.47
1976	10.06	9.94	5.04	16.64	18.04	5.47
1980	10.25	9.76	4.86	16.03	18.96	5.74
1984	9.99	10.01	5.11	16.87	18.85	5.71
1988	9.92	10.08	5.18	17.71	19.77	5.99
1992	9.96	10.04	5.14	16.97	19.02	5.76

Chapter 1 Assessment

In this chapter you studied physics as it relates to track and field events. Now you have the knowledge to write a physics manual about track and field training for your high school team to help improve its performance. The manual should:

- **help students compare themselves with the competition;**
- **include a description of physics principles as they relate to track events;**
- **provide specific techniques to improve performance.**

Review and remind yourself of the grading criteria that was agreed on by the class at the beginning of the chapter. You may have decided that some or all of the following qualities should be graded in your track and field manual:

- **physics principles**
- **inclusion of charts**
- **past records**
- **relevant equations**
- **definitions**
- **specific techniques**

Any advice you give should be understandable to athletes who have not studied physics. You can describe any activities you have done to explain how you know that the technique works, but you should not tell so much about each activity that the reader becomes bored.

Physics You Learned

Distance, Time, Speed

Speed vs. Time graphs

Histograms

Average speed, Instantaneous speed

Distance vs. Time graphs

Acceleration

Accelerometers

Speed = Frequency × Stride Length

Centripetal acceleration

Leaning during acceleration

Trajectory motion

Horizontal and Vertical motion

Optimum angle

Energy

Potential, Kinetic

S 50

Alternative Chapter Assessment

1. A sprinter runs the 100-m race in 9.80 s. Calculate her average speed for the race.

2. At the 50-meter mark in the above race, do you think that the elapsed time was less than, equal to, or more than 4.90 s? Explain your answer.

3. Who wins the race?

   a) the runner with the highest finishing speed

   b) the runner with the best average speed

   c) the runner with the greatest top speed

4. The following table gives the split times for a 100-m race run by Leroy Burrell in the 1991 World Championships:

Distance(m)	Time(s)
10	1.83
20	2.89
30	3.79
40	4.68
50	5.55
60	6.41
70	7.28
80	8.12
90	9.01
100	9.88

   a) Fill in the table below by calculating the time it takes Leroy Burrell to run each interval and his average speed for each interval.

Interval (m)	Time(s)	Speed(m/s)
0-10	_____	_____
10-20	_____	_____
20-30	_____	_____
30-40	_____	_____
40-50	_____	_____

50-60	_____	_____
60-70	_____	_____
70-80	_____	_____
80-90	_____	_____
90-100	_____	_____

b) At what distance did Leroy stop accelerating?

c) What is Leroy's average speed for the entire race?

5. A runner is holding a cork accelerometer. Draw the position of the accelerometer for each phase of the race.

   a) The runner is accelerating to the right.

   b) The runner is running at constant speed.

6. A runner has a stride length of 1.0 m. During the race, he takes 26 strides in 13 s. What is his average speed?

7. A person standing on a 50.0 m-high cliff throws a ball out horizontally with an initial speed of 10 m/s. At the same instant, she drops her mitt. Which hits the ground first? Why?

8. What launching angle produces the greatest range if the launch and landing heights are the same?

9. At what angle with respect to the horizontal should the shot put be launched if the starting height is 2.0 m, to have the greatest range?

10. The following chart shows the times for three runners to run each of the five 20.0 m intervals in a 100-m race.

Runner	0-20.0 m	20.0-40.0 m	40.0-60.0 m	60.0-80.0 m	80.0-100 m
A	2.62 s	2.60 s	2.53 s	2.52 s	2.49 s
B	2.60 s	2.58 s	2.53 s	2.45 s	2.52 s
C	2.62 s	2.61 s	2.53 s	2.48 s	2.50 s

Answer the following questions based on this information:

a) Which runner had the fastest speed in any one time interval? How fast was that runner moving in that interval?

b) Which runner was running the fastest at the end of the race? How fast was that runner moving?

c) Which runner wins the race? Explain your answer using the concept of speed in your explanation.

11. Determine the maximum possible height that a pole vaulter could clear when a speed of 6.0 m/s is reached prior to planting the pole. How will the possible height change if the maximum speed is 12 m/s? Assume all the kinetic energy is converted to potential enery and that the vaulter does no work while rising.

12. Explain the effect on a runner's speed if attempting to double the stride length also cuts the frequency in half.

13. An accelerometer indicates that an acceleration to the left has occurred. Discuss and explain two possible motions that would result in this observation.

14 a) A ball is thrown into the air. Describe the speed and acceleration at the following places:

i) on its way up.

ii) at the peak of its path.

iii) while falling back down.

b) How and where would the speed and acceleration change if a human were jumping into the air and coming back down?

Alternative Chapter Assessment Answers

1. $v = d/t = 100$ m/9.80 s = 10.2 m/s

2. Since it takes 50 m or more to reach maximum speed when starting from rest, the elasped time will be more than the time at the end of the race.

3. b)

4. a)

| Leroy Burrell | | 1991 | |
Distance(m)	Time(s)	Interval time	Speed(m/s)
10	1.83	1.83	5.46
20	2.89	1.06	9.43
30	-3.79	0.9	11.11
40	4.68	0.89	11.24
50	5.55	0.87	11.49
60	6.41	0.86	11.63
70	7.28	0.87	11.49
80	8.12	0.84	11.90
90	9.01	0.89	11.24
100	9.88	0.87	11.49

   b) 60-m mark

   c) v= 10.1 m/s

5. Student draw accelerometer with cork

   a) leaning to the left

   b) straight up and down

6. $v = 2.0$ m/s

7. They hit at the same time because the horizontal motion is independent of the vertical motion.

8. 45°.

9. 41°.

10.a) runner B   speed = 8.16 m/s during 60.0-80.0 interval

 b) runner A   speed = 8.03 m/s during last interval

 c) runner B

The total time for B is 12.68 s. Average speed for B = 100 m/12.68 s = 7.89 m/s.

The total time for C is 12.74 s. Average Speed = 7.85 m/s.

The total time for A is 12.76 s. Average Speed = 7.84 m/s.

The runner with the greatest average speed will win the race.

11. Using $1/2\ mv^2$=mgh and solving for h the speed of 6.0 m/s will result in h = 1.8 m, if v is doubled h will quadruple giving a new height of 7.2 m.

12. Doubling the frequency alone would double the speed, cutting the frequency in half would cut the speed in half, since both are changed, the overall speed will remain exactly the same as what it was.

13. An acceleration to the left could occur under the following conditions:

 object is moving to the left and speeding up

 object is moving to the right and slowing down

14. a) Going up — speed is up and decreasing in magnitude, acceleration is constant and directed down.

At peak of path — speed is zero and acceleration is still constant and down.

Going down — speed is down and increasing in magnitude, acceleration is still constant and directed down.

 b) A human once in the air behaves exactly like the ball in terms of velocity and acceleration.

©1999 American Association of Physics Teachers

2

Sports Chapter 2-Physics in Action
National Science Education Standards

Chapter Challenge

PBS has decided to televise sporting events in a program that has educational value. Students are challenged to provide the voice-over on a sports video that explains the physics of the action as an audition for the job of sports broadcaster.

Chapter Summary

To gain knowledge and understanding of physics principles necessary to meet this challenge, students work collaboratively on activities that investigate a variety of the forces that affect the changes in motion that are commonly observed in sports. These experiences engage students in the following content identified in the *National Science Education Standards*.

Content Standards

Unifying Concepts

• Systems, order, and organization

• Evidence, models and explanations

• Constancy, change and measurement

Science as Inquiry

• Identify questions and concepts that guide scientific investigations

• Use technology and mathematics to improve investigations

• Communicate and defend a scientific argument

• Formulate & revise scientific explanations & models using logic and evidence

Physical Science

• Motions and Forces

• Conservation of energy & increase in disorder

History and Nature of Science

• Science as a human endeavor

• Nature of scientific knowledge

• Historical perspectives

Key Physics Concepts and Skills

Activity Summaries	Physics Principles

Activity One: A Running Start

Students measure the motion of a ball rolling down then up the sides of a bowl and find the ratio of the "running start" to the vertical distance. From this, they are introduced to the concept of inertia.

- **Acceleration**
- **Gravity**
- **Galileo's Principle of Inertia**
- **Newton's First Law of Motion**

Activity Two: Push or Pull

Students construct, calibrate, and use a simple force meter to explore the variables involved in throwing a shot put. They then connect their observations and data to a study of the laws of motion.

- **Newton's Second Law of Motion**
- **Relationship of mass and force to acceleration**
- **Gravity**

Activity Three: Center Of Mass

By finding the balance points on objects with a variety of shapes, students are introduced to the affect motion of the athlete's center of mass has on balance and performance.

- **Center of Mass**
- **Gravity**

Activity Four: Defy Gravity

Students learn to measure hang time and analyze vertical jumps of athletes using slow-motion videos. This introduces the concept that work when jumping is force applied against gravity.

- **Gravity**
- **Potential and kinetic energy**
- **Work**
- **Vertical accelerated motion**

Activity Five: Run and Jump

Thinking about the direction in which they apply force to move in a desired way introduces students to the concept that a force has an equal and opposite force. They test this concept, then apply it to a variety of motions observed in sports.

- **Force vectors**
- **Weight and gravity as forces**
- **Newton's Third Law of Motion**

Activity Six: The Mu of the Shoe

Students measure the amount of force necessary to slide athletic shoes on a variety of surfaces. From this and the weight of the shoe, they learn to calculate friction coefficients. They then consider the affect of friction on an athlete's performance.

- **Gravity**
- **Frictional force**
- **Normal force**
- **Coefficient of Sliding friction**

Activity Seven: Concentrating on Collisions

Students investigate the affect of a ball's velocity on its motion after a collision. They then apply these observations and what they now know about opposing forces in motion to describe collisions of balls and athletes in sporting events.

- **Newton's Third Law of Motion**
- **Mass**
- **Velocity**
- **Momentum**

Activity Eight: Conservation of Momentum

Additional collisions between objects allow students to investigate what happens when the objects stay together or "stick" after the collision.

- **Newton's Third Law of Motion**
- **Momentum = Mass x Velocity**
- **Velocity**
- **Law of Conservation of Momentum**

Activity Nine: Circular Motion

Students use an accelerometer to test the direction of acceleration when spinning in a chair. From this, they investigate the forces involved in the movement of turning objects and athletes.

- **Inertia**
- **Centripetal acceleration**
- **Centripetal force**

Equipment List For Chapter Two

QTY	TO SERVE	ACTIVITY	ITEM	COMMENT
1	Class	4	*Active Physics Sports* Content Video	Segments: Ice skaters, basketball player.
1	Group	3	Adhesive dot or patch of tape	For making spot on Objects A, B, C, D.
1	Group	8	Apparatus for monitoring speed	See suggestions in Teacher's Edition for Activity Eight.
1	Group	6	Athletic shoes	Student's shoe may be used.
1	Class	8	Balance or spring scale	To measure masses of colliding carts or gliders.
1	Class	7	Balance or spring scale	To measure masses of golf & tennis ball.
2	Group	7,9	Ball, bowling	
1	Group	7	Ball, golf	
1	Group	7,9	Ball, soccer	Substitute: Volleyball.
1	Group	2,7	Ball, tennis	
1	Group	6	Ballast to double weight of athletic shoe	Metal washers or sand in bag will serve.
3	Group	2	Balls or lab carts having different masses	To be pushed using force meter.
1	Individual	All	Calculator, basic	One per student best; one per group minimum.
1	Group	2	Clamp(s) for holding plastic strip	Kind depends on furniture.
1	Group	9	Cork Accelerometer	
2	Group	8	Dynamics cart	Substitute: Air track gliders.
1	Group	2,9	Flexible plastic ruler or plastic strip	Such as a 30-cm "floppy" acetate ruler.
1	Group	9	Force meter	The same flexible plastic strip used in Activity Two.
1	Class	3	Hammer & catch box	For demonstration.
1	Class	1	Low friction devices, assorted examples	See Teacher's Edition, Activity One A for suggestions..
1	Group	1	Marking pen, washable ink	Marks able to be washed from salad bowl surfaces.
1	Group	1,3,4,5	Meter stick	
1	Group	2	Metric ruler, mm marking	Meter stick will serve.
1	Individual	4	Patch of tape	For marking level of body's center mass.
1	Group	5	Penny coin or metal washer	
4	Group	2	Penny coins or metal washers	
1	Group	3	Pin or nail for suspension	For suspending shapes and plumb bobs.
1	Group	3	Plumb bob	Simple weight (such as a washer) on strings.
1	Group	6	Rough horizontal surface	Such as carpet sample or rough wood.
1	Group	1	Ruler, flexible plastic	Able to bend to conform to curve of bowl.
1	Class	5	Safety helmet, knee and elbow pads	Use with skateboard or wheeled chair.
1	Group	1	Salad bowl, large diameter	Large bowl having washable, gently sloped inner surface.
1	Group	3	Set of Shapes A, B, C, D (2-d shapes)	See Background Materials for specifications.
1	Class	5	Skateboard or wheeled chair	Use safety helmet and pads listed separately.
1	Group	6	Smooth horizontal surface	Could use classroom floor or tabletop.
1	Group	6	Spring scale, 0-5 newton range	To convert from gram calibration, 100 g = 1.0 N.
2	Group	7	Starting ramp for bowling ball	Wood boards joined to have "V" cross section will serve.
1	Group	1	Super Ball or equivalent, approx. 1" diameter	Ball which will not roll, not slide on salad bowl surface.
2	Group	1	Tape, short lengths	For making positions on ramps.
1	Group	1	Track with adjustable outrun slope	2 m of Hot Wheels™ track or board 0.5 m, 0.5 m and 1.5 m long.
1	Class	4	VCR & TV Monitor	
1	Group	8	Weight for loading cart or glider	To provide 2:1 mass ratio.
10	Group	5	Weights, 1-newton (100 g masses)	Substitute: Large steel washers, mass approx. 100 g each.

Organizer for Materials Available in Teacher's Edition

Activity in Student Text	Additional Material	Alternative / Optional Activities
ACTIVITY ONE: A Running Start, p. S54	Performance Assessment Rubrics, pgs. 138-139	Activity One A: Can Objects Move Forever?, p. 137
ACTIVITY TWO: Push or Pull, p. S61		
ACTIVITY THREE: Center of Mass, p. S68	Templates for Shapes A, B, C, and D, p. 158-159	Activity Three A: Alternative Method for Determining Center of Gravity, p. 157
ACTIVITY FOUR: Defy Gravity, p. S73	Calculating Hang Time and Force During a Vertical Jump (Worksheet) pgs. 170-171	Activity Four A: High-Tech Alternative for Monitoring Vertical Jump Height, p. 169
ACTIVITY FIVE: Run and Jump, p. S81		Activity Five A: Using a Bathroom Scale to Measure Forces, pgs. 180-181
ACTIVITY SIX: The Mu of the Shoe, p. S86	Assessment Rubric, Physics to Go Question 8, p. 190 Backround Information for Activity Six A: Alternative Activity for Measuring the Mu of the Shoe, pgs. 193-194	Activity Six A: Alternative Activity for Measuring the Mu of the Shoe, p. 192
ACTIVITY SEVEN: Concentrating on Collisions, p. S92		
ACTIVITY EIGHT: Conservation of Momentum, p. S97		
ACTIVITY NINE: Circular Motion, p. S103		

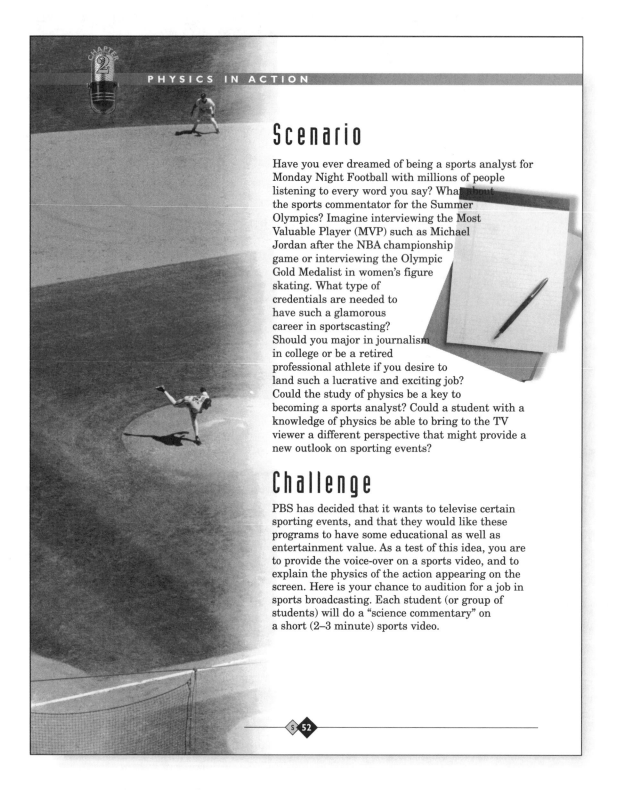

PHYSICS IN ACTION

Scenario

Have you ever dreamed of being a sports analyst for Monday Night Football with millions of people listening to every word you say? What about the sports commentator for the Summer Olympics? Imagine interviewing the Most Valuable Player (MVP) such as Michael Jordan after the NBA championship game or interviewing the Olympic Gold Medalist in women's figure skating. What type of credentials are needed to have such a glamorous career in sportscasting? Should you major in journalism in college or be a retired professional athlete if you desire to land such a lucrative and exciting job? Could the study of physics be a key to becoming a sports analyst? Could a student with a knowledge of physics be able to bring to the TV viewer a different perspective that might provide a new outlook on sporting events?

Challenge

PBS has decided that it wants to televise certain sporting events, and that they would like these programs to have some educational as well as entertainment value. As a test of this idea, you are to provide the voice-over on a sports video, and to explain the physics of the action appearing on the screen. Here is your chance to audition for a job in sports broadcasting. Each student (or group of students) will do a "science commentary" on a short (2–3 minute) sports video.

S 52

Chapter and Challenge Overview

In this chapter Newton's Laws of Motion and also the concepts of force, inertia (mass), momentum, and the physics of rotation are introduced.

Students will be asked to produce a voice-over (or a script for a voice-over) to explain the physics behind a short sports video. You will select the video, but if the students are very ambitious, they can either find some footage themselves, shoot some scenes with a camcorder, or tape some sporting events from TV. The entire chapter will build toward this end, and the final evaluation of the student's progress will be based on the video voice-over.

There are a number of objectives in this chapter, one of which is to show the students that the laws of physics hold true not only in their science class and lab, but out in the world as well. The students should be able to look at a sporting event and realize what physical principle is involved. Hopefully this will carry over to everyday life, and the student will then be able to see the physics in the world around them.

Each class might start with a short video segment showing sports bloopers. They are commercially available and many of the students may have their own. After the class has covered some of the material, it is increasingly appropriate to discuss the physics that is being displayed in the blooper. Many of these bloopers are very humorous and the students look forward to the beginning of the class.

As you review the Challenge assignments, reassure the students that while they may feel incompetent now, by the end of the chapter they will have the necessary skills and vocabulary to respond adequately.

On the following pages of the Teacher's Edition there are suggestions on how to evaluate students on this material. It is very important at this time that the students be made aware of the method you are going to use and how you will evaluate their work. Have the students actively participate in deciding the criteria for evaluation.

The Physics To Go at the end of each section often contains more questions than should ever be assigned for homework. This section has been written in such a way as to give you a choice as to how much work, and the nature of the work the students will be expected to do each day out of class.

As you work with *Active Physics*, be aware that the same physics concepts appear repeatedly in different contexts. It is not necessary for the students to achieve total understanding the first time that they encounter Newton's Laws of Motion, and the physics of rotation.

As the assessment of how well you understand this material, you (or your group) are to:

- **submit a written script; or**
- **narrate live; or**
- **dub onto the video soundtrack; or**
- **record on an audio cassette.**

Your task is not to give a play-by-play description of the sporting event or give the rules of the game, but rather to go a step beyond and educate the audience by describing to them the rules of nature that govern the event. This approach will give the viewer (and you) a different perspective of both sports and physics. The laws of physics cover not only obscure phenomena in the lab, but everyday events in the real world as well.

Criteria

What criteria should be used to evaluate a voice-over dialogue or script of a sporting event? Since the intention is to provide an analysis of and interest in the physics of sports, the voice-over should include the use of physics terms and physics principles. All of these terms and principles should be used correctly. How many of these terms and principles would constitute an excellent job? Would it be enough to use one physics term correctly and explain how one physics principle is illustrated in the sport? Should use of one physics term and one physics principle be a minimum standard to get minimal credit for this assessment? Discuss in your small groups and your class and decide on reasonable expectations for the physics criteria for the assessment.

Assessment Rubric for Voice-Over Dialogue or Script

Meets the standard of excellence. **5**	• A significant number of physics principles are consistently and correctly addressed. • Physics concepts from the chapter are repeatedly integrated in the appropriate places. • Physics terminology and equations are consistently incorporated as applicable. • Correct estimates of the magnitude of physical quantities are frequently used. • Additional research, beyond basic concepts presented in the chapter, is evident. Knowledge of the rules of the game are evident. • The voice-over has great entertainment value. It contains humor and excitement.
Approaches the standard of excellence. **4**	• A significant number of physics principles are correctly addressed often. • Physics concepts from the chapter are integrated in the appropriate places. • Physics terminology and equations are incorporated as applicable. • Correct estimates of the magnitude of physical quantities are frequently used. • Knowledge of the rules of the game are evident. • The voice-over has entertainment value. It contains some humor and excitement.
Meets an acceptable standard. **3**	• A sufficient number of physics principles are correctly addressed. • Physics concepts from the chapter are integrated in the appropriate places. • A limited amount of physics terminology and equations are incorporated as applicable. • Correct estimates of the magnitude of physical quantities are occasionaly used. • Knowledge of the rules of the game are general. • The voice-over has some entertainment value.
Below acceptable standard and requires remedial help. **2**	• Very few physics principles are addressed. • Physics concepts from the chapter are not always integrated in the appropriate places. • A limited amount of physics terminology is incorporated as applicable. • Estimates of the magnitude of physical quantities are seldom used. • Knowledge of the rules of the game is weak. • The voice-over has a limited entertainment value.
Basic level that requires remedial help or demonstrates a lack of effort. **1**	• Physics principles are not addressed correctly. • Physics concepts from the chapter are not integrated in the appropriate places. • Physics terminology is not used. • No attempt is made to include the magnitude of physical quantities. • Knowledge of the rules of the game is lacking. • The voice-over is difficult to follow and portions are missing.

2

For use with *Sports*, Chapter 2
©1999 American Association of Physics Teachers

Assessment Rubric for Voice-Over Dialogue or Script

Meets the standard of excellence. **5**	• Scientific vocabulary is used consistently and precisely. • Sentence structure is consistently controlled. • Spelling, punctuation, and grammar are consistently used in an effective manner. • Scientific symbols for units of measurement are used appropriately in all cases.
Approaches the standard of excellence. **4**	• Scientific vocabulary is used appropriately in most situations. • Sentence structure is usually consistently controlled. • Spelling, punctuation, and grammar are generally used in an effective manner. • Scientific symbols for units of measurement are used appropriately in most cases.
Meets an acceptable standard. **3**	• Some evidence that the student has used scientific vocabulary although usage is not consistent or precise. • Sentence structure is generally controlled. • Spelling, punctuation, and grammar do not impede the meaning. • Some scientific symbols for units of measurement are used. Generally, the usage is appropriate.
Below acceptable standard and requires remedial help. **2**	• Limited evidence that the student has used scientific vocabulary. Generally, the usage is not consistent or precise. • Sentence structure is poorly controlled. • Spelling, punctuation, and grammar impedes the meaning. • Some scientific symbols for units of measurement are used, but most often, the usage is inappropriate.
Basic level that requires remedial help or demonstrates a lack of effort. **1**	• Limited evidence that the student has used scientific vocabulary and usage is not consistent or precise. • Sentence structure is poorly controlled. • Spelling, punctuation, and grammar impedes the meaning. • No attention to using scientific symbols for units of measurement.

For use with *Sports*, Chapter 2   **Maximum = 10 points**

What is in the Physics InfoMall for Chapter 2?

Chapter 2 deals with the physics of sports, which was already introduced in Chapter 1. If you are beginning *Sports* with this chapter, part of the following introduction is also provided in Chapter 1.

If you have had much experience with the *Physics InfoMall CD-ROM*, you have probably done a few searches, and no doubt some of the searches have resulted in "Too many hits." Surprisingly, searching the entire CD-ROM with the keyword "sport*" does not give "too many" hits, but provides some interesting hits. Note that the asterisk is a wild character; this searches for any word beginning with "sport."

If you do the search just mentioned, the first hit is a resource letter

("Resource letter PS-1: Physics of sports," *American Journal of Physics, vol. 54, issue 7*) that discusses the published discussions on the physics of sports. According to this letter, "there is surprisingly little published information about the basic physics underlying most sports, even though the relevant physics is all classical." Included is a list of places you might find such information, including journals and books. The letter contains a list of specific references grouped by sport, such as Physics of basketball, *American Journal of Physics, vol. 49, issue 4*. Another interesting article is "Students do not think physics is 'relevant.' What can we do about it?," in the *American Journal of Physics, vol. 36, issue 12*.

Given that the physics in sports is classical, you might search for student difficulties learning classical physics in general. One article you might find is "Factors influencing the learning of classical mechanics," *American Journal of Physics 48, issue 12*. Knowledge of such factors affecting learning can be a valuable tool. Perform other searches that meet your needs, and the InfoMall is very likely to provide good information. And we have not even opened the Textbook Trove yet!

2

ACTIVITY ONE
A Running Start

Background Information

Two major ideas are introduced in this activity:

• Galileo's Principle of Inertia

• Newton's Second Law of Motion

Before attempting to identify causes and effects for generating, sustaining, and arresting motion, a pivotal question first must be answered: What kinds of motion require explanation?

Two distinct kinds of motion along a straight line often are encountered in nature: (1) motion with constant speed and (2) motion with uniform, or constant, acceleration.

Since the contributions of Galileo, physics has operated from the perspective that the first of these kinds of motion, constant speed, has no cause. Galileo devised a number of arguments and demonstrations, some of which are replicated in this activity, to support this notion.

The cause of all accelerated motion is force; some agent(s) must be pushing or pulling—exerting a force—on any object observed to be accelerating. Sources or kinds of forces abound. Every situation that involves acceleration has an associated net force. Observation: If an orange is dropped, it accelerates; assigned cause: the downward force due to gravity. When the orange hits the floor it stops; another acceleration, another force. The force which stops the orange is provided by the floor, upward. A magnet brought near another magnet will cause an acceleration; therefore, there must be a magnetic force.

Sometimes, we can also discover forces hiding in constant-speed linear motion. Drop a coffee filter: it accelerates downward for a bit, but the amount of acceleration falls off to zero, so that the coffee filter falls most of the way at constant speed. Did the force of gravity decrease or disappear? No, a coffee filter seems to weigh (a measure of the force of gravity) the same at every point in the descent path. Conclusion: there must be another force, the force of air resistance, acting in the opposite direction to gravity. The force of air resistance eventually balances out the gravitational force. It is possible for a combination of forces to have a net effect of zero.

So, it is the net force on an object that imparts the acceleration. Newton's First Law of Motion states the case: An object at rest tends to remain at rest, and an object in motion (in a straight line) tends to remain in motion unless acted upon by an outside (net, non-zero) force. This statement is more complete than the one provided to the students in Activity One. Whenever speed, direction, or both speed and direction, are observed to change, a net force is the cause.

The First Law does not attempt to quantify the relationship between accelerations and the forces that cause them. Establishing the quantitative relationship requires experimental evidence which is the purpose of the next activity.

Active-ating the Physics InfoMall

Note that this activity has students perform a simple experiment, and gradually leads them to concentrate on one aspect of the motion, then leads to predictions and generalizations. The importance of the prediction should not be overlooked; indeed, predictions force students to examine their understanding of a phenomena and actively engage thought. If you were to search the InfoMall to find more about the importance of predictions in learning, you would find that you need to limit your search. For example, a search for "prediction*" AND "inertia" resulted in several hits; the first hit is from *A Guide to Introductory Physics Teaching: Elementary Dynamics,* Arnold B. Arons' Book Basement entry. Here is a quote from that book: "Because of the obvious conceptual importance of the subject matter, the preconceptions students bring with them when starting the study of dynamics, and the difficulties they encounter with the Law of Inertia and the concept of force, have attracted extensive investigation and generated a substantial literature. A sampling of useful papers, giving far more extensive detail than can be incorporated here, is cited in the bibliography [Champagne, Klopfer, and Anderson (1980); Clement (1982); di Sessa (1982); Gunstone, Champagne, and Klopfer (1981); Halloun and Hestenes (1985); McCloskey, Camarazza, and Green (1980); McCloskey (1983); McDermott (1984); Minstrell (1982); Viennot (1979); White (1983), (1984)]." Note that students' preconceptions can have a large effect on how they learn something. It is important that they are forced to consciously acknowledge their preconceptions by making predictions.

Not surprisingly, among the list of hits from the search just mentioned is an article on Galileo, "Galileo, yesterday and today," *American Journal of*

Physics vol. 33, issue 9, 1965. This article provides an interesting insight into Galileo and his work, as well as several historical accounts of his work. Check it out; it might provide interesting additional reading for your students.

Of course, Newton had something to say about inertia, and another hit from the same search provides *Physics for Science and Engineering* in the Textbook Trove. See Chapter 4, Newton's Principles of Motion.

The search above was conducted initially to explore the importance of predictions, especially as related to the concept of inertia. As we can see, additional information was provided that was easily relevant to this topic. This is not unusual when searching the InfoMall — you will often find many interesting bits of information that may take you on unexpected, but enlightening, tangents.

In Physics To Go, question 2, you are encouraged to find something about "curling." Sadly, the InfoMall has only one reference to this sport, and it is a short passage indicating that Lord Kelvin broke his leg while curling, and limped badly thereafter. While not directly related to anything in this section, it is another of those interesting articles one can find on the InfoMall.

Planning for the Activity

Time Requirements

- One class period.

Materials Needed

For each group:
- salad bowl, large diameter
- super ball or equivalent, approximately one inch diameter
- ruler, flexible plastic
- marking pen, washable ink
- track with adjustable outrun slope

Advance Preparation and Setup

You may wish to consider using ball bearings or glass marbles rolling within flexible, transparent plastic tubing for the second part of For You To Do instead of a ball rolling on an adjustable track. If so, you may wish to procure the tubing and bearings in

advance. However, "Hot Wheels™" track, or an equivalent, used during Chapter 1 may work for you if students will be able to adjust the slope easily enough.

Teaching Notes

Active Physics uses a modified constructivist model. By confronting students' misconceptions and by having them do hands-on exploration of ideas, we seek to replace their misconceptions with correct perceptions of reality. In order to do this, a consistent scheme is integrated into the course activities to elicit the students' misconceptions early in any activity.

Students' current mental models are sampled by one or more What Do You Think? questions. Students are not expected to know a "right" answer. These questions are supposed to elicit from students their beliefs regarding a very specific prediction or outcome, and students should commit to a written specific answer in their logs.

When students have completed For You To Do, convene the entire class for a demonstration of objects moving at constant speed on low-friction surfaces. A demonstration is included at the end of this activity. Possible materials for demonstrating motion at constant speed for an object given a push start on a low-friction, flat surface such as a smooth counter top or the glass surface include:

- balloon puck on a smooth, hard surface
- a piece of dry ice on a smooth, hard surface
- glider on air track
- puck on air table
- puck on raw rice in a ripple tank
- puck on plastic bead bearings in a ripple tank.

Then direct students to read the For You To Read and Physics Talk sections. Reserve some time for closure after students have completed the reading.

You may wish to direct the students' attention to the fact that several sports involve motion for which an initial speed does not involve running in the literal sense, but may involve an object, such as a shot put ball, being given an initial speed by an athlete.

2

Activity Overview

Student Objectives

Students will:

- understand and apply Galileo's Principle of Inertia.

- understand and apply Newton's First Law of Motion.

- recognize inertial mass as a physical property of matter.

ANSWERS FOR THE TEACHER ONLY

What Do You Think?

The horizontal distance a basketball player travels while "hanging" is determined by the speed upon jumping; since the speed often is high, the trajectory is quite flat near the peak of flight, giving the illusion that the player "hangs" in the air.

Skaters maintain speed on ice due to very low friction between the blades and the ice.

PHYSICS IN ACTION

Activity One
A Running Start

WHAT DO YOU THINK?

Many things that happen in athletics are affected by the amount of "running start" speed an athlete can produce.

- **What determines the amount of horizontal distance a basketball player travels while "hanging" to do a "slam-dunk" during a fast break?**

- **How do figure skaters keep moving across the ice at high speeds for long times while seldom "pumping" their skates?**

Record your ideas about these questions in your *Active Physics log*. Be prepared to discuss your responses with your small group and the class.

FOR YOU TO DO

1. Use a salad bowl and a ball to explore the question, "When a ball is released to roll down the inside surface of a salad bowl, is the motion of the ball up the far side of the bowl the 'mirror image' of the ball's downward motion?" Use a non-permanent pen to mark a starting position for the ball near the top edge of the bowl. Use a flexible ruler to measure, in centimeters, the distance along the bowl's curved surface from the bottom-center of the bowl to the mark.

SPORTS S 54

ANSWERS

For You To Do

1. a) Students copy tables into their logs.
   b) Students record data.

✎ a) Make a table similar to the one below in your log.

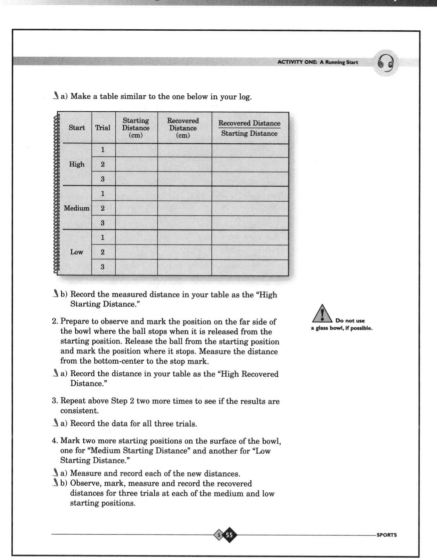

Start	Trial	Starting Distance (cm)	Recovered Distance (cm)	Recovered Distance / Starting Distance
High	1			
	2			
	3			
Medium	1			
	2			
	3			
Low	1			
	2			
	3			

✎ b) Record the measured distance in your table as the "High Starting Distance."

2. Prepare to observe and mark the position on the far side of the bowl where the ball stops when it is released from the starting position. Release the ball from the starting position and mark the position where it stops. Measure the distance from the bottom-center to the stop mark.

✎ a) Record the distance in your table as the "High Recovered Distance."

3. Repeat above Step 2 two more times to see if the results are consistent.

✎ a) Record the data for all three trials.

4. Mark two more starting positions on the surface of the bowl, one for "Medium Starting Distance" and another for "Low Starting Distance."

✎ a) Measure and record each of the new distances.
✎ b) Observe, mark, measure and record the recovered distances for three trials at each of the medium and low starting positions.

⚠ **Do not use a glass bowl, if possible.**

ANSWERS

For You To Do (continued)

2. - 4. Students record data. When rolling the ball within the salad bowl, students should find the Recovered Distance to be very nearly equal to the Starting Distance.

PHYSICS IN ACTION

❧ c) Complete the table by calculating and recording the value of the ratio of the Recovered Distance to the Starting Distance for each trial. (The ratio is the Recovered Distance divided by the Starting Distance.)

> **Example**
> If the Recovered Distance is 6 cm for a Starting Distance of 10 cm, the value of the ratio is $\frac{6\ cm}{10\ cm} = 0.6$.

❧ d) For each of the three starting distances, to what extent is the motion of the ball up the far side of the bowl the "mirror image" of the downward motion? Use data as evidence for your answer.

❧ e) Does the fraction of the starting distance "recovered" when going up the far side of the bowl depend on the amount of starting distance? Describe any pattern of data which supports your answer.

5. Repeat the activity but roll the ball along varying slopes during its upward motion. Make a track which has the same slope on both sides, as shown below. Your teacher will suggest how high the ends of the track sections should be elevated. This time, concentrate on comparing the vertical height of the ball's release position to the vertical height of the position where the ball stops.

❧ a) Measure and record the vertical height (not the distance along the track) from which the ball will be released at the top end of the left-hand section of track.

❧ b) Prepare to observe and mark the position on the right-hand section of track where the ball stops when it is released from the starting position. Release the ball from the top end of the left-hand section of track and mark the position where it stops. Measure and record the vertical height of the position where the ball stops.

❧ c) Calculate the ratio of the recovered height to the starting height. How is this case, and the result, similar to what you did when using the salad bowl? How is it different?

ANSWERS

For You To Do *(continued)*

4. c-e) The ratio of Recovered Distance to Starting Distance should be only slightly less than 1.00 and typical 0.90 or more. The actual value will, of course, depend on the coefficient of friction for the particular kind of ball and bowl used. You may expect that the error of measurement will be nearly as much as the observable difference in distances, indicating nearly complete "conservation" of distance. The ratio should remain essentially constant regardless of the starting height.

5. a-c) Students record data and calculate the ratios of Recovered to Starting Distance.

6. Leave the left-hand starting section of track unchanged, but change the right-hand section of track so that it has less slope and is at least long enough to allow the ball to recover the starting height. The track should be arranged approximately as shown below.

🖋 a) Predict the position where the ball will stop on the right-hand track if it is released from the same height as before on the left-hand track. Mark the position of your guess on the right-hand track and explain the basis for your prediction in your log.

7. Release the ball from the same height on the left-hand section of track as before and mark the position where the ball stops on the right-hand section of track.

🖋 a) How well did you guess the position? Why do you think your guess was "on," or "off?"

🖋 b) Measure the vertical height of the position where the ball stopped and again calculate the ratio of the recovered height to the starting height. Did the ratio change? Why, do you think, the ratio either did or did not change?

8. Imagine what would happen if you again did not change the left-hand starting section of track, but changed the right-hand section of track so that it would be horizontal, as shown below.

🖋 a) How far along the horizontal track would the ball need to roll to recover its starting height (or most of it)? How far do you think the ball would roll?

🖋 b) When rolling on the horizontal track, what would "keep the ball going?"

ANSWERS

For You To Do *(continued)*

6. a) Make certain that students record their predictions.

7. a-b) When rolling the ball down the adjustable track, students must shift attention from comparing distances traveled along the "down" and "up" paths to vertical distances "down" and "up." For symmetrical slopes, the former measurement—distance along either track—would serve, but, as the "up" slope is made less, the ball will roll farther on the "up" slope to gain nearly the same vertical height as the height from which it was released on the "down" slope.

8. a-b) The intent is for students to realize, in accord with Galileo's ideal, that when the "up" track has no slope, the ball will roll "forever" in its attempt to gain the height from which it was released.

FOR YOU TO READ

Inertia

Italian philosopher Galileo Galilei (1564–1642), who can be said to have introduced science to the world, noticed that a ball rolled down one ramp seems to seek the same height when it rolls up another ramp. He also did a "thought experiment" in which he imagined a ball made of extremely hard material set into motion on a horizontal, smooth surface, similar to the final track in For You To Do. He concluded that the ball would continue its motion on the horizontal surface with constant speed along a straight line "to the horizon" (forever). From this, and from his observation that an object at rest remains at rest unless something causes it to move, Galileo, formed the "Principle of Inertia:"

Inertia is the natural tendency of an object to remain at rest or to remain moving with constant speed in a straight line.

Isaac Newton, born in England on Christmas day in 1642 (the year that Galileo died) used Galileo's Principle of Inertia as the basis for developing his First Law of Motion, presented in Physics Talk. Crediting Galileo and others for their contributions to his thinking, Newton said, "If I have seen farther than others, it is because I have stood on the shoulders of giants."

Running Starts

Running starts take place in many sporting activities. Since there seems to be this prior motion in many sports, there must be some advantage to it.

In sports where the objective is to maximize the speed of an object or the distance traveled in air, the prior motion may be essential. When a javelin is thrown, at the instant of release it has the same speed as the hand that is propelling it.

- The hand has a forward speed relative to the elbow, the elbow has a forward speed relative to the shoulder (because the arm is rotating around the elbow and shoulder joints), and the shoulder has a forward speed relative to the ground because the body is rotating and the body is also moving forward.

- The javelin speed then is the sum of each of the above speeds. If the thrower is not running forward, that speed does not add into the equation.

You can write a velocity equation to show the speeds involved.

$$v_{javelin} = v_{hand} + v_{elbow} + v_{shoulder} + v_{ground}$$

Motion captures everyone's attention in sports. Starting, stopping, and changing direction (accelerations) are part of the motion story, and they are exciting components of many sports. Ordinary, straight-line motion is just as important but is easily overlooked.

PHYSICS TALK

Newton's First Law of Motion

Isaac Newton included Galileo's Principle of Inertia as part of his First Law of Motion:

In the absence of an unbalanced force, an object at rest remains at rest, and an object already in motion remains in motion with constant speed on a straight line path.

Newton also explained that an object's mass is a measure of its inertia, or tendency to resist a change in motion.

Here is an example of how Newton's First Law of Motion works:

Inertia is expressed in kilograms of mass. If an empty grocery cart has a mass of 10 kg and a cart full of groceries has a mass of 100 kg, which cart would be the most difficult to move (have the greatest tendency to remain at rest)? If both carts already were moving at equal speeds, which cart would be the most difficult to stop (would have the greatest tendency to keep moving)? Obviously in both cases, the answer is the more massive cart.

REFLECTING ON THE ACTIVITY AND THE CHALLENGE

Running starts can be observed in many sports. Many observers may not realize the important role that inertia plays to preserve the speed already established when an athlete engages in activities such as jumping, throwing, or skating from a "running start." "Immovable objects" such as football linemen which illustrate the tendency of highly massive objects to remain at rest also can be observed in many sports. You should have no problem finding a great variety of video segments which illustrate Newton's First Law presented in this activity.

Physics To Go

1. Answers will vary.
Possible examples:
An outfielder diving for a line drive. The outfielder continues in motion, sliding along the ground, his hat also continues in motion.

A slap shot in hockey. The puck continues to move in a constant horizontal motion, once it has been set in motion by the player.

2. Curling, an Olympic competition, is similar to shuffleboard and involves sliding "stones" on ice.

3. A skater has either nearly constant speed, or very small uniform deceleration.

4. A hockey puck on ice has either nearly constant speed, or very small uniform deceleration.

5. A baseball player slides into second or third to decelerate to a stop at the base because if the base is overrun, the player would be "out" if tagged; at first base, a player can overrun the base without danger of being tagged out and the fastest way of beating a throw to first base is to run without sliding into the base.

6. It does not seem possible to eliminate friction to arrive at perpetual motion in the real world.

PHYSICS IN ACTION

PHYSICS TO GO

1. Provide three illustrations of Newton's First Law in sporting events. Describe the sporting event and which object when at rest, stays at rest, or which object when in motion, stays in motion. Describe these same three illustrations in the manner of an entertaining sportscaster.

2. Find out about a sport called "curling" (it is an Olympic competition which involves some of the oldest Olympians) and how this sport could be used to illustrate Newton's First Law of Motion.

3. When a skater glides across the ice on only one skate, what kind of motion does the skater have? Use principles of physics as evidence for your answer.

4. Use what you have learned in this activity to describe the motion of a hockey puck between the instant the puck leaves a player's stick and the instant it hits something. (No "slap shot" allowed; the puck must remain in contact with the ice.)

5. Why do baseball players often slide into second base and third base, but never slide into first base after hitting the ball? (The answer depends on both the rules of baseball and the laws of physics.)

6. Do you think it is possible to arrange conditions in the "real world" to have an object move, unassisted, in a straight line at constant speed forever? Explain why or why not.

S 60

Activity One A

Can Objects Move Forever?

FOR YOU TO DO

In this exercise, you will observe the motion of various objects on a variety of surfaces to see whether an object might be able to move forever as Galileo concluded. Since you do not have an infinite time to work, nor an infinitely large room to work in, this question can only be approached by looking at limited examples and by trying to imagine the limitless consequences of what you see. Also, since space is limited, it is important to use as little space as possible for the process of getting the object started on the motion that is to be investigated. That leaves more room to see what happens when the object is "on its own."

1. The first object to start in motion is an eraser on a table top or on the floor. The spine of the eraser is to be in contact with the floor.

a) Does the eraser sustain its motion on its own? Describe what happens.

2. Replace the table top with smooth glass or plastic.

a) Describe the eraser's motion.

3. Put about ⅛ cup of raw rice on the same surface and try the eraser again.

a) Describe the eraser's motion.

4. Remove the rice, and replace it with dried peas. Try the eraser again.

a) Describe the eraser's motion.

5. Remove the peas, and pour on a thin, sparse layer of minute plastic beads. Be careful not to spill the beads on the floor, as they are very difficult to pick up and clean up. Try the eraser again.

a) Describe the eraser's motion.

b) Try other objects on the beaded surface (coins, small blocks of wood, objects with weights on them, etc.). Report the results.

6. Set a non-inflated air puck in motion along a table top.

a) Describe what happens.

7. Fill the air reservoir and set the puck in motion again.

a) Describe and explain what happens in each case.

Performance Assessment Rubrics

Part 1 = maximum 4
Part 2 = maximum 2
Part 3 = maximum 5
Part 4 = maximum 6

1. **Student records experimental data demonstrating that the distance a ball travels down a salad bowl is nearly the same distance that the ball travels up the same salad bowl.**

Descriptor	Task accomplished	Task not accomplished
a) Measurement is taken with ruler.		
b) Units of measurement are recorded in centimeters.		
c) Three trials are used for high, medium, and low distances.		
d) Release height is compared with recovery height.		

Maximum 4 marks if each of the sub tasks is accomplished.

Total marks: _____

2. **Student uses deductive reasoning to make a generalization about how the downward motion of the ball mirrors the upward motion of the ball. The greater the release height, the greater the recovery distance.**

Descriptor	Task accomplished	Task not accomplished
a) Student explains that the recovery distance is dependent upon the start distance.		
b) Student notes the constant ratio by comparing start distance of three places (high, medium, and low) to recovery distance.		

Maximum 2 marks if each of the sub tasks is accomplished.

Total marks: _____

For use with *Sports*, Chapter 2, ACTIVITY ONE: A Running Start
©1999 American Association of Physics Teachers

3. Student uses experimental data to find a constant ratio of release distance to recovery distance.

Descriptor	Task accomplished	Task not accomplished
a) Vertical measurement is taken with a ruler.		
b) Units of measurement are recorded in centimeters.		
c) Release height is compared with recovery height.		
d) Ratio of the recovery distance to start distance is calculated correctly.		
e) Comparison is made between the ramp and salad bowl. Similarities and differences are identified.		

Maximum 5 marks if each of the sub tasks is accomplished.

Total marks: _____

4. Student correctly decreases the slope on the right of the incline and notes that the release height affects the distance that the ball rolls along a vertical plane.

Descriptor	Task accomplished	Task not accomplished
a) Prediction of recovery distance.		
b) Student notes that the ball travels a greater distance along a horizontal plane.		
c) Student measures the vertical distance that the ball traveled up the ramp.		
d) The ratio of the vertical start height to horizontal recovery height is calculated correctly. The ratio remains constant.		
e) Gravity is identified as the force that caused the ball to move downward and slowed the ball as it moved upward.		
f) Student concludes that the force of gravity remains constant for both downward and upward movement of the ball.		

Maximum 6 marks if each of the sub tasks is accomplished.

Total marks: _____

For use with *Sports*, Chapter 2, ACTIVITY ONE: A Running Start
©1999 American Association of Physics Teachers

ACTIVITY TWO
Push or Pull

Background Information

It is suggested that the "Physics Talk: Newton's Second Law of Motion" and "For You To Read: Weight and Newton's Second Law" sections of the student text for Activity Two be read before proceeding in this section.

The unit of mass, or quantity of matter, in the International System of Units is the kilogram. One of seven base units from which all other units are derived, the kilogram originally was conceived as the quantity of matter represented by 1 liter of water at the temperature of maximum density, 4 °C; today, the kilogram is defined by a carefully protected metal standard called the International Prototype Kilogram. When a balance which employs the force of gravity is used to measure the mass of an object by comparison to prototype masses, the resulting measurement is known as the "gravitational mass" of the object. Mass also is internationally recognized as a measure of the inertial resistance of an object to acceleration. When a standard force is used to compare an object's acceleration to the acceleration of a prototype mass as a means of measuring the mass of the object, the resulting measurement is known as the "inertial mass" of the object. It can be shown that 1 kilogram of gravitationally determined mass is equivalent to 1 kilogram of inertial mass.

A derived unit of force, the newton, is defined in terms of base units of mass, length and time using Newton's Second Law of Motion, $F = ma$. 1 newton (N) is the force which will cause 1 kilogram to accelerate at 1m/s², or 1 N = 1 (kg)m/s².

The word "weight" denotes a force; the weight of an object is the product of its mass and the acceleration due to gravity, 9.81 m/s². Since weight is the force due to gravity, weight is measured in newtons. One newton is roughly a quarter pound, prompting the identification of the familiar "quarter-pounder" as a "newton burger".

In summary, matter seems to have two distinct properties:
1. It exhibits a resistance to acceleration, property called "inertia".

2. It has the property of gravitation; matter is attracted to other matter.

It is clear why it is that all objects, irrespective of mass, have the same free fall acceleration at a given location. The more mass, the more gravitational force; but the more mass, the more difficult it is to accelerate the object. These two factors exactly compensate to produce the same acceleration for every freely falling object at a given location.

Active-ating the Physics InfoMall

A big concept in this activity is the concept of force. Students' understanding of this concept has been studied extensively. An InfoMall search using "force" AND "misconception*" in only the Articles and Abstracts Attic produced many great references. The first such hit is the article containing the Force Concept Inventory. The second is "Common sense concepts about motion," *American Journal of Physics,* vol. 53, issue 11, 1985. in which it is mentioned that "(a) On the pretest (post-test), 47% (20%) of the students showed, at least once, a belief that under no net force, an object slows down. However, only 1% (0%) maintained that belief across similar tasks. (b) About 66% (54%) of the students held, at least once, the belief that under a constant force an object moves at constant speed. However, only 2% (1%) held that belief consistently." More results are reported in this article.

The third hit in this search is "Physics that textbook writers usually get wrong," in *The Physics Teacher,* vol. 30, issue 7, 1992. This article is good reading for any introductory physics teacher. The list of hits from this search is long. In fact, it had to be limited to just the Articles and Abstracts Attic to prevent the "Too many hits" warning. If you search the rest of the CD-ROM, you will find many other great hits, such as this quote from Chapter 3 of Arons' *A Guide to Introductory Physics Teaching: Elementary Dynamics*: "In the study of physics, the Law of Inertia and the concept of force have, historically, been two of the most formidable stumbling blocks for students, and, as of the present time, more cognitive research has been done in this area than in any other."

Newton's Second Law is discussed in virtually every physics textbook in existence, not to mention the InfoMall. Depending on the level at which you wish to present this Law, you may wish to examine the conceptual-level texts, the algebra-based texts, or even the calculus-bases textbooks on the InfoMall.

If you want more exercises to give to your students, searching the InfoMall is a bad idea — there are too many problems on the CD-ROM. Searching with keywords "force" AND "acceleration" AND "mass" in the Problems Place alone produces "Too many hits." However, you will find more than enough by simply going to the Problems Place and browsing a few of the resources you will find there. For example, *Schaum's 3000 Solved Problems in Physics* has a section on Newton's Laws of Motion. You will surely find enough problems there to keep any student busy for some time!

Planning for the Activity

Time Requirements

• One class period.

Materials Needed

For each group:
• flexible plastic ruler or plastic strip
• clamp(s) for holding plastic strip
• penny coins or metal washers (4)
• balls or laboratory carts having different masses (3)

Advance Preparation and Setup

Identify the particular combination of flexible rulers (or plastic strips) and weights (coins or metal washers) which will serve as force meters; the same strips used in Chapter 1, Activity Ten probably will work. Identify a means of preventing the weights from slipping off the bent plastic strip, such as a lightweight cardboard "lip" taped to the plastic strip.

Also identify the set of objects to be accelerated; either balls of about the same diameter but having different masses (such as a bowling ball, basketball and inflated beach ball) or laboratory carts which can be loaded to vary the mass would work. If possible, have at least three objects of different masses available for each group.

Try for yourself the calibration procedure and the use of the force meter to accelerate objects in advance of class. You may need to try different sizes of coins or metal washers to find a kind which will produce a reasonable

amount of bend in the ruler for a 4-coin load while at the same time providing a reasonable amount of acceleration when the smallest and largest objects are pushed using the smallest and largest forces.

Teaching Notes

Students can be expected to need practice to exert constant amounts of force on moving objects. Only semiquantitative comparisons of the amounts of acceleration (e.g., low, higher, even higher) which result from varying the amount of force (while mass is held constant) and from varying the amount of mass (when force is held constant) are intended.

Direct all students to silently read the For You To Read section. Then conduct a brief discussion of the assumption presented in the section. You may wish to point out that assumptions represent beliefs which may be argued, but not proven as "right" or "wrong." Another example of an assumption which could be used for the discussion is "There is a tooth fairy."

You may wish to see if students really believe that gravity treats all athletes equally by probing students about the "hang time" of basketball stars.

2

NOTES

Activity Two
Push or Pull

WHAT DO YOU THINK?

Moving a football one yard to score a touchdown requires strategy, timing, and many forces.

• **What is a force?**

• **Can the same force move a bowling ball and a ping pong ball?**

Record your ideas about these questions in your *Active Physics log*. Be prepared to discuss your responses with your small group and the class.

FOR YOU TO DO

1. Make a crude "force meter" from a strip of plastic. Use coins to make a scale of measurement for (calibrate) the meter in pennyweights. The force you are using to calibrate the meter is gravity the force with which Earth pulls downward on every object near its surface. **Carefully** clamp the plastic strip into position as shown in the diagram.

S 61   SPORTS

ANSWERS

For You To Do

1. Student activity.

Activity Overview

In this activity students calibrate a crude "force meter" by deforming (bending) a plastic strip using pennies. Students then use the force meter to accelerate the same object using different forces, and different objects using the same force.

Student Objectives

Students will:

• recognize that a force is a push or a pull.

• identify the forces acting on an object.

• determine when the forces on an object are either balanced or unbalanced.

• calibrate a force meter in arbitrary units.

• use a force meter to apply measured amounts of force to objects.

• compare amounts of acceleration semi-quantitatively.

• understand and apply Newton's Second Law of Motion, *F=ma*.

• understand and apply the definition of the newton as a unit of force,
$$1\text{ N} = 1\text{ (kg)m/s}^2.$$

• understand weight as a special application of Newton's Second Law,
$$\text{Weight} = mg.$$

2

ANSWERS FOR THE TEACHER ONLY

What Do You Think?

In simple terms, a force is a push or a pull. Some forces, such as gravitational and magnetic forces can act on objects without having to be in contact with them. Many other forces, called mechanical forces, act when particles or objects contact each other Forces are very important in physics because they determine how matter interacts with other matter.

The same force could be used to move both a bowling ball and a ping pong ball. The difference would be the amount by which each is accelerated by the force. The greater the mass, the less the acceleration experienced by an object when the same force is applied to it. Mass affects acceleration.

PHYSICS IN ACTION

2. Draw a line on a piece of paper. Hold the paper next to the plastic strip so that the line is even with the edge of the strip. Mark the position of the end of the strip on the reference line and label the position as the "zero" mark.

3. Place one coin on the top surface of the strip near the strip's outside end. Notice that the strip bends downward and then stops. Hold the paper in the original position, and mark the new position of the end of the strip and label the mark as "1-pennyweight."

4. Repeat step 3 for 2, 3, and 4 coins placed on the strip. In each case mark and label the new position of the end of the strip.

   a) Copy the reference line and the calibration marks from the piece of paper into your log.

5. Practice holding one end of the "force meter" (plastic strip) in your hand and pushing the free end against an object until you can bend the strip by forces of 1, 2, 3, and 4-pennyweight amounts. To become good at this, you will need to check the amount of bend in the strip against your calibration marks as you practice.

6. Use the force meter to push an object such as a tennis ball with a continuous 1-pennyweight force. You will need to keep up with the object as it moves, and keep the proper bend in the force meter. You may need to practice a few times to be able to do this.

   a) In your log, record the amount of force used, a description of the object, and the kind of motion the object seemed to have.

7. Repeat step 6 three more times, pushing on the same object with steady (constant) 2, 3, and 4-pennyweight amounts of force.

   a) Record the results in your log for each amount of force.

ANSWERS

For You To Do (continued)

2. - 3. Student activity.

4. a) Students record calibration in their logs. As each penny is added, the ruler deflects more. The force due to gravity of each penny is responsible for bending the ruler.

5. Student activity.

6. a) Students record observations in their logs.

7. a) The greater the force applied to the tennis ball, the greater its acceleration, as demonstrated by the increasing difficulty in keeping up with the ball to maintain the force on it.

8. Based on your observations, complete the statement:
"The greater the constant, unbalanced force pushing on
an object,..."

◣ a) Write the completed statement in your log.

9. Select an object which has a small mass. Use the force
meter to push on the object with a rather large, steady force
such as 3 or 4-pennyweights.

◣ a) Record the amount of force used, a description of the
object pushed (especially including its mass, compared
to the other objects to be pushed) and the kind of motion
the object seemed to have.

10. Repeat step 9 using the same amount of force to push
objects of greater and greater mass.

◣ a) Record the results in your log for each object.

11. Based on your observations, complete the statement:
When equal amounts of constant, unbalanced force are
used to push objects having different masses, the more
massive object..."

◣ a) Write the completed statement in your log.

ANSWERS

For You To Do (continued)

8. a) The greater the constant, unbalanced force pushing on an
object, the greater the acceleration of that object.

9. a) Students record results in their logs.

10. a) As the mass of the objects increases, the acceleration decreases.

11. a) When equal amounts of constant, unbalanced forces are used to
push objects having different masses, the more massive objects
are accelerated less.

PHYSICS TALK

Newton's Second Law of Motion

Based on observations from experiments similar to yours, Isaac Newton wrote his Second Law of Motion:

The acceleration of an object is directly proportional to the unbalanced force acting on it and is inversely proportional to the object's mass. The direction of the acceleration is the same as the direction of the unbalanced force.

If 1 newton is defined as the amount of unbalanced force which will cause a 1-kilogram mass to accelerate at 1 meter per second every second, the law can be written as an equation:

$$f = ma$$

where f is expressed in newtons (symbol N), mass is expressed in kilograms (kg), and acceleration is expressed in meters per second every second (m/s^2).

By definition the unit "newton" can be written in its equivalent form: (kg)m/s^2

Newton's Second Law can be arranged in three possible forms:

$$f = ma \qquad a = f/m \qquad m = f/a$$

Example

How much unbalanced force is needed to cause an 8.0 kg object to accelerate at 3.0 m/s^2?

$$f = ma$$
$$= 8 \text{ kg} \times 3 \text{ m/s}^2$$
$$= 24 \text{ (kg)m/s}^2 \text{ or } 24 \text{ N}$$

FOR YOU TO READ

Weight and Newton's Second Law

Newton's Second Law explains what "weight" means, and how to measure it. If an object having a mass of 1 kg is dropped, its free fall acceleration is approximately 10 m/s^2.

Using Newton's Second Law,

$$f = ma$$

the force acting on the falling mass can be calculated as

$$f = ma$$
$$= 1 \text{ kg} \times 10 \text{ m/s}^2 \text{ or } 10 \text{ N.}$$

The 10 N force causing the acceleration is known to be the gravitational pull of Earth on the 1 kg object. This gravitational force is given the special name "weight." Therefore, it is correct to say, "The weight of a one-kilogram mass is ten newtons."

What is the weight of a 2-kg mass? If dropped, a 2-kg mass also would accelerate due to gravity (as do all objects in free fall) at 10 m/s^2. Therefore, according to Newton's Second Law, the weight of a 2-kg mass is equal to

$$2 \text{ kg} \times 10 \text{ m/s}^2 \text{ or } 20 \text{ N.}$$

In general, to calculate the numerical value of an object's weight in newtons, it is necessary only to multiply the numerical value of its mass by the numerical value of the "g," which is 10.

$$\text{Weight} = mg$$

The above equation is the "special case" of Newton's Second Law which must be applied to any situation in which the force causing an object to accelerate is Earth's gravitational pull.

REFLECTING ON THE ACTIVITY AND THE CHALLENGE

What you learned in this activity really increases the possibilities for interpreting sports events in terms of physics. Now you can explain why accelerations occur in terms of the masses and forces involved. You know that forces produce accelerations. Therefore, if you see an acceleration occur, you know to look for the forces involved. You can apply this to the sport which you will describe.

Also, you can explain, in terms of mass and weight, why gravity has no "favorite" athletes; in every case of free fall in sports, g has the same value, about 10 m/s^2.

 65   SPORTS

ANSWERS

Physics To Go

1. See chart below.

2. a) The long jumper and the shot put ball both are cases of free fall; therefore the acceleration is *g*, the acceleration due to gravity.

   b) The negative sign is used to denote that the force and acceleration are in a direction opposite the motion.

   c) Since acceleration occurs in the direction of the causal force, yes the force should be shown as negative.

   d) Students should be able to provide a plausible "voice-over" narration for an imagined video clip showing each event in the table.

3. 4.2 N / 0.30 kg = ~~14~~ m/s²
   140

PHYSICS IN ACTION

PHYSICS TO GO

1. Copy and complete the following table using Newton's Second Law of Motion. Be sure to include the unit of measurement for each missing item.

Newton's 2nd Law:	*f*	=	*m*	×	*a*
Sprinter beginning 100-meter dash	?		70 kg		5m/s²
Long jumper in flight	800 N		?		10 m/s²
Shot put ball in flight	70 N		7 kg		?
Ski jumper going down hill before jumping	400 N		?		5 m/s²
Hockey player "shaving ice" while stopping	−1,500 N		100 kg		?
Running back being tackled	?		100 kg		−30 m/s²

2. The following items refer to the table in question 1.

   a) In which cases in the table does the acceleration match "*g*," the acceleration due to gravity 10 m/s²? Are the matches to *g* coincidences, or not? Explain.

   b) The force on the hockey player stopping is given in the table as a negative value. Should the player's acceleration also be negative? What do you think it means for a force or an acceleration to be negative?

   c) The acceleration of the running back being tackled also is given as negative. Should the unbalanced force acting on him also be negative? Explain.

   d) In your mind, "play" an imagined video clip which illustrates the event represented by each horizontal row of the above table. Write a brief "voice-over" script for each video clip which explains how Newton's Second Law of Motion is operating in the event. Use appropriate physics terms, equations, numbers and units of measurement in the scripts.

3. What is the acceleration of a 0.30 kg volleyball when a player uses a force of 42 N to spike the ball?

Newton's Second Law:	*f*	=	*m*	×	*a*
Sprinter beginning 100-meter dash	350 N		70 kg		5m/s²
Long jumper in flight	800 N		80 kg		10 m/s²
Shot put ball in flight	70 N		7 kg		10 m/s²
Ski jumper going down hill before jumping	400 N		80 kg		5 m/s²
Hockey player "shaving ice" while stopping	—1,500N		100 kg		-15 m/s²
Running back being tackled	—3000 N		100 kg		—30m/s²

4. What force would be needed to accelerate a 0.040 kg golf ball at 20 m/s²?

5. Most people can throw a baseball farther than a bowling ball, and most people would find it less painful to catch a flying baseball than a bowling ball flying at the same speed as the baseball. Explain these two apparent facts in terms of:

   a) Newton's First Law of Motion.
   b) Newton's Second Law of Motion.

6. Calculate the weight of a new fast food sandwich which has a mass of 0.1 kg. Think of a clever name for the sandwich which would incorporate its weight.

7. In the USA, people measure body weight in "pounds." Write down the weight, in pounds, of a person which is known to you (this could be your weight, or someone else's).

   a) Convert the person's weight in the British unit of force, pounds, to the international unit of force, newtons. To do so, use the below conversion equation:
   Weight in newtons = Weight in pounds × 4.38 newtons/pound
   b) Use the person's body weight, in newtons, and the equation
   Weight = mg
   to calculate the person's body mass, in kilograms.

8. Imagine a sled (such as a bobsled or luge used in Olympic competitions) sliding down a 45-degree slope of extremely slippery ice. Assume there is no friction or air resistance (not really possible). Even under such ideal conditions, it is a fact that gravity could cause the sled to accelerate at a maximum of only 7.1 m/s². Why would the "ideal" acceleration of the sled not be "g," 10 m/s²? Your answer is expected only to suggest reasons why, on a 45-degree hill, the ideal "free fall" acceleration is "diluted" from 10 m/s² to about 7 m/s²; you are not expected to give a complete explanation of why the "dilution" occurs.

9. If you were doing the voice-over for a tug-a-war, how would you explain what was happening? Write a few sentences as if you were the science narrator of that athletic event.

10. You throw a ball. When the ball is many meters away from you, is the force of your hand still acting on the ball?

Physics To Go
(continued)

4. 0.040 kg x 20 m/s² = 0.8 N

5. a) A bowling ball has greater inertia (mass) than a baseball; therefore, a bowling ball has a greater tendency to either remain at rest or remain in motion than does a baseball.

   b) More force is required to cause a bowling ball to accelerate than a baseball; therefore, throwing (accelerating) or catching (decelerating) a bowling ball involves much greater forces than throwing or catching a baseball when equal speeds are involved.

6. The sandwich would weigh 0.1 kg x 10 m/s² = 1 N.

7. Example: Weight
   = 150 lb. x 4.38 N/lb.
   = 657 N

   Mass
   = 657 N ÷ 10 m/s² = 65.7 kg

8. The component, or effectiveness, of the weight in the downhill direction, parallel to the slope of the hill, is 0.71 times the downward force of gravity (weight), or 7.1 N/kg; therefore, the acceleration is 7.1 N/kg = 7.1 m/s². This can be analyzed using either a scale drawing or trigonometry and should not be expected of all students at this level.

9. Students provide voice-over for tug-of-war.

10. No.

ACTIVITY THREE
Center of Mass

Background Information

The center of mass of an object is the only idea introduced in this activity.

Definition: The center of mass is the point at which the entire mass of an object may be thought of as being concentrated for purposes of analyzing the translational motion (motion along a path) or rotational motion (spinning motion) of the object.

For practical purposes, the location of the center of mass of an object having only one significant dimension—such as a straight stick, loaded teeter totter, twirler's baton, screwdriver or wrench—corresponds to the object's balance point. For a two-dimensional object—such as a sheet of plywood cut into any shape—the location of the center of mass corresponds to the balance point located on either of the two large, flat surfaces of the object; to the extent that a two-dimensional object—such as a triangle cut from a sheet of plywood—may have significant thickness and, therefore, actually be three-dimensional, the center of mass would be located within the object, "in line" with the balance point, at the center of the thickness dimension.

For objects having simple three-dimensional shapes—such as homogeneous or symmetrically layered spheres (examples, in respective order: bowling ball, basketball), cubes, rectangular solids and cylinders—the center of mass is located within the object, at its center.

An alternative to balancing an object to locate the center of mass is to suspend the object from any point which is not the center of mass. When suspended, gravity serves to orient the object so that its center of mass is located directly below the point of suspension (this is an example that the Earth "views" an object near it as a "point mass" (located at the object's center of mass) and pulls the point mass as close to Earth as possible). A line extended straight downward from the point of suspension passes through the object's center of mass. The intersection of two such lines, corresponding to two points of suspension, locates the object's center of mass.

It is possible that the center of mass may not be located within the material of the object for some shapes. The "boomerang" shape is an example of such an object.

For purposes of applying Newton's Laws of Motion, an object is treated as if all of its mass is concentrated at the center of mass. The fact that objects behave this way in nature is verified by the observation that when a baton is thrown through the air as a twirling projectile, the baton's center of mass, if marked for high visibility, is seen to trace the familiar parabolic trajectory of a projectile. A twirling baton brings up another aspect of center of mass: when a force acting on an object is aligned with the object's center of mass, the object accelerates in accordance with Newton's second law; however, if the applied force is not aligned with the center of mass, the object also will rotate, or spin. The latter kind of case is not treated in *Active Physics*.

Considerable emphasis in future activities will be placed on the center of mass of the human body. Except for contorted positions of body parts (e.g., the arched "Fosbury Flop" position in the high jump), the normal location of the body's center of mass is within the body at about the level of the navel.

Active-ating the Physics InfoMall

The methods outlined in the *Active Physics* text are standard for finding the center of mass for objects. However, you may want demonstrations. A search of the Demo & Lab Shop produces many great, and tested, demonstrations. Just use keywords "Center of mass" and search only the Demo & Lab Shop. Of course, you can also find many problems in the Problems Place, if you wish, using the same keywords.

Planning for the Activity

Time Requirements

• One class period.

Materials Needed

For the class:

• hammer and catch box
  (demonstration, step 8)

For each group:

• set of shapes A, B, C, and D

• adhesive dot or patch of tape

• pin or nail for suspension

• plumb bob (weight on string)

• meter stick

Advance Preparation and Setup

Cutouts of shapes A, B, C, and D need to be made for each group in the shapes of templates provided in the Additional Materials for this activity. (The templates are provided only for your convenience. Other shapes, in greater variety, may be used and the size may be scaled differently, if desired. If you depart from the shapes provided, be sure to include a "boomerang" shape for which the C of M will be outside the object.) The shapes may be cut from any thin, flat material such as corrugated cardboard or (more durable) plywood, plastic or metal; it would also be convenient to cut the shapes from a sheet of pegboard material to avoid need to drill holes for suspension.

Drill holes to serve to suspend the shapes, and plan how the shapes will be suspended from pins or nails from areas such as a bulletin board or pieces of wood mounted on laboratory table rods.

Teaching Notes

Prepare a demonstration of one or more objects having complex shapes moving as spinning projectiles. For example, the centers of mass of shapes A, B, C, and D could be brightly marked and observed from a distance as two persons play catch with each of the objects. Even if an object spins, the center of mass will trace a parabolic trajectory. A baton having the center of mass marked with bright tape also could be used. When all students have completed steps 1 to 5 of For You To Do, convene the entire class to observe the motion of the center of mass as two persons play catch with the objects planned for the demonstration.

You may wish to recommend the opening montage on the *Active Physics* video as a possibility for tracing the motion of the center of mass moving as a projectile.

2

Activity Overview

In this activity students locate the center of mass of various shaped objects by locating the center of gravity. This is done by balancing the object on a finger as well as by suspending the object and using a plumb bob. Students also estimate the location of their own center of mass.

Student Objectives

Students will:

• locate the center of mass of oddly shaped two-dimensional objects.

• infer the location of the center of mass of symmetrical three-dimensional objects.

• measure the approximate location of the center of mass of the student's body.

• understand that the entire mass of an object may be thought of as being located at the object's center of mass.

ANSWERS FOR THE TEACHER ONLY

What Do You Think?

The center of mass is the point at which all of the mass of an object may be thought of as being concentrated. (As defined above: The center of mass is the pint at which the entire mass of an object may be thought of as being concentrated for purposes of analyzing the translational motion (motion along a path) or rotational motion (spinning motion) of the object.)

The normal location of the body's center of mass is within the body at about the level of the navel.

PHYSICS IN ACTION

Activity Three
Center of Mass

WHAT DO YOU THINK?

The center of mass of a high jumper using the "Fosbury Flop" (arched back) technique passes below the bar as the jumper's body successfully passes over the bar.

• **What is "center of mass?" What does it mean?**

• **Where is your body's center of mass?**

Record your ideas about these questions in your *Active Physics log*. Be prepared to discuss your responses with your small group and the class.

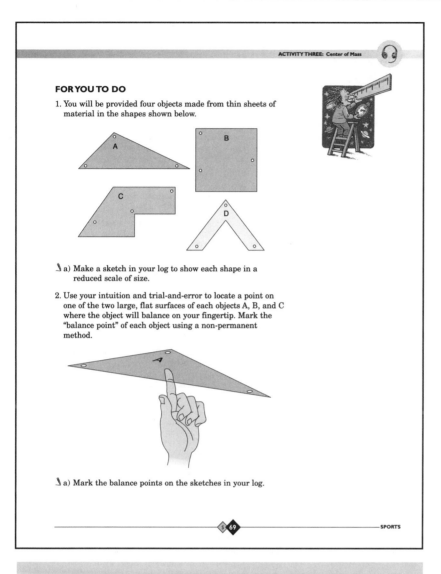

ANSWERS

For You To Do

1. a) You may wish to provide students with a copy of the templates at the end of this activity, rather than have them redraw each in their log.

2. a) Students should be able to locate the balance points for each shape.

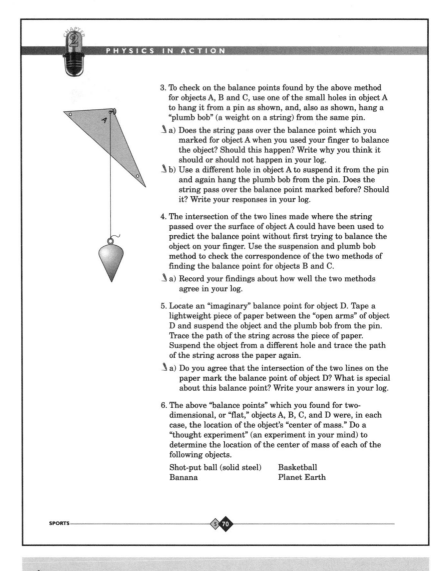

PHYSICS IN ACTION

3. To check on the balance points found by the above method for objects A, B and C, use one of the small holes in object A to hang it from a pin as shown, and, also as shown, hang a "plumb bob" (a weight on a string) from the same pin.

   a) Does the string pass over the balance point which you marked for object A when you used your finger to balance the object? Should this happen? Write why you think it should or should not happen in your log.

   b) Use a different hole in object A to suspend it from the pin and again hang the plumb bob from the pin. Does the string pass over the balance point marked before? Should it? Write your responses in your log.

4. The intersection of the two lines made where the string passed over the surface of object A could have been used to predict the balance point without first trying to balance the object on your finger. Use the suspension and plumb bob method to check the correspondence of the two methods of finding the balance point for objects B and C.

   a) Record your findings about how well the two methods agree in your log.

5. Locate an "imaginary" balance point for object D. Tape a lightweight piece of paper between the "open arms" of object D and suspend the object and the plumb bob from the pin. Trace the path of the string across the piece of paper. Suspend the object from a different hole and trace the path of the string across the paper again.

   a) Do you agree that the intersection of the two lines on the paper mark the balance point of object D? What is special about this balance point? Write your answers in your log.

6. The above "balance points" which you found for two-dimensional, or "flat," objects A, B, C, and D were, in each case, the location of the object's "center of mass." Do a "thought experiment" (an experiment in your mind) to determine the location of the center of mass of each of the following objects.

   Shot-put ball (solid steel)   Basketball
   Banana   Planet Earth

ANSWERS

For You To Do (continued)

3. a) Yes, the plumb bob should pass over the balance point. When suspended, gravity serves to orient the object so that its center of mass is located directly below the point of suspension. A line extended straight downward from the point of suspension passes though the object's center of mass.

   b) The string will pass over the balance point. The spot where the lines from a) and b) intersect represents the C of M.

4. a) Students will find that the two methods will produce similar results.

5. a) The C of M is not located on shape D. Students will find that the two lines cross as a point outside the shape.

6. a) Students answer will vary.

ACTIVITY THREE: Center of Mass

➘a) For each object, describe in your log how you decided upon the location of the center of mass.

7. The technique that was used to find the center of mass (C of M) relied on the fact that the C of M always lies beneath the point of support when an object is hanging. Similarly, when an object is balanced, the C of M is always above the point of support. To find your C of M carefully balance on one foot and then the other. Your C of M is located where a vertical meter stick from one foot and the other intersect. Locate this point. The actual C of M is inside your body, since nobody has zero thickness.

➘a) Record the location of your C of M.

8. Your teacher will balance a hammer on a finger to locate the hammer's C of M. As your teacher drops the hammer into the catch box on the floor, and it twists and turns, notice the movement of the C of M.

➘a) How does the movement of the C of M compare to the motion of the entire hammer?

**REFLECTING ON THE ACTIVITY
AND THE CHALLENGE**

The center of mass is an important concept in any sports activity. The motion of the center of mass of a diver or gymnast is much easier to observe than the movements of the entire body. The sure-fire way of having a football player fall is to move his center of mass away from his support.

Think about the possibilities for using a transparent plastic cover on a TV monitor and using a pen to trace the motion of the center of mass of an athlete executing a "free fall" jump or dive. This could be used to simulate the light pen technique used by TV commentators when they comment on football replays. This would seem a good way to add an interesting feature to your TV sports commentary.

2

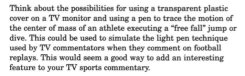

ANSWERS

For You To Do (continued)

7. a) The center of mass of a body is located inside the body at about the level of the navel.

8. a) The C of M moves directly down in a straight line, whereas the hammer twists and turns as it falls.

ANSWERS

Physics To Go

1. If not directed toward the center of mass, part of the force will be used to make the object rotate, not accelerate along a line.

2. Referring to the above answer to question 1, a player having a low center of gravity must be "hit" low, at the level of the center of mass, to have his state of rest or motion changed.

3. The body's center of mass has no support directly beneath it, so it falls.

4. Fosbury Flop: the center of mass is located behind the back, in the air outside the body.

5. The pushoff force is directed at an angle to the intended path of travel.

6. If the car were suspended from a crane twice, each time from a different point of attachment of the cable to the car, the intersection of lines representing, in each case of suspension, an extension of the cable through the car would locate the center of mass.

7. Students will probably find the center of mass by balancing the bat on their finger. Ask students to record what they did, and any problems they may have encountered.

8. When the support is moved away from the center of mass, the book will fall. By tackling below the center of mass, the support is moved away from under the center of mass, and the player will fall.

PHYSICS IN ACTION

PHYSICS TO GO

1. When applying a force to make an object move, why is it most effective to have the applied force "aimed" directly at the object's center of mass?

2. "Center of gravity" means essentially the same thing as "center of mass." Why is it often said to be desirable for football players to have a low center of gravity?

3. Stand next to a wall facing parallel to the wall. With your right arm at your side pushing against the wall and with the right edge of your right foot against the wall at floor level, try to remain standing as you lift your left foot. Why is this impossible to do?

4. Think of positions for the human body for which the center of mass might be located outside the body. Describe each position and where you think the center of mass would be located relative to the body for each position.

5. An object tends to rotate (spin) if it is pushed on by a force which is not aimed at the center of mass. How do athletes use this fact to initiate spins before they fly through the air as in gymnastics, skating, and diving events?

6. Could the suspension technique for finding the center of mass used in For You To Do be adapted to locate the center of mass of a 3-dimensional object? If you had a crane which you could use to suspend an automobile from various points of attachment, how could you locate the auto's center of mass?

7. Find the center of mass of a baseball bat using the technique that you learned in class.

8. Carefully balance a light object (not too massive) over a table or catch box. Notice that the C of M is directly over the point of support. Move the support a little bit. Explain how this technique can be adapted to tackling in football.

S 72

Activity Three A
Alternative Method for Determining Center of Gravity

FOR YOU TO DO

1. Locate the center of mass (often abbreviated C of M) of your body. For this you will need an equal arm "teeter-toter," meter stick, and two assistants. Your teacher will give you safety precautions.

Lay on your back on the teeter-toter as an assistant stabilizes each end to prevent extreme tipping. Adjust your position until balance is achieved without the assistants touching the system.

2. When at balance, have an assistant measure the distance, in meters, from the bottom of the heel of your shoe to the fulcrum (middle support) of the teeter-toter.

3. As the assistants again stabilize the teeter-toter, get off the teeter-toter.

🖎 a) Record the distance measured by the assistant.

      Distance from heel to C of M = _____ m

4. Standing erect, measure the above distance from the floor upward to locate the height of your body's center of mass relative to a constant reference point such as your navel.

🖎 a) Record the location in your log so that you will be able to recover it easily (Example: three fingerwidths below navel). Actually, your center of mass is located inside your body (within your belly) when your body is in most positions. Usually you will need to know only how high above floor level it is located.

Template for Shapes A and B

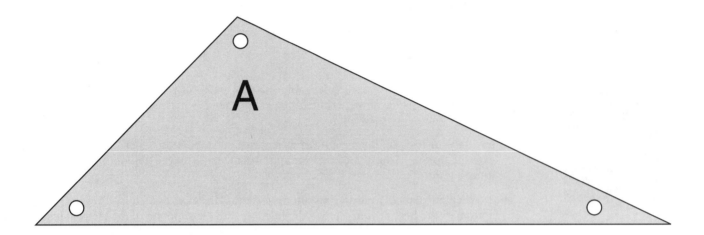

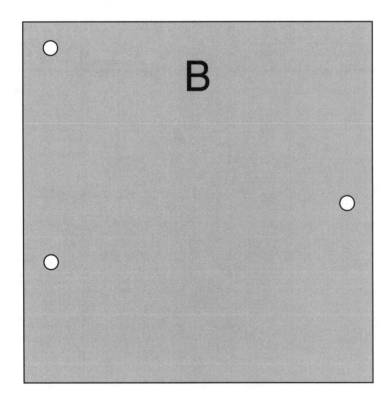

For use with *Sports*, Chapter 2, ACTIVITY THREE: Center of Mass
©1999 American Association of Physics Teachers

Template for Shapes C and D

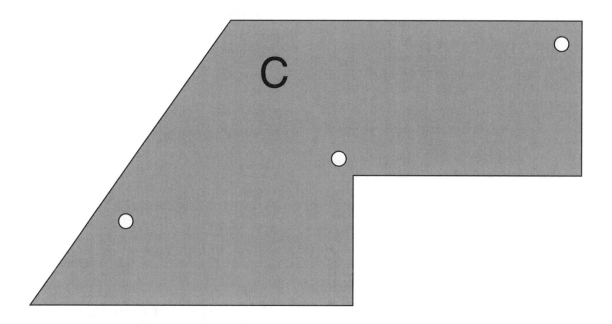

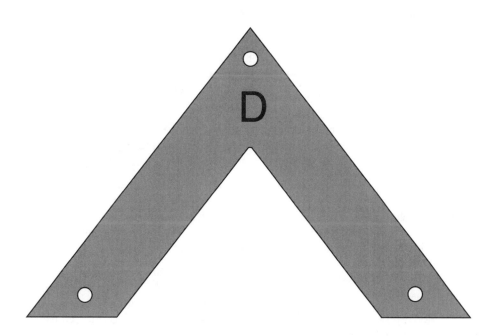

For use with *Sports*, Chapter 2, ACTIVITY THREE: Center of Mass

©1999 American Association of Physics Teachers

ACTIVITY FOUR
Defy Gravity

Background Information

It is suggested that you read the "Physics Talk" and "Example Analysis" sections in the student text for this activity before proceeding in this section. It is also suggested that you review the teacher background and student text for Chapter 1, Activity Ten, "Energy in the Pole Vault."

Work, the product force x distance, is expressed in joules. Work is equivalent to energy and, indeed, is transformed into kinetic energy and gravitational potential energy in the vertical jump.

Research has shown that the location of the center of mass within the jumper's body varies only slightly for the body positions assumed during the process of the vertical jump.

The force which lifts and accelerates the body's center of mass during a vertical jump is provided by muscles of the leg, ankles, and feet. The method of analysis used for this activity assumes that the muscular force is constant as the body rises from "ready" to "launch" positions; this is not entirely accurate—in a real jump, the force varies—but is a reasonable approximation of reality.

Active-ating the Physics InfoMall

While "hang time" is discussed on the InfoMall, it is in the sense of how long a football stays in the air during a punt, and not how long a basketball player stays (or seems to stay) in the air.

Note that gravitational potential energy is mentioned in this activity, a topic we encountered at the end of Chapter 1. And, like at the end of Chapter 1, you are encouraged to browse the textbooks. If you perform a search of the InfoMall for Work, Potential Energy, or Kinetic Energy, you will want to limit your search to only one or two stores at a time, or use additional keywords to restrict your search.

Should you desire additional problems for your students to work on, consult the Problems Place. For example, *Progressive Problems in Physics* has 16 problems on Work, and 25 on Energy.

Planning for the Activity

Time Requirements

• One class period.

Materials Needed

For the class:

• *Active Physics Sports* content video

(Segments: ice skater performing triple axel jump, basketball player "hanging" during slam dunk)

• VCR & TV monitor

For each group:

• meter stick

• patch of tape

Advance Preparation and Setup

Reserve a VCR and TV monitor for showing segments of the *Active Physics Sports* video.

Teaching Notes

View the slow-motion sequences of the jumping figure skater and the jumping basketball player to examine "hang time" and to see if either athlete remains suspended at the peak of flight. These are clear cases of free fall; since the basketball player has high horizontal running speed at take-off, the top of the trajectory is quite flat, giving illusion, when viewed in "real time" that the player "hangs" in the air.

Students may be expected to need help when applying their own data to replicate the calculations presented as an example in Physics Talk.

If a sonic ranger is available, monitor a jump from above and analyze the graphs of distance, speed, and acceleration versus time with the entire class. Directions for using a sonic ranger are provided at the end of this activity.

Suggest to students who have access to VCRs with slow-motion playback capability that they could record jumps during athletic contests and perform analysis similar to those conducted using the *Active Physics* Video.

Activity Four
Defy Gravity

WHAT DO YOU THINK?

No athlete can escape the pull of gravity.

• **Does the "hang time" of some athletes defy the above fact?**

• **Does a world class skater defy gravity to remain in the air long enough to do a triple axel?**

Record your ideas about these questions in your *Active Physics log*. Be prepared to discuss your responses with your small group and the class.

FOR YOU TO DO

1. Your teacher will show you a slow-motion video of a world-class figure skater doing a triple axel jump. The image of the skater will appear to "jerk" because a video camera completes one "frame," or one complete picture, every $\frac{1}{60}$ * second. When the video is played at normal speed, the human mind and eye perceive the action as continuous; played at slow motion, the individual frames can be detected and counted. The duration of each frame is $\frac{1}{60}$ s. *

 73 ──────────────── **SPORTS**

* Incorrect. Please change to $\frac{1}{30}$ s.

Activity Overview

In this activity students measure the positions of the C of M of a student during a vertical jump, and then analyze the amount of force and energy required by the student to perform the jump.

Student Objectives

Students will:

• measure changes in height of the body's center of mass during a vertical jump.

• calculate changes in the gravitational potential energy of the body's center of mass during a vertical jump.

• understand and apply the definition of work, Work = *fd*

• recognize that work is equivalent to energy.

• understand and apply the joule as a unit of work and energy using equivalent forms of the joule:

$1 \text{ J} = 1 \text{ Nm} = 1 \text{ (kg)m/s}^2 \times \text{(m)} = 1 \text{ (kg)m}^2/\text{s}^2$

• apply conservation of work and energy to analysis of a vertical jump, including weight, force, height, and time of flight.

ANSWERS FOR THE TEACHER ONLY

What Do You Think?

The answer to both questions is the same. There is no evidence that athletes are able to defy gravity.

2

ANSWERS

For You To Do

1. a) The skater is in the air for 15 frames.

   b) Time in air (s) = Number of frames x $^1/_{30}$ s

   $$=15 \quad \times {^1/_{30}} \text{ s}$$
   $$={^{15}/_{30}} \text{ s} = {^1/_2} \text{ s}$$

   c) During the time frame as viewed on the video, the skater's position is constantly changing. There is no "hang" time.

2. a) The basketball player is in the air for 31 frames.

   Time in air (s) = Number of frames x $^1/_{30}$ s

   $$=31 \quad \times {^1/_{30}} \text{ s}$$
   $$={^{31}/_{30}} \text{ s} = 1 \, {^1/_{30}} \text{ s}$$

   b) During the time frame as viewed on the video, the basketball player's position is constantly changing. There is no "hang" time. (Since the ball is moving upward before the player leaves the ground, and since on the way down his arms are extending and lifting the ball into the net, the illusion of hanging in the air may be created.)

3. a) Students' answers will vary according to their weight in pounds.

The following equations were presented on page S67:

Weight in newtons = Weight in pounds x 4.38 N/lb.

Weight (N) = Weight (lbs) x 4.38 N/lb

Weight = mg

Weight (N) = m(kg) x g(m/s²)

$$m \text{ (kg)} = \frac{\text{Weight (N or kg. m/s}^2)}{g \text{ (m/s}^2)}$$

---

a) Count and record in your log the number of frames during which the skater is in the air.

b) Calculate the skater's "hang time." (Show your calculation in your log.)

Time in air (s) = Number of frames $\times \frac{1}{60}$s *

c) Did the skater "hang" in the air during any part of the jump, appearing to "defy gravity?" If necessary, view the slow-motion sequence again to make the observations necessary to answer this question in your log. If your observations indicate that hanging did occur, be sure to indicate the exact frames during which it happened.

2. Your teacher will show you a similar slow-motion video of a basketball player whose "hang time" is believed by many fans clearly to "defy gravity."

a) Using the same method as above for the skater, show in your log the data and calculations used to determine the player's hang time during the "slam dunk."

b) Did the player "hang?" Cite evidence from the video in your answer.

3. How much force and energy does a person use to do a vertical jump? A person uses body muscles to "launch" the body into the air, and, primarily, it is leg muscles which provide the force. First, analyze only the part of jumping which happens before the feet leave the ground. Find your body mass, in kilograms, and your body weight, in newtons, for later calculations. If you wish not to use data for your own body, you may use the data for another person who is willing to share the information with you. (See Activity 2, Physics to Go, question 7, page S67, for how to convert your body weight in pounds to weight in newtons and mass in kilograms.)

a) Record your weight, in newtons, and mass, in kilograms, in your log.

SPORTS S 74

* Incorrect. Please change to $\frac{1}{30}$ s.

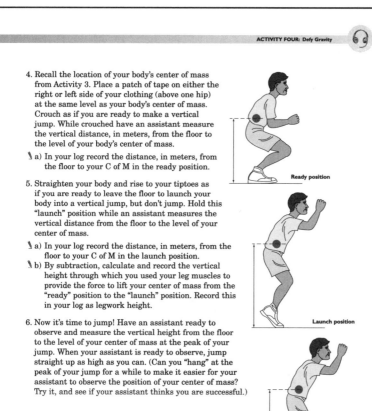

4. Recall the location of your body's center of mass from Activity 3. Place a patch of tape on either the right or left side of your clothing (above one hip) at the same level as your body's center of mass. Crouch as if you are ready to make a vertical jump. While crouched have an assistant measure the vertical distance, in meters, from the floor to the level of your body's center of mass.

a) In your log record the distance, in meters, from the floor to your C M in the ready position.

Ready position

5. Straighten your body and rise to your tiptoes as if you are ready to leave the floor to launch your body into a vertical jump, but don't jump. Hold this "launch" position while an assistant measures the vertical distance from the floor to the level of your center of mass.

a) In your log record the distance, in meters, from the floor to your C of M in the launch position.

b) By subtraction, calculate and record the vertical height through which you used your leg muscles to provide the force to lift your center of mass from the "ready" position to the "launch" position. Record this in your log as legwork height.

Launch position

6. Now it's time to jump! Have an assistant ready to observe and measure the vertical height from the floor to the level of your center of mass at the peak of your jump. When your assistant is ready to observe, jump straight up as high as you can. (Can you "hang" at the peak of your jump for a while to make it easier for your assistant to observe the position of your center of mass? Try it, and see if your assistant thinks you are successful.)

⚠️ **Make sure the floor is dry and the area in which you are jumping is clear of obstructions.**

Peak position

S 75 ——— SPORTS

ANSWERS

For You To Do
(continued)

4.-6. You may wish to provide the students with a copy of the Calculating Hang Time and Force during a Vertical Jump Worksheet provided after this activity. Expect to help students when applying their own data to replicate the calculations presented as an example in Physics Talk.

2

PHYSICS IN ACTION

a) In your log record the distance from the floor to C of M at peak position.

b) By subtraction, calculate and record the vertical height through which your center of mass moved during the jump.

Jump height = Peak position – Launch position

7. Carefully read the Physics Talk and the example given on page S77 before proceeding to the next step of this activity.

The information needed to analyze the muscular force and energy used to accomplish your jump—and an example of how to use sample data from a student's jump to perform the analysis—is presented in Physics Talk and the Example Analysis.

a) Use the information presented in the Physics Talk and Example Analysis sections and the data collected during above steps 4 through 6 to calculate the hang time and the total force provided by **your** leg muscles during your vertical jump. Show as much detail in your log as is shown in the Example Analysis.

8. An ultrasonic ranging device coupled to a computer or graphing calculator which can be used to monitor position, speed, acceleration, and time for moving objects may be available at your school. If so, it could be used to monitor a person doing a vertical jump. This would provide interesting information to compare to the data and analysis which you already have for the vertical jump. Check with your teacher to see if this would be possible.

SPORTS

 S 76

ANSWERS

For You To Do (continued)

6. a) Answers will vary.

   b) Answers will vary.

7. a) Answers will vary.

8. Student activity.

PHYSICS TALK

Work

When you lifted your body from the "ready" (crouched) position to the "launch" (standing on tiptoes) position before "takeoff" during the vertical jump activity, you performed what physicists call "work." In the context of physics, the word "work" is defined as:

The work done when a constant force is applied to move an object is equal to the amount of applied force multiplied by the distance through which the object moves in the direction of the force.

Symbols can be used to write the definition of work as:

$$W = fd$$

where f is the applied force in newtons, d is the distance the object moves in meters, and work is expressed in joules (symbol, J). At any time it is desired, the unit "joule" can be written in its equivalent form as force times distance, "(N)(m)."

The unit "newton" can be written in the equivalent form "(kg)m/s^2." Therefore, the unit "joule" also can be written in the equivalent form (kg)m^2/s^2. In summary, the units for expressing work are:

$$1 \text{ J} = 1 \text{ (N)(m)} = 1 \text{ (kg)m}^2/\text{s}^2$$

It is very common in sports that work is transformed into kinetic energy, and then, in turn, the kinetic energy is transformed into gravitational potential energy. This "chain" of transformations can be written as:

$$\text{Work} = \text{KE} = \text{PE}$$

$$fd = \frac{1}{2}mv^2 = mgh$$

These transformations are used in the analysis of data for a vertical jump.

PHYSICS IN ACTION

EXAMPLE:
Calculation of Hang Time and Force During Vertical Jump

DATA: Body Weight = 100 pounds = 440 N
       Body Mass = 44 kg
       Legwork Height = 0.35 m
       Jump Height = 0.60 m

ANALYSIS:
Work done to lift the center of mass from ready position to launch position without jumping ($W_{\text{R to L}}$):
$$W_{\text{R to L}} = fd = (\text{Body Weight}) \times (\text{Legwork Height})$$
$$= 440 \text{ N} \times 0.35 \text{ m} = 150 \text{ J}$$

Gravitational Potential Energy gained from jumping from launch position to peak position (PE_J):
$$PE_J = mgh = (\text{Body Mass}) \times (g) \times (\text{Jump Height})$$
$$= 44 \text{ kg} \times 10 \text{ m/s}^2 \times 0.60 \text{ m}$$
$$= 260 \text{ (kg)(m)}^2/\text{s}^2 = 260 \text{ (N)(m)} = 260 \text{ J}$$

The jumper's kinetic energy at takeoff was transformed to increase the potential energy of the jumper's center of mass by 260 J from launch position to peak position. Conservation of energy demands that the kinetic energy at launch was 260 J:
$$KE = \tfrac{1}{2}mv^2 = 260 \text{ J}$$

This allows calculation of the jumper's launch speed:
$$v = \sqrt{2(KE)/m} = \sqrt{2(260 \text{ J})/(44 \text{ kg})} = 3.4 \text{ m/s}$$

From the definition of acceleration, $a = \Delta v/\Delta t$, the jumper's time of flight "one way" during the jump was:
$$\Delta t = \Delta v/a = (3.4 \text{ m/s}) / (10 \text{ m/s}^2) = 0.34 \text{ s}$$
Therefore, the total time in the air (hang time) was
$2 \times 0.34 \text{ s} = 0.68 \text{ s}$

The total work done by the jumper's leg muscles before launch, W_T, was the work done to lift the center of mass from ready position to launch position without jumping, $W_{R \text{ to } L} = 150$ J, plus the amount of work done to provide the center of mass with 260 J of kinetic energy at launch, a total of 150 J + 260 J = 410 J. Rearranging the equation $W = fd$ into the form $f = W/d$, the total force provided by the jumper's leg muscles, f_T was:

$$f_T = \frac{W_T}{(\text{Legwork Height})}$$

$$= 410 \text{ J} / 0.35 \text{ m}$$

$$= 1{,}200 \text{ N}$$

Approximately one-third of the total force exerted by the jumper's leg muscles was used to lift the jumper's center of mass to the launch position, and approximately two-thirds of the force was used to accelerate the jumper's center of mass to the launch speed.

REFLECTING ON THE ACTIVITY AND THE CHALLENGE

Work, the force applied by an athlete to cause an object to move (including the athlete's own body as the object in some cases) multiplied by the distance the object moves while the athlete is applying the force, explains many things in sports. For example, the vertical speed of any jumper's take-off (which determines height and "hang time") is determined by the amount of work done against gravity by the jumper's muscles before take-off. You will be able to find many other examples of work in action in sports videos, and now you will be able to explain them.

PHYSICS TO GO

1. How much work does a male figure skater do when lifting a 50 kg female skating partner's body a vertical distance of 1 m in a pairs competition?

Physics To Go

1. Work = $fd = (mg)d$
   = 50 kg x 10 m/s^2 x 1 m = 50 j.

ANSWERS

Physics To Go
(continued)

2. Team members do work while running and pushing the sled to give it and their bodies kinetic energy before jumping on the sled, $fd = 1/2mv^2$. After the team has jumped on the sled, the total energy of the team + sled is equal to the kinetic energy gained during the pushing phase plus the gravitational potential energy $= mgh$, where h is the vertical distance to the bottom of the hill. At the bottom of the hill, the kinetic energy of the sled should be equal to the kinetic energy gained during the pushing phase plus the loss in potential energy due to coming down the hill, $1/2mv^2 + mgh$. The brake must do enough work to cause the sled to lose all of its kinetic energy by exerting a force in the direction opposite the sled's motion.

3. It is apparent that the person wants to believe that the player can defy gravity and is attempting to justify that belief by rejecting scientific evidence. It could be said that the person is not reflecting open-mindedness, a desirable attribute in scientific pursuits.

4. The burden of proof rests with the person making the claim.

5. Increase the force the athlete is able to exert using muscles, lose weight without decreasing muscular force.

6. a) 1.0 N x 1.0 m = 1 j

   b) 1.0 N x 10 m = 10 j

   c) 10 N x 1.0 m = 10 j

   d) 0.10 N x 100 m = 10 j

   e) 100 N x 0.10 m = 10 j

7. All answers are the same as for #6 above.

8. All answers are the same as for #6 above.

PHYSICS IN ACTION

2. Describe the energy transformations during a bobsled run, beginning with team members pushing to start the sled and ending when the brake is applied to stop the sled after crossing the finish line. Include work as one form of energy in your answer.

3. Suppose that a person who saw the video of the basketball player used in For You To Do said, "He really can hang in the air. I've seen him do it. Maybe he was just having a 'bad hang day' when the video was taken, or maybe the speed of the camera or VCR was 'off.' How do I know that the player in the video wasn't a 'look alike' who can't hang?" Do you think these are legitimate statements and questions? Why or why not?

4. If someone claims that a law of physics can be defied or violated, should the person making the claim need to provide observable evidence that the claim is true, or should someone else need to prove that the claim is not true? Who do you think should have the burden of proof? Discuss this issue within your group and write your own personal opinion in your log.

5. Identify and discuss two ways in which an athlete can increase his or her maximum vertical jump height.

6. Calculate the amount of work, in joules, done when:
   a) a 1.0 N weight is lifted a vertical distance of 1.0 m.
   b) a 1.0 N weight is lifted a vertical distance of 10 m.
   c) a 10 N weight is lifted a vertical distance of 1.0 m.
   d) a 0.10 N weight is lifted a vertical distance of 100 m.
   e) a 100 N weight is lifted a distance of 0.10 m.

7. List how much gravitational potential energy, in joules, each of the weights in question 6 above would have when lifted to the height listed for it.

8. List how much kinetic energy, in joules, each of the weights in questions 6 and 7 above would have at the instant before striking the ground if each weight were dropped from the height listed for it.

 80

Activity Four A
High-Tech Alternative for Monitoring Vertical Jump Height

FOR YOU TO DO

1. Place a computer motion sensor near the ceiling, pointing straight down.

2. Adjust the software so that the duration of the time axis is 5 s or less.

3. Activate the Distance versus Time graph.

4. Click the start button and, as soon as you hear the motion sensor clicking, jump. Try not to get closer than 50 cm to the sensor or you will get erroneous results.

5. Look at the resulting graph and try to find the following parts of the motion:

   • The initial bending of your knees in preparation for the jump.

   • The part of the motion when you were in the air.

   • The bending of your knees upon landing.

   a) Describe what each part of the jump looks like on the graph.

6. Use the software to zoom in on the part of the graph that contains the above-mentioned parts.

7. Switch to a Velocity versus Time graph.

   a) Describe your velocity while in the air.

8. Repeat the experiment for a higher and a lower jump. Compare and contrast your results.

For use with *Sports*, Chapter 2, ACTIVITY FOUR: Defy Gravity
©1999 American Association of Physics Teachers

Calculating Hang Time and Force During a Vertical Jump

Use a calculator to complete the following analysis of a vertical jump.

DATA:

Calculate body weight.

Weight (N) = Weight (lb.) x 4.38 N/lb.

= _____x 4.38 N/lb.

= _____

Body Weight = _____

Calculate body mass.

Weight (N) = mg

m (kg) = $\dfrac{\text{weight (kg.m/s}^2)}{g\text{(m/s}^2)}$

= $\dfrac{\text{_____}}{10 \text{ m/s}^2}$

= _____

Body Mass = _____

Calculate your legwork height.

Legwork Height = Launch position – Ready position

= _____ – _____

= _____

Legwork Height = _____

Calculate your jump height.

Jump Height   = Peak position – Launch postion

= _____ – _____

= _____

Jump Height = _____

Calculate the work done to lift the center of mass from ready position to launch position without jumping ($W_{R \text{ to } L}$)

$W_{R \text{ to } L} = fd$ = (Body Weight) x (Legwork Height)

= _____N x _____ m

= _____ N.m or J (joules)

Calculate the gravitational potential energy gained from jumping from launch position to peak position (PE_J).

$PE_J = mgh$ = (Body Mass) x (g) x (Jump Height)

= _____ kg x 10 m/s^2 x _____ m

= _____ kg.m^2/s^2

= _____ N.m or J (joules)

Conservation of energy demands that the kinetic energy at launch is equal to the gravitational potential energy at peak position.

$KE = PE$

= _____ J

(Insert the figure you calculated for PE above.)

For use with *Sports*, Chapter 2, ACTIVITY FOUR: Defy Gravity
©1999 American Association of Physics Teachers

Calculate the jumper's launch speed by writing the following equation in a different form.

$$KE = 1/2mv^2$$

$$v = \sqrt{2(KE)/m}$$

$$= \sqrt{2 \underline{\hspace{4cm}} / \underline{\hspace{5cm}}}$$

(Insert the KE from above)   (Insert the figure for mass from data above)

$$= \underline{\hspace{5cm}} \text{ m/s}$$

Calculate the jumper's time of flight "one way" during the jump by using the following equation (acceleration is change in velocity divided by the time). The change in velocity is the jumper's final velocity subtracted from the jumper's launch velocity. In this case, the final velocity at the top of the jump will be zero, the velocity before the jumper begins to come down. The acceleration is the acceleration due to gravity ($a = g$).

$$a = \Delta v / \Delta t$$

$$g = \Delta v / \Delta t$$

This equation can be rearranged in the following form to find the jumper's time of flight one way.

$$\Delta t = \Delta v / g$$

$$\Delta t = \underline{\hspace{6cm}} /10 \text{ m/s}^2$$

(Insert the value of jumper's launch speed)

$$= \underline{\hspace{5cm}} \text{ s}$$

Calculate the total time in the air (hang time) by multiplying by two, to account for the time going up and the time coming down.

Hang time   = 2 x jumper's flight one way

$$= 2 \times \underline{\hspace{5cm}}$$

(Insert the value of time for one-way trip.)

$$= \underline{\hspace{6cm}}$$

Calculate the total work done by the jumper's leg muscles before launch, W_T. This was the work done to lift the center of mass from ready position to launch position without jumping ,$W_{R\,to\,L}$, plus the amount of work done to provide the center of mass with kinetic energy (KE) at launch.

$$W_T = W_{R\,to\,L}, + KE$$

$$= \underline{\hspace{3cm}} + \underline{\hspace{3cm}}$$

$$= \underline{\hspace{4cm}} \text{ J}$$

Calculate the total force provided by the jumper's leg muscles, f_T by rearranging the following equation

$$W = fd$$

$$f = W/d,$$

$$f_T = W_T \underline{\hspace{2cm}}$$

(Legwork Height)

$$= \underline{\hspace{3cm}}$$

$$= \underline{\hspace{3cm}}$$

ACTIVITY FIVE
Run and Jump

Background Information

It is recommended that you read the section Physics Talk: Newton's Third Law of Motion in the student text for this activity before proceeding in this section.

The explanation of forces involved in walking given in the teacher's background information for Activity Four will serve to explain the forces involved with walking and running brought up in this activity. You may wish to review the background information for Activity Four before proceeding.

The pairs of equal and opposite forces identified during earlier activities to explain friction and walking are examples of Newton's Third Law of Motion, often stated as: "For every action there is an equal and opposite reaction." Another equal and opposite pair of forces arises during this activity when a student standing on a skateboard sets himself into motion by using a leg and foot to push off from the wall.

Inevitably, forces exist in equal and opposite pairs, and often the force which we identify as the force responsible for motion is not the correct one. For example, a person who says, "I pushed down on the trampoline with a mighty force, and my force launched me upward in a high jump," is mistaken; it was the equal and opposite reaction force provided by the trampoline that launched the person upward.

Active-ating the Physics InfoMall

In addition to looking for information on Newton's Third Law (look at problem 4.14 in *Schaum's 3000 Solved Problems in Physics*, in the Problems Place), perform a search using "force diagrams" as the keywords, and the first hit is a great one! It is, again, from Arons' *A Guide to Introductory Physics Teaching: Elementary Dynamics*, Chapter 3. Section 3.12 is on Newton's Third Law and Free Body Diagrams. Arons mentions common problems and suggests solutions, including suggestions of what not to do.

Arons also notes that "Students do not really begin to understand the concept of force until they

become able to apply the third law correctly and draw proper, isolated force diagrams of interacting objects," in his article "Thinking, reasoning, and understanding in introductory physics courses," in *The Physics Teacher*, vol. 19, issue 3, 1981. Check out this article.

This same search produces the warning that "Introductory textbooks are liberally decorated with diagrams, but they fail to convey to students the essential role of diagrams in problem solving or, indeed, to distinguish the roles of different kinds of diagrams" from "Toward a modeling theory of physics instruction," in the *American Journal of Physics*, vol. 55, issue 5, 1987. It is clear that the practice and ability to draw force diagrams are important.

Stretching Exercise: Add the word "elevator" to the search above (so now it is "force diagrams" AND "elevator*") for some nice discussions related to the Stretching Exercise.

Planning for the Activity

Time Requirements
- One class period.

Materials Needed
For each group:
- skateboard or wheeled chair
- safety helmet, knee and elbow pads
- meter stick
- penny coin or metal washer
- 1-newton weights or 100-gram masses (10)

Advance Preparation and Setup

If you do not have access to a skateboard and safety equipment ask the students if you may borrow theirs for use in class.

Teaching Notes

SAFETY PRECAUTIONS: Close supervision is needed to prevent injury or damage to equipment and surrounding items when students set themselves into motion by pushing off from the wall.

You may wish to emphasize that the reaction forces which we "feel" throughout a day are not limited to the reaction forces felt while running or walking, but include the reaction force which occurs whenever we touch something.

You may wish to do Activity Five A, Using a Bathroom Scale to Measure Forces, following this activity in the Teacher's Edition. If you do, be sensitive to the fact that many students will not wish to step on scales in front of their peers. Be sure that the students are given the opportunity to volunteer for this activity. Also, forewarn students that the reading of the bathroom scale "goes crazy" when the applied force increases or decreases by great amounts in short time intervals because the scales inertia causes it to "overshoot" maximum or minimum reading. Therefore, the scale reading must be observed very soon after a dramatic change is made in the force applied to the scale. Also point out to students that a specific value of force is not expected to be read on the scale during the push-off phase of the vertical jump; only the nature of the change in force—e.g., greater, no change, less—needs to be observed.

2

NOTES

Activity Five
Run and Jump

WHAT DO YOU THINK?

The men's high jump record is over 8 feet.

• **Pretend that you have just met somebody who has never jumped before. What instructions could you provide to get the person to jump up (that is, which way do you apply the force)?**

Record your ideas about this question in your *Active Physics log.* Be prepared to discuss your responses with your small group and the class.

FOR YOU TO DO

1. Carefully stand on a skateboard or sit on a wheeled chair near a wall. By touching only the wall, not the floor, cause yourself to move away from the wall to "coast" across the floor. Use words and diagrams to record answers to the following questions in your log.

   a) When is your motion accelerated? For what distance does the accelerated motion last? In what direction do you accelerate?

S 81   — SPORTS

Activity Overview

In this activity students analyze the forces involved in running, stopping, and jumping. First they push against a wall while standing on a skateboard. Then they perform a thought experiment about the forces involved.

Student Objectives

Students will:

• understand the definition of acceleration.

• understand meters per second per second as the unit of acceleration.

• use an accelerometer to detect acceleration.

• use an accelerometer to make semi-quantitative comparisons of accelerations.

• distinguish between acceleration and deceleration.

ANSWERS FOR THE TEACHER ONLY

2

What Do You Think?

Jumping involves the downward force exerted by the feet on the jumping surface and the equal and opposite reaction force exerted by the jumping surface on the feet and, in turn, the body.

ANSWERS

For You To Do

1. a) Your motion is accelerated when you push away from the wall. The acceleration lasts for a short distance after which you move at a constant velocity, and then slow down. The direction of acceleration is away from the wall.

ANSWERS

For You To Do
(continued)

1. b) Motion is at a constant speed just after the initial acceleration, when the force is acting on you. If you neglected friction you would keep moving until another force acted on you to slow you down or stop you.

c) The force is supplied by the wall. The force must be acting in the direction of motion, away from the wall.

d) You push on the wall, in a direction towards the wall.

e) The two forces are equal, but opposite in direction.

2. a-b) You push your foot on the ground backwards from yourself. How much you can push your foot parallel to the surface of the sidewalk depends on how much frictional force can be sustained by the interaction of the sole of the shoe and the sidewalk's surface. If the shoe does not slip on the surface, the sidewalk surface's equal and opposite reaction to the rearward force of friction causes your body to accelerate forward.

c) On the slippery ice surface, the force of friction is greatly reduced. You are unable to apply much of a backwards force on the ice surface because your shoe will slide on the surface. In turn the opposite reaction of the sidewalk will also be minimal with the result that you go nowhere, there is no force to push you forward.

3. a-b) Students generate a force diagram similar to the one shown.

PHYSICS IN ACTION

b) When is your motion at constant speed? Neglecting effects of friction, how far should you travel? (Remember Galileo's Principle of Inertia when answering this question)

c) Newton's Second Law, $f = ma$, says that a force must be active when acceleration occurs. What is the source of the force, the push or pull, that causes you to accelerate in this case? Identify the object that does the pushing on your mass (body plus skateboard) to cause the acceleration. Also identify the direction of the push which causes you to accelerate.

d) Obviously, you do some pushing, too. On what object do you push? In what direction?

e) How do you think, on the basis of both amount and direction, the following two forces compare?
  • the force exerted by you on the wall,
  • the force exerted by the wall on you

2. Do a "thought experiment" about the forces involved when you are running or walking on a horizontal surface. Use words and sketches to answer the following questions in your log.

a) With each step, you push the bottom surface of your shoe, the sole, horizontally backward. The force acts parallel to the surface of the ground, trying to scrape the ground in the direction opposite your motion. Usually, friction is enough to prevent your shoe from sliding across the ground surface. How is this similar to you pushing against the wall in the above activity? How is it different?

b) Since you move forward, not rearward, there must be a force in the forward direction which causes you to move. Identify where the forward force comes from, and compare its amount and direction to the rearward force exerted by your shoe with each step.

c) Would it be possible to walk or run on an extremely slippery skating rink when wearing ordinary shoes? Discuss why or why not in terms of forces.

3. Think about the vertical forces acting on you while you are standing on the floor.

a) Copy the diagram of a person in your log.

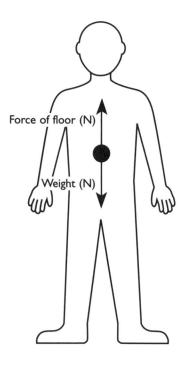

Force of floor (N)

Weight (N)

b) Identify all the vertical forces. Use an arrow to designate the size and direction of the force. Draw the forces from the dot.

c) How can you be sure that the force with which you push on the floor and the floor pushes on you are equal?

4. Set up a meter stick with a few books for support as shown.

5. Place a penny in the center of the meter stick.

a) In your log record what happens.

6. Remove the penny and replace it with 100 g (weight of 100 g = 1.0 N). Continue to place 1.0 N weights on the center of the meter stick. Note what happens as you place each weight on the stick.

⚠️ Do not exceed 10 N of weight.

a) Measure the deflection for each 1.0 N of weight and record the values for these deflections.

b) How does the deflection of the meter stick compare to the weight it is supporting? In your log sketch a graph to show this relationship.

c) Write a concluding statement concerning the penny and the deflection of the meter stick.

PHYSICS TALK

Newton's Third Law of Motion

Newton's Third Law of Motion can be stated as:

For every applied force, there is an equal and opposite force.

If you push or pull on something, that something pushes or pulls back on you with an equal amount of force in the opposite direction. This is an inescapable fact; it happens every time.

For You To Do
(continued)

3. c) The forces must be equal because you are not moving.

4. Student activity.

5. a) Nothing happens.

6. a-c) The more weight that is added to the meter stick, the greater the deflection of the stick. As the meter stick is deflected, the restoring forces in the wood build up until they exert an upward force equal to the downward force of the weight.

2

ANSWERS

Physics To Go

1. Yes, the forces are equal and opposite.

2. The restoring forces within the material from which the chair is made build up until the upward force exerted by the chair equals the downward force caused by your weight; if someone sits in your lap, the chair "bends" (or is otherwise deformed) more, resulting in a higher equal and opposite reaction force.

3. The forces on the ball and the bat are equal and opposite; sometimes the force exerted by the ball on the bat is enough to break the wood of the bat.

4. The forces on the players are equal and opposite, but the smaller player experiences a greater acceleration, which can have more harmful effects on the human body than a lesser acceleration.

5. The forces are equal and opposite; the hockey player is more likely than the boards to complain about the pain involved.

6. Gloves having padding which compresses and/or webbing which deforms when the ball hits the glove. The "softness" of a glove reduces the force which the glove exerts on the ball to a lower amount than a stationary hand would need to exert to stop the ball. The lower force causes the ball to decelerate at a lower rate, also reducing the reaction force which the ball exerts on the glove during stopping. A sure way to reduce the forces during a collision is to increase the amount of time that the objects exert forces on each other during the collision; that is one reason why airbags reduce injuries in automobile collisions.

PHYSICS IN ACTION

REFLECTING ON THE ACTIVITY AND THE CHALLENGE

According to Newton's Third Law, each time an athlete acts to exert a force on something, an equal and opposite force happens in return. Countless examples of this exist as possibilities to include in your video production. When you kick a soccer ball the soccer ball exerts a force on your foot. When you push backwards on the ground, the ground pushes forward on you (and you move). When a boxer's fist exerts a force on another boxer's body, the body exerts an equal force on the fist. Indeed, it should be rather easy to find a video sequence of a sport which illustrates all three of Newton's Laws of Motion.

PHYSICS TO GO

1. When preparing to throw a shot put ball, does the ball exert a force on the athlete's hand equal and opposite to the force which the hand exerts on the ball?

2. When you sit on a chair, the seat of the chair pushes up on your body with a force equal and opposite to your weight. How does the chair know exactly how hard to push up on you—are chairs intelligent?

3. For a hit in baseball, compare the force exerted by the bat on the ball to the force exerted by the ball on the bat. Why do bats sometimes break?

4. Compare the amount of force experienced by each football player when a big linebacker tackles a small running back.

5. Identify the forces active when a hockey player "hits the boards" at the side of the rink at high speed.

6. Newton's Second Law, $f = ma$, suggests that when catching a baseball in your hand, a great amount of force is required to stop a high speed baseball in a very short time interval. The great amount of force is needed to provide the great amount of deceleration required. Use Newton's Third Law to explain why baseball players prefer to wear gloves for catching high speed baseballs. Use a pair of forces in your explanation.

SPORTS S 84

ACTIVITY FIVE: Run and Jump

7. Write a sentence or two explaining the physics of an imaginary sports clip using Newton's Third Law. How can you make this description more exciting so that it can be used as part of your sports voice-over?

8. Write a sentence or two explaining the concept that a deflection of the ground can produce a force. How can you make this description more exciting so that it can be used as part of your sports voice-over?

STRETCHING EXERCISE

Ask the manager of a building which has an elevator for permission to use the elevator for a physics experiment. Your teacher may be able to help you make the necessary arrangements.

1. Stand on a bathroom scale in the elevator and record the force indicated by the scale while the elevator is:

   a) at rest.
   b) beginning to move (accelerating) upward.
   c) seeming to move upward at constant speed.
   d) beginning to stop (decelerating) while moving upward.
   e) beginning to move (accelerating) downward.
   f) seeming to move downward at constant speed.
   g) beginning to stop (decelerating) while moving downward.

2. For each of the above conditions of the elevator's motion, the earth's downward force of gravity is the same. If you are accelerating up, the floor must be pushing up with a force larger than the acceleration due to gravity.

   a) Make a sketch which shows the vertical forces acting on your body.
   b) Use Newton's laws of motion to explain how the forces acting on your body are responsible for the kind of motion—at rest, constant speed, acceleration or deceleration—that your body has.

Stretching Exercise

1. For the answers below, m is the person's mass, g is the acceleration due to gravity, mg is the person's weight, and a is the acceleration of the elevator.

   a) mg

   b) $mg + ma$

   c) mg

   d) $mg - ma$

   e) $mg - ma$

   f) mg

   g) $mg + ma$

2

ANSWERS

Physics To Go *(continued)*

7. - 8. Students may wish to use parts of the voice-overs generated for these questions for their Chapter Challenge.

Activity Five A
Using a Bathroom Scale to Measure Forces

FOR YOU TO DO

1. Stand on a bathroom scale. Preferably, the scale should be calibrated in newtons; if is not, see Chapter 2, Activity Two, Physics to Go, Problem #5 for how to convert the calibration of the scale to newtons.

   a) Reproduce the sketch of you standing on a bathroom scale in your log. Use a dot to show the approximate location of your center of mass.

   b) Add an arrow pointing downward from your center of mass to show your weight, in newtons. Label the arrow: Weight = ___ N (enter the amount).

   c) Since, when standing on the scale, you are not moving, the scale must be pushing upward on you with a force which balances your weight. Decide upon the amount and direction of the second force which must be acting on you. Starting from your center of mass in the sketch, draw the arrow to represent the force. Label the arrow: Force of Scale = ___ N (enter the amount).

2. Stand on the bathroom scale again, but stand in a crouched position.

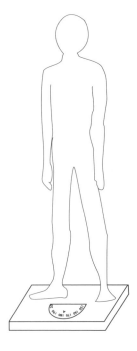

   a) Reproduce the position of you crouching on a bathroom scale sketch in your journal.

   b) Without moving any part of your body while in a crouched position are you able to "bear down" to increase the scale reading to more than when you are standing upright on the scale? Explain why or why not, and show the forces acting on your center of mass when crouched on the scale.

3. Crouch on the scale and rise to full standing position at very low, constant speed. Accelerate as little as possible as you move; you will need to accelerate somewhat to start moving upward, but, after that, try to move your center of mass at low, constant speed upward.

   a) Describe in your log that you are moving your center of mass upward at constant speed from the positions used in the above steps.

   b) Even though the scale reading may "wiggle" somewhat due to small, not-on-purpose accelerations as you move your center of mass upward at constant speed, how does the scale reading compare to when you were at rest in both crouched and standing positions? In your sketch, draw arrows to represent the forces acting on your center of mass as it moves upward at constant speed.

4. Explain how Newton's First Law of Motion (see Chapter 2, Activity One, Physics Talk) applies to the above situations when you were:

a) at rest, standing upright on the scale.

b) at rest, crouched on the scale.

c) standing on the scale while moving your center of mass upward at constant speed.

5. Again stand on the scale in crouched position. This time, you will accelerate your center of mass to jump upward. Design your jump to go slightly forward so that you land on the floor, not the scale, when you come back down – this is to prevent damage to the scale by landing on it.

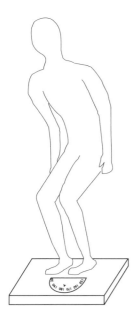

Have an assistant ready to observe how the scale reading changes as you are accelerating your center of mass upward, but before you "launch" into the air. This observation will be "tricky" because bathroom scales cannot rapidly respond to changing forces. Also, inertia causes the scale to "overshoot" the maximum reading when the force on the scale is suddenly increased. Therefore, observation of how the scale reading changes will need to be made a fraction of a second after you begin rising from crouched position.

a) Sketch the situation in your log.

b) Jump as your assistant observes the scale. Compared to when your center of mass is at rest or moving with constant upward speed, is the downward force you exert on the scale when accelerating your center of mass upward more or less? Since the scale couldn't "keep up" with the action to allow reading a specific value of force, report in your journal whether the force was "going up" or "going down" as you were pushing on the scale before launching into your jump.

c) In your sketch, draw arrows to represent the forces acting on your center of mass as it accelerates upward before launch, including (i) the downward pull of gravity on your body (your weight) and (ii) the upward push of the scale. You do not have a value, in newtons, to determine how long to make the arrow representing the upward push of the scale, so indicate, according to how the scale reading was changing, that the amount of force is either greater or less than your weight.

d) Which of Newton's Laws of Motion best applies to this situation, the First Law or the Second Law? How does the law which you choose apply?

ACTIVITY SIX
The Mu of the Shoe

Background Information

When it is desired to accelerate an object (change its speed, its direction, or both), a net force must act on the object. If you are standing still on a sidewalk and want to get moving, you must somehow cause a force to be exerted on your body's mass. You accomplish application of a force by pushing your foot in the backward direction parallel to the surface of the sidewalk. How much you can push your foot parallel to the surface of the sidewalk depends on how much frictional force can be sustained by the interaction of the sole of the shoe and the sidewalk's surface. If the shoe does not slip on the surface, the sidewalk surface's equal and opposite reaction to the rearward force of friction causes your body to accelerate forward; if there is ice on the sidewalk, the available force will be reduced, and your shoes may just slide on the surface with the result that you go nowhere.

The maximum frictional force, F, that can be generated between the surfaces of two materials in contact is expressed by the equation $F = \mu N$ where N is the "normal force" (the word "normal" in this context means "perpendicular") that is perpendicularly pushing the two surfaces together, and μ is the "coefficient of friction" for the pair of materials from which the surfaces are made. In the above example, the normal force, N, would be equal to your weight. The value of μ depends on the quality of the two materials in contact. For a given pair of materials, such as leather shoe soles on a concrete sidewalk surface, there are two kinds of μ, the "coefficient of static (starting) friction" and the "coefficient of sliding friction" (sometimes called the coefficient of "kinetic"—meaning "moving"— friction). The larger of the two, the "coefficient of static (stationary) friction" applies when the surfaces are at rest with respect to each other; the value of F resulting from calculations using the coefficient of static friction is the minimum force required to "tear the surfaces loose" to cause them to begin sliding across one another. The second kind of μ, the smaller of the two kinds, applies when the surfaces are moving with respect to each other in a sliding mode; the value of F resulting from calculations using the coefficient of sliding friction is the minimum force required to cause the surface to slide across one another at constant speed. It is the second kind, the coefficient of sliding friction, that is measured in this activity

The frictional force generated between the shoe and the sidewalk—due to pushing the foot rearward as gravity and the upward restoring force of the sidewalk squeeze the sole of the shoe and the surface of the sidewalk together—is answered by a corresponding equal forward push by the sidewalk on the foot. The latter force, the forward push by the sidewalk, is the push you "feel" and which causes you to accelerate forward.

It is not always the case that the normal force, N, is equal to the weight of the object. In the case of an object on a sloped surface, the normal force is less than the weight, equaling the component, or effectiveness, of the object's weight in the direction perpendicular to the sloped surface. Other examples of a cases where N is not equal to the weight of an object bearing on a surface would include the frictional force between belts riding on pulleys in machines; in such cases, tensioned springs usually are used to force the surfaces together to provide sufficient N to prevent sliding, or, intentionally as when stopping a machine, to reduce tension to cause N to be reduced to an amount where a belt will slide on a pulley.

Active-ating the Physics InfoMall

Discussions of friction can be found throughout the InfoMall. If you choose to do a search using the keyword "friction," you will need to limit your search to only a few stores at a time. If you look in the Articles and Abstracts Attic, one of the titles that may interest you is "Twas the class before Christmas," from *The Physics Teacher*, vol. 24, issue 9, 1986. At the very least, the problems involving friction can be amusing.

Try searching the Demo & Lab Shop with the keyword "friction." You will want to look at these yourself, so no examples are included here. Choose the demonstration that best suits your style and situation.

And you will not be surprised to find that there are many, many problems you can find involving friction in the Problems Place.

Planning for the Activity

Time Requirements

• One class period.

Materials Needed

For each group:

• spring scale, 0-5 newton range

• athletic shoe

• rough horizontal surface

• smooth horizontal surface

• ballast (to approximately double weight of shoe)

Advance Preparation and Setup

You will need a variety of athletic shoes and samples of floor materials. If enough students wear athletic shoes to school, use their shoes as samples; if not, arrange to have one shoe per group available. Floor materials may include your classroom floor, a table top to simulate a floor, samples of floor materials from a retail store, or, if they are different from the floor in your classroom, floors in other areas of your school such as the gymnasium. It is desired to have contrasting degrees of "roughness" represented in the samples. Surfaces which would be of most interest to your students would be best to use.

Teaching Notes

It is recommended that a brief discussion of the symbolism used in physics be conducted with the class after students have read the section "What is Mu?" and before beginning "For You To Do." Students may feel it is "cool" to ask other students not taking physics about the "μ" of their athletic shoes.

Students may wish to measure μ for more kinds of shoes and floor materials called for in the instructions. The instructions include only the minimum number of samples needed to acquire meaningful data; encourage students to test more samples of shoes and/or floor materials if time allows.

Coefficients of friction for many pairs of materials are listed in the *Handbook of Chemistry and Physics* and in many traditional physics textbooks. You may wish to have listings available for students to examine.

In addition to athletic shoes which are specialized for particular sports, the variety of waxes used by skiers

to control friction in various conditions may be of high interest to some students.

This activity can also be done with a smart pulley, computer, and weights hanging over the smart pulley.

You may wish to do the Activity Six A : Alternative Activity for Measuring the Mu of the Shoe presented after this activity in the Teacher's Edition as a Stretching Exercise with the class, after you have completed the activity in the textbook. The explanation for this activity is also presented following the activity.

2

Activity Overview

In this activity students investigate the effect of different surfaces, and different weights on the coefficient of friction.

Student Objectives

Students will:

- understand and apply the definition of the coefficient of sliding friction, μ.

- measure the coefficient of sliding friction between the soles of athletic shoes and a variety of floor surface materials.

- calculate the effects of frictional forces on the motion of objects.

ANSWERS FOR THE TEACHER ONLY

What Do You Think?

Vast amounts of engineering knowledge and research about friction are applied to the design of athletic footgear.

PHYSICS IN ACTION

Activity Six
The Mu of the Shoe

WHAT DO YOU THINK?

A shoe store may sell as many as one hundred different sport shoes.

- **Why do some sports require special shoes?**

Record your ideas about this question in your *Active Physics log*. Be prepared to discuss your responses with your small group and the class.

FOR YOU TO DO

1. Take an athletic shoe. Use the spring scale to measure the weight of the shoe, in newtons.

   a) Record a description of the shoe (such as its brand) and the shoe's weight in your log.

SPORTS ————————◆ S 86 ◆————————

ANSWERS

For You To Do

1. a) Answers will vary depending on the brand of shoe used and the size of the shoe.

2. Place the shoe on one of the two horizontal surfaces (either rough or smooth) designated by your teacher to be used for testing. Attach the spring scale to the shoe as shown below so that the spring scale can be used to slide the shoe across the surface while, at the same time, the amount of force indicated by the scale can be read.

✎ a) Record in your log a description of the surface on which the shoe is to slide.

✎ b) Measure and record the amount of force, in newtons, needed to cause the shoe to slide on the surface at constant speed. Do not measure the force needed to start, or "tear the shoe loose," from rest. Measure the force needed, after the shoe has started moving, to keep it sliding at low, constant speed. Also, be careful to pull horizontally so that the applied force neither tends to lift the shoe nor pull downward on the shoe.

✎ c) Use the data you have gathered to calculate μ, the coefficient of sliding friction for this particular kind of shoe on the particular kind of surface used. Show your calculations in your log.

The coefficient of sliding friction, symbolized by μ, is calculated using the following equation.

$$\mu = \frac{\text{force required to slide object on surface at constant speed}}{\text{perpendicular force exerted by the surface on the object}}$$

Example:

Brand X athletic shoe has a weight of 5 N. If 1.5 N of applied horizontal force is required to cause the shoe to slide with constant speed on a smooth concrete floor, what is the coefficient of sliding friction?

$$\mu_x \text{ on concrete} = \frac{1.5\ N}{5.0\ N} = 0.30$$

ANSWERS

For You To Do (continued)

2. a-c) Students' answers will vary depending on the surface and the shoe used.

ANSWERS

For You To Do
(continued)

3. a-c) Taking into account possible errors in measurement, students should find that the value of μ is not affected by the weight of the shoe.

4. a) Student sketch (see bottom right).

b-c) Students should recognize that the value of μ will be different for different surfaces. The "rougher" the surface, the greater the coefficient of friction. For example, the coefficient of friction for rubber on dry concrete is 140 times greater than rubber on ice.

d) The previous step in the activity indicates that the weight of the shoe should not make a difference in the coefficient of friction.

PHYSICS IN ACTION

3. Add "filler" to the shoe to approximately double its weight and repeat the above procedure for measuring the μ of the shoe.

a) Calculate μ for this surface, showing your work in your log.

b) Taking into account possible errors of measurement, does the weight of the shoe seem to affect μ? Use data to answer the question in your log.

c) How do you think the weight of an athlete wearing the shoe would affect μ? Why?

4. Place the shoe on the second surface designated by your teacher and repeat the procedure again.

a) Make another sketch to show the forces acting on the shoe.

b) Calculate μ.

c) How does the value of μ for this surface compare to μ for the first surface used? Try to explain any difference in μ.

d) Would it make any difference if you used the empty shoe, or the shoe with the filler to calculate μ in this activity? Explain your answer.

REFLECTING ON THE ACTIVITY AND THE CHALLENGE

Many athletes seem more concerned about their shoes than most other items of equipment, and for good reason. Small differences in the shoes (or skates or skis) athletes wear can affect performance. As everyone knows, athletic shoes have become a major industry because people in all "walks" of life have discovered that athletic shoes are great to wear, not only on a track but, as well, just about anywhere. Now that you have studied friction, a major aspect of what makes shoes function well when need exists to be "sure-footed," you are prepared to do "physics commentary" on athletic footgear and other effects of friction in sports. Your sports commentary may discuss the μ of the shoe, the change in friction when a playing field gets wet, and the need for friction when running.

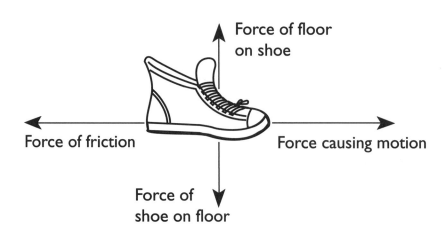

PHYSICS TALK

Coefficient of Sliding Friction, μ

There are not enough letters in the English alphabet to provide the number of symbols needed in physics, so letters from another alphabet, the Greek alphabet, also are used as symbols. The letter μ, pronounced "mu," traditionally is used in physics as the symbol for the "coefficient of sliding friction."

The coefficient of sliding friction, symbolized by μ, is defined as the ratio of two forces:

$$\mu = \frac{\text{force required to slide object on surface at constant speed}}{\text{perpendicular force exerted by the surface on the object}}$$

Facts about the coefficient of sliding friction:

- μ does not have any units because it is a force divided by a force; it has no unit of measurement.

- μ usually is expressed in decimal form such as, 0.85 for rubber on dry concrete (0.60 on wet concrete).

- μ is valid only for the pair of surfaces in contact when the value is measured; any significant change in either of the surfaces (such as the kind of material, surface texture, moisture or lubrication on a surface, etc.) may cause the value of μ to change.

- only when sliding occurs on a horizontal surface is the perpendicular force which the sliding object exerts on the surface over which sliding occurs equal to the weight of the object.

ANSWERS

Physics To Go

1. In football, players change the length of shoe cleats or sometimes wear shoes without cleats to improve footing in bad weather.

2. Downhill skiers use wax to reduce friction.

3. A common misconception is that the coefficient of friction—and, therefore, the force of friction—depends only on the shoe; it depends as much as on the nature of the surface beneath the shoe. No, the athlete cannot be assured that the same amount of frictional force will be present when the same shoe is used on a court having a different surface.

4. $F = 0.03 \times 600$ N = 18 N

5. Normal force, F_N = Weight of vehicle = mg = 1000 kg x 10 m/s^2 = 10,000 N

Frictional (stopping) force = μN = 0.55 x 10,000 N = 5,500 N

Work to stop vehicle = fd = 5,500 N x 100 m = 550,000 joules

Work to stop vehicle = KE of vehicle before brakes were applied:

550,000J = 1/2 mv^2

Therefore, $v = \sqrt{2\ (550{,}000\ J)/m} = \sqrt{1{,}100{,}000\ J/1{,}000\ kg}$

$= \sqrt{1{,}100\ m^2/s^2} = 33$ m/s = 75 miles/hr

The driver has a problem because the laws of physics will prevail

PHYSICS TO GO

1. Identify a sport and changing weather conditions which probably would cause an athlete to want to increase friction to have better "footing." Name the sport, describe the change in conditions and explain what the athlete might do to increase friction between the shoes and ground surface.

2. Identify a sport in which athletes desire to have frictional forces as small as possible and describe what the athletes do to reduce friction.

3. If a basketball player's shoes provide an amount of friction which is "just right" when she plays on her home court, can she be sure the same shoes will provide the same amount of friction when playing on another court? Explain why or why not.

4. A cross-country skier who weighs 600 N has chosen ski wax which provides μ = 0.03. What is the minimum amount of horizontal force which would keep the skier moving at constant speed across level snow?

5. A race car having a mass of 1,000 kg was traveling at high speed on a wet concrete road under foggy conditions. The tires on the vehicle later were measured to have μ = 0.55 on that road surface. Before colliding with the guard rail, the driver locked the brakes and skidded 100 m, leaving visible marks on the road. The driver claimed not to have been exceeding 65 miles per hour (29 m/s). Use the equation:

Work = Kinetic Energy

to estimate the driver's speed upon hitting the brakes. (Hint: In this case, the force which did the work to stop the car was the frictional force; calculate the frictional force using the weight of the vehicle, in newtons, and use the frictional force as the force for calculating work)

6. Identify at least three examples of sports in which air or water have limiting effects on motion similar to sliding friction. Do you think forces of "air resistance" and "water resistance" remain constant or do they change as the speeds of objects (such as athletes, bobsleds or rowing sculls) moving through them change? Use examples from your own experience with these forms of resistance as a basis for your answer.

7. If there is a maximum frictional force between your shoe and the track, does that set a limit on how fast you can start (accelerate) in a sprint? Does that mean you cannot have more than a certain acceleration even if you have incredibly strong leg muscles? What is done to solve this problem?

8. How might an athletic shoe company use the results of your experiment to "sell" a shoe? Write copy for such an advertisement.

9. Explain why friction is important to running. Why are cleats used in football, soccer, and other sports?

10. Choose a sport and describe an event in which friction with the ground or with the air plays a significant part. Create a voice-over or a script which uses physics to explain the action.

Physics To Go

(continued)

6. Any sports which involve objects moving through air or water will involve fluid resistance (air resistance will be treated in detail in Chapter 3, Activity Nine).

7. Yes, the maximum frictional force between your shoe and the track does place a limit on acceleration. Sprinters use blocks when beginning a race, thereby providing a surface that is not horizontal, and does not depend on the weight of the runner.

8. See the assessment rubric following this activity in the Teacher's Edition.

9. Without friction it would be impossible to walk or run. Cleats increase the friction between the shoe and the ground by increasing the amount of surface in contact.

10. Students provide a voice-over related to friction.

Assessment Rubric: Physics To Go Question 8

How might an athletic shoe company use the results of your experiment to sell a shoe?
Write a copy for an advertisement.

Descriptors	Levels of Attainment			
	poor	average	good	excellent
1. Physics concepts are accurately presented. • Newton's Third Law of Motion is explained in terms of running: For every applied force, there is an equal and opposite force. • The force required to cause the shoe to move on different surfaces is explained: *Coefficient of friction is expressed as a ratio of the force needed to move the shoe at a constant speed, by the force exerted by the surface of the shoe.*	1	2	3	4
2. Physics concepts explained in everyday language. • Examples are provided for each concept. • Presentation style does not talk down to the consumer.	1	2	3	4
3. The need for buying specialized sports shoes is established. • Examples are provided illustrating why different shoes are used for different sporting events. *Presentation identifies the amount of friction provided by different surfaces and the advantages of solid traction for different sports.*	1	2	3	4
4. Design and appeal. • Organization of information is short and snappy. • Message is clearly identified with a target audience. • Presentation designed for a specific media: i.e., visual images utilize color effectively and enhance message for television presentation.	1	2	3	4

For use with *Sports*, Chapter 2, ACTIVITY SIX: The Mu of the Shoe

NOTES

2

Activity Six A

Alternative Activity for Measuring the Mu of the Shoe

FOR YOU TO DO

"Step right up. Have the Mu of your shoe measured right here! Takes only a minute." Can the coefficient of sliding friction be measured in a quicker, easier way than the method used in For You To Do? Well, it can.

1. For a quick-and-easy way to measure, all you need is a sample of "floor" material (such as a wooden board) which can be arranged into a ramp, a carpenter's square (if not available, a ruler can be substituted), a calculator, and, of course, a shoe. Use a stack of books (or some other method) to raise one end of the floor sample to form a sloped ramp and place the shoe on the ramp facing "downhill."

2. Adjust the slope of the ramp to find the slope at which the shoe, given only a "nudge" to start it sliding, continues sliding down the ramp "on its own" (without further pushing) at low, constant speed:

   • If the shoe does not slide down the ramp on its own after it is tapped, increase the slope of the ramp.

   • If the shoe accelerates while sliding down the ramp after it is tapped, decrease the slope of the ramp.

3. When you have found the proper slope for the ramp, use a carpenter's square (or a ruler) to measure the vertical rise and corresponding horizontal run of the ramp.

   ✎ a) Record the rise and the run in your log.

4. The coefficient of sliding friction is:

$$\mu = \frac{\text{Length of vertical rise of ramp}}{\text{Length of horizontal run of ramp}}$$

   ✎ a) Calculate the coefficient of sliding friction of the shoe.
   This method gives the same result as the method used in For You To Do. Your teacher may be able to help you understand why the two methods are equivalent.

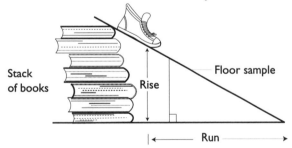

Stack of books   Rise   Floor sample   Run

For use with *Sports*, Chapter 2, ACTIVITY SIX: The Mu of the Shoe
©1999 American Association of Physics Teachers

Background Information for Activity Six A: Alternative Activity for Measuring the Mu of the Shoe

Below is the theoretical basis for the method of measuring μ presented in the Alternative Activity Six A. Highly able students who have had experience with geometry may be able to understand the principles underlying the method. It is recommended that the procedure presented in the alternative activity be read carefully before proceeding to read the below explanation.

The purpose of this explanation is to prove that μ = (Rise) ÷ (Run) when the slope of the inclined plane made from a sample of floor material is adjusted so that the shoe, or other object represented by mass m, whose coefficient of friction with the floor material is desired to be measured—slides down the slope at low, constant speed when given a "tap" start. (A tap start is required to overcome the force of "static," or starting, friction which is greater than the force of sliding friction.) Referring to the below diagram, side lengths A and B of triangle ABC correspond, respectively, to the Rise and Run distances. Therefore, it is to be proven that μ = A/B.

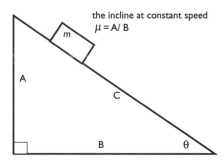

the incline at constant speed
$\mu = A/B$

The weight, W, of mass m is a force which acts straight downward, as shown in the below diagram. The amount of force due to the object's weight, $W = mg$, is represented by the length of vector W.

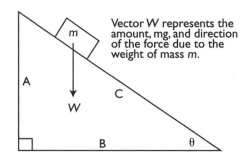

Vector W represents the amount, mg, and direction of the force due to the weight of mass m.

The weight vector, W, can be resolved into, or thought of as having the same effect as, two vectors called "components" of vector W: (1) a vector R which has a direction normal, or perpendicular, to the surface represented by side C of triangle ABC and (2) a vector P which has a direction parallel to side C. Vector R is the force which mass m exerts perpendicular to the surface on which sliding occurs, and vector P is the force which causes mass m to slide on the surface at constant speed.

The below diagram shows vectors R and P as components of vector W.

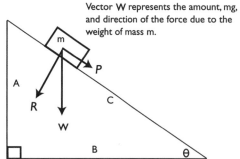

Vector W represents the amount, mg, and direction of the force due to the weight of mass m.

Vector mathematics requires that component vectors, in this case vectors P and R, must, when added together as vectors, equal the vector of which they are components, in this case vector W. On the right-hand side of the above diagram, vectors P and R are shown repositioned, but not altered in length or direction, for head-to-tail vector addition. As shown, when the component vectors are placed head-to-tail with one of the component vectors having its tail end corresponding to the tail end of vector W, the head end of the second component vector drawn in the head-to-tail vector addition process arrives at the head end of vector W. The fact that the component vectors "close" on the head end of W when placed head-to-tail shows, according to vector algebra, that the components add together to equal vector W. Since the directions of the component vectors P and R were specified as, respectively, parallel and normal to triangle side C, only the lengths of P and R shown in the diagram would satisfy the requirement of the vectors to close on the head end of vector W.

Referring to the above diagram, triangle ABC is similar to triangle PRW formed for the vector addition process. This is true by the theorem that two triangles which each contain a right angle and which have two mutually perpendicular sides are similar. Therefore, the sides of the two triangles exist in equal proportions, and:

$$A/B = P/R$$

Since, by definition, μ is the ratio of the force required to slide mass m at constant speed on the surface to the force which m exerts perpendicular to the surface:

$\mu = P/R = A/B$

Therefore, it is proven: the Rise divided by the Run, A/B, when mass m slides at constant speed is equal to μ.

For those familiar with trigonometry, A/B in triangle ABC is the tangent of angle θ.

Therefore, as an alternative to measuring the Rise and Run to determine μ, θ can be measured instead, and μ can be determined using a calculator or table by the relation:

$\mu = \tan\theta$

All of the forces active as mass m slides on the incline at constant speed are shown in the below diagram:

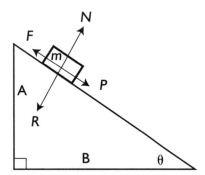

Where the new forces introduced to the diagram, F and N, are, respectively, the frictional force and the restoring force of the material represented by side C of triangle ABC. It is important to observe that, since mass m moves at constant speed, the sum—vector sum, that is—of the four forces acting on mass m is zero, or, in other words, all of the forces acting on mass m are balanced.

For students who wish to do the acitivity , samples of floor materials will need to be able to be arranged on an adjustable slope; boards 2 or 3 feet in length could be used as a base for samples to be arranged on a slope.

NOTES

2

ACTIVITY SEVEN
Concentrating on Collisions

Background Information

This background information will serve for both this and the next activity because both involve the same topic, momentum. Momentum is introduced in this activity, and conservation of momentum is implied; the next activity, Activity Eight, focuses on the Law of Conservation of Momentum.

Before proceeding in this section it is recommended that you read the following sections in this chapter of the student text: "Physics Talk: Mass, Velocity and Momentum" in Activity Seven, and "Physics Talk: Conservation of Momentum" in Activity Eight. Both selections are assumed as background for the below discussion of momentum.

A corollary to Newton's Third Law is that action/reaction pairs of forces always act for the same amount of time as each other. In any interaction, forces may vary in complicated ways during the time of interaction; nevertheless, the average force during the interaction often can be used to provide an accurate, but greatly simplified, analysis of the net result of an interaction such as a collision.

If F is used to represent the average force on one object during a collision and Δt is used to represent the duration of the collision, one can produce the following mathematical path to the quantity of ultimate interest for this discussion, the quantity called momentum. Newton's Second Law, written in rearranged form, may be applied to the object involved in the collision as:

$$F\Delta t = m\Delta v = m(v_f - v_i) = (mv_f - mv_i) = \Delta(mv)$$

$$F\Delta t = \Delta(mv)$$

where v_i is the speed with which the object entered into the collision and v_f is the speed of the object after the collision.

In the above equation $F\Delta t = \Delta(mv)$, the product of force and time, $F\Delta t$, is called the "impulse," and the quantity mv is called the "momentum." The equation can be read literally as, "When during a collision an object of mass m experiences an impulse $F\Delta t$, the object experiences a change in momentum $\Delta(mv)$. Indeed, it is the impulse that causes the object's change in momentum."

The force-time product, impulse, is expressed in the unit newton-seconds; dimensional analysis shows that Ns is equivalent to the unit of momentum, which, as a mass-velocity product, must be (kg)m/s:

Ns = [(kg)(m/s²)] (s) = (kg)m/s   (True, the unit of impulse equals the unit of momentum.)

But, according to Newton's Third Law, whatever happens to one of a pair of objects during a collision should, in terms of force, happen equally, but in the opposite direction, to the other object. Also, the objects involved in a collision touch each other for equal amounts of time, the duration of mutual contact. Thus, whatever change in momentum one object experiences, the other object must experience an equal and opposite change in momentum. For example, if one object gains momentum in a head-on collision, the other object must lose an equal amount of momentum. It must always be true that the combined momenta of the two objects before the collision are preserved, or conserved, after the collision. It is a far-reaching law of physics that momentum is conserved in all interactions, no matter how complex, convoluted, and intense the interactions may be.

This chapter considers only collisions that are "head on," so the motions all are along the same line before and after the collision. However, the caution is made that when dealing with momentum in collision events, one must take care to keep track of the directions of travel of the objects both before and after the collision. For example, for a collision in which an object of mass m leaves the collision traveling at the same rate but in an opposite direction is not one for which the momentum change is zero. In this case, the momentum change is $2mv$ or $-2mv$, depending on how positive and negative signs are attributed to direction of travel.

Similarly, if two objects enter a collision with equal magnitudes of momentum, the opposite signs mean that the sum of momenta is zero. As a result, when the collision is completed, one possible outcome is to have both objects at rest at the point of impact. The only other possibility is that both objects will rebound with equal and opposite momenta, but not necessarily equal and opposite speeds because the masses may be unequal.

Depending on the nature of the objects involved in a collision, kinetic energy may be conserved to extents ranging from not at all (as when the objects stop upon colliding) to almost completely (as when extremely hard objects such as ball bearings or gas molecules collide and rebound). Whether or not kinetic energy is conserved, momentum is conserved, which makes the Law of Conservation of Momentum a very powerful tool.

Active-ating the Physics InfoMall

The physics of collisions is yet another of those topics discussed in almost every textbook around. To keep InfoMall searches interesting and related to this *Active Physics* book, search the Articles and Abstracts Attics using keywords "collisions" AND "sport*". Several articles result, including "Batting the ball," from the *American Journal of Physics*, vol. 31, issue 8, 1963. You will also find plenty on billiard balls, kicking footballs, and using tennis rackets.

You may wish to investigate the known misconceptions students have regarding momentum. Try a search with "momentum" AND "misconcept*". One of the hits is "Verification of fundamental principles of mechanics in the computerized student laboratory," in *American Journal of Physics,* vol. 58, issue 10, 1990. This article is also great for using computers for teaching.

Planning for the Activity

Time Requirements

- One class period.

Materials Needed

For the class:
- balance or spring scale

For each group:
- bowling balls (2)
- starting ramp for bowling ball (2)
- tape
- soccer ball
- tennis ball
- golf ball

Advance Preparation and Setup

A pair of (nearly) matched bowling balls is recommended for each group. Contacting bowling establishments in your area perhaps will result in donations of balls, and putting out a call to your school community—parents, faculty, etc.—also should be productive. If you simply cannot obtain bowling balls, hard balls, such as croquet balls, could be substituted.

Two safe, sturdy ramps also will be needed for each group to give the balls controlled starts. A "V-shaped" trough can be made by fastening two pieces of inexpensive 1" x 4" lumber edge-to-edge. Some provision will be needed to elevate one end so that the slope of the ramp will be about 20°. It may also be desirable to cut the low end of the trough at an angle to match the floor surface, allowing the ball to have a smooth transition from the ramp to the floor. Some of your students may be able to design and construct ramps, perhaps with help from a technical education teacher or a parent.

You also will need, for each group, a soccer ball, a golf ball and a tennis ball. You may be able to borrow the balls from your school's athletic department.

Teaching Notes

Discuss with the students what is happening in the humorous illustration on page S92. What will happen to the soccer ball? Students who enjoy drawing, may wish to draw the "next frame" in this action. You may wish to return to the illustration at the end of the activity to compare students responses to their original predictions.

Bowling balls, with their large inertias, are less sensitive to annoyances like small imperfections in the floor or a little grit in the path. They also have the advantage of a large diameter, so it is easier to create collisions that are nearly head on. They also demand a lot of attention. Their disadvantage is their large mass!

SAFETY PRECAUTION: Moving bowling balls can hurt people (especially toes and fingers) and property. Instruct students to use extreme care and provide close supervision when bowling balls are in use.

Contact sports are, of course, rich sources of collisions to be analyzed in terms of momentum.

Need to assign positive and negative values to velocity and momentum is implied in the discussion in Physics Talk. The unit of momentum, (kg)m/s is new, but is sufficiently straightforward that it should not be a problem for students; somehow, the unit of momentum seems to have escaped being collected up under someone's name.

Activity Overview

In this activity students investigate the principle of momentum by rolling balls down ramps and staging collisions. They will infer the relative masses of two balls by staging and observing collisions between them.

Student Objectives

Students will:

• understand and apply the definition of momentum: Momentum = mv

• conduct semi-quantitative analyses of the momentum of pairs of objects involved in one-dimensional collisions.

• infer the relative masses of two objects by staging and observing collisions between the objects.

ANSWERS FOR THE TEACHER ONLY

What Do You Think?

Both mass and speed have roles in collisions between two objects. The mass of each player and their speeds must be considered.

PHYSICS IN ACTION

Activity Seven

Concentrating on Collisions

WHAT DO YOU THINK?

In contact sports, very large forces happen during short time intervals.

• **A football player runs toward the goal line and a defensive player tries to stop him with a head-on collision. What factors determine whether the offensive player scores?**

Record your ideas about this question in your *Active Physics log*. Be prepared to discuss your responses with your small group and the class.

FOR YOU TO DO

1. You will stage a head-on collision between two matched bowling balls. Set up a launch ramp for one ball, and find a level area clear of obstructions nearby where the other bowling ball can be at rest.

⚠ **Moving bowling balls can cause injury to people and property. Be careful!**

SPORTS ——— ◆ S 92

ANSWERS

For You To Do

1. Student activity.

---

2. Temporarily remove the "target" bowling ball. Find a point of release within the first one-fourth of the ramp's total length that gives the ball a slow, steady speed across the floor. Mark the point of release on the ramp with a piece of tape.

3. Replace the target ball. Adjust the aim until a good approximation of a head-on collision is obtained. Stage the collision.

   a) Record the results in your log. Use a diagram and words to describe what happened to each ball.

4. Repeat the above type of collision, but this time move the release point up the ramp to at least double the distance.

   a) Describe the results in your log.

   b) How did the results of the collision change from the first time?

   c) Identify a real-life situation which this collision could represent.

5. Arrange another head-on collision between the balls, but this time have both balls moving at equal speeds before the collision. Using a second, identical ramp, aim the second ramp so that the second ball's path is aligned with the first ball's path. Mark a release point on the second ramp at a height equal to the mark already made on the first ramp. This should assure that the balls will have low, approximately equal speeds. On a signal, two persons should release the balls simultaneously from equal ramp heights.

   a) Describe the results in your log.

   b) Identify a real-life situation which this collision could represent.

ANSWERS

For You To Do
(continued)

2. Student activity.

3. a) Assume the bowling balls have reasonably well-matched masses. A moving ball striking a stationary ball of equal mass should result in the moving ball stopping upon colliding and the stationary ball moving away from the collision at about the same speed that the incoming ball had before the collision. Bowling balls have a high coefficient of restitution, and the first ball can be expected to come nearly to a stop, while the other leaves at something close to the speed of the incident ball.

4. a-b) As the speed of the released ball increases, the speed at which the stationary ball moves away will increase.

   c) This type collision may occur when a forward collides with a goal tender, or the head of a golf club collides with a golf ball.

5. a) Balls of equal mass and equal speed colliding head-on should result in both balls rebounding at speeds less than or equal to their speeds before the collision.

   b) This type of collision may occur between two football players. However, since people are less elastic than bowling balls, they won't bounce off each other so far. However, you can expect both players to bounce in such a manner that each finds himself on his back.

2

ANSWERS

For You To Do
(continued)

6. a) In collisions between a soccer ball and a bowling ball it is generally the case that the soccer ball "loses," or undergoes the greatest change in velocity. When a stationary soccer ball is hit head-on by a moving bowling ball, the soccer ball will leave the collision at a speed higher than the incoming speed of the bowling ball, and the bowling ball will keep moving in its original direction of motion after the collision, but with a slightly reduced speed.

7. a) When a stationary bowling ball is hit head-on by a moving soccer ball, it should occur, for ordinary incoming speeds of the soccer ball, that the bowling ball will move away from the collision at a relatively low speed and the soccer ball will rebound from the collision.

8. a) Students will stage a collision between a golf ball and a tennis ball and determine from observing the balls after the collision that the golf ball is much more massive.

b) Students determine mass of both balls. Their results from the previous method will be verified.

PHYSICS IN ACTION

6. Repeat steps 1, 2, and 3, but replace the stationary bowling ball with a soccer ball.

　a) Be sure to write all responses, including identification of a similar situation in real life, in your log.

7. Repeat steps 1, 2, and 3, but in this case have the soccer ball roll down the ramp to strike a stationary bowling ball.

　a) Be sure to write all responses, including identification of a similar situation in real life, in your log.

8. Using your observations, determine the relative mass of a golf ball compared to a tennis ball by staging collisions between them.

　a) Which ball has the greater mass? How many times more massive is it than the other ball? Describe what you did to decide upon your answer.

　b) Use a scale or balance to check your result. Comment on how well observing collisions between the balls worked as a method of comparing their masses.

FOR YOU TO READ
Momentum

Taken alone, neither the masses nor the velocities of the objects were important in determining the collisions you observed in this activity. The crucial quantity is momentum (mass × velocity). A soccer ball has less mass than a bowling ball, but a soccer ball can have the same momentum as a bowling ball if the soccer ball is moving fast. A soccer ball moving very fast can affect a stationary bowling ball more than a soccer ball moving very slowly. This is similar to the damage small pieces of sand moving at very high speeds can cause (such as when a sand blaster is used to clean various surfaces).

Sportscasters often use the term "momentum" in a different way. When a team is doing well, or "on a roll," that team has momentum. A team can gain or lose momentum, depending on how things are going. This momentum clearly does not refer to the mass of the entire team multiplied by the team's velocity.

Other times, sportscasters use the term momentum to mean exactly how it is defined in the activity (mass × velocity), when they say things such as, "Her momentum carried her out of bounds."

REFLECTING ON THE ACTIVITY AND THE CHALLENGE

You already have identified several real-life situations which involve collisions, and many such situations happen in sports. Some involve athletes colliding with one another as in hockey and football. Others cases include athletes colliding with objects such as when kicking a ball. Still others include collisions between objects such as a golf club, bat, or racquet and a ball. Some spectacular collisions in sports provide fun opportunities for demonstrating your knowledge about collisions during voice-over commentaries. Use the concept of momentum when describing collisions in your sports video.

PHYSICS TO GO

1. Sports commentators often say that a team has "momentum" when things are going well for the team. Explain the difference between that meaning of the word "momentum" and its specific meaning in physics.

2. Suppose a running back collides with a defending linebacker who has just come to a stop. If both players weigh the same, what do you expect to see happen in the resulting collision?

3. Describe the collision of a running back and a linebacker of equal mass running toward each other at equal speeds.

Physics To Go

1. To say that a team or a candidate for election has momentum is to say that things are going well, are "on the rise;" momentum in physics is defined as mass times velocity.

2. You would expect the defending linebacker to be set in motion, and the speed of the running back to be greatly reduced.

3. The two players will bounce back from each other, at least to the extent that they will end up on their backsides.

ANSWERS

Physics To Go
(continued)

4. a) The heavier bat has, for the same speed, more momentum than the lighter bat and will transfer more momentum to the ball, hitting the ball farther.

   b) For the same effect, the lighter bat would need to be swung at a speed 38/30 = 1.3 times faster than the heavy bat.

5. It is difficult to change the momentum of a massive person.

6. The relative speeds of the players determines who gets knocked backward; if the small player moves fast enough, the big player can get knocked backward.

7. 100 kg x 10 m/s
   =(0.10 kg) v,
   v = 10,000 m/s

8. Before the collision one puck is moving and the other is stationary; after the collision, the pucks have "traded" conditions, with the puck which originally was moving being stationary and the puck which originally was stationary moving at about the same speed which the other puck had before the collision.

PHYSICS IN ACTION

4. Suppose that you have two baseball bats, a heavy (38-ounce) bat and a light (30-ounce) bat.

   a) If you were able to swing both bats at the same speed, which bat would allow you to hit the ball the farthest distance? Explain your answer.

   b) How fast would you need to swing the light bat to produce the same hitting effect as the heavy bat? Explain your answer.

5. Why do football teams prefer defensive linemen who weigh about 300 pounds?

6. What determines who will get knocked backward when a big hockey player checks a small player in a head-on collision?

7. A 100-kg athlete is running at 10 m/s. At what speed would a 0.10-kg ball need to travel in the same direction so that the momentum of the athlete and the momentum of the ball would be equal?

8. Use the words mass, velocity, and momentum to write a paragraph which gives a detailed "before and after" description of what happens when a moving shuffleboard puck hits a stationary puck of equal mass in a head-on collision.

9. Describe a collision in some sport by using the term momentum. Adapt this description to a 15-second dialogue that could be used as part of the voice-over for a video.

NOTES

2

ACTIVITY EIGHT
Conservation of Momentum

Background Information

The background information for this activity is presented in the Background Information for Activity Seven, "Concentrating on Collisions" because the same topic, momentum, is involved in both activities.

Active-ating the Physics InfoMall

In addition to the references to Activity Seven, you may wish to examine the Problems Place for even more exercises in momentum conservation. Remember, *Schaum's 3000 Solved Problems in Physics* has the problem and the solution. It can be a source for you, as well as a way to provide your students with solved problems for them to study!

Planning for the Activity

Time Requirements

• One class period.

Materials Needed

For the class:
• balance

For each group:
• apparatus for measuring speed

Advance Preparation and Setup

Two matters need to be considered in advance: (1) the pairs of objects to be used by each group in the collisions, including how the masses will be varied and how the masses will be caused to stick together upon colliding, and (2) how the velocity will be measured before and after the collision.

Regarding the objects to be collided, two possibilities seem to exist, air track gliders or laboratory carts. At a minimum, three collisions are desired,

involving mass ratios of 1:1, 2:1 and 1:2 (the moving mass is listed first in the ratios listed – see the data table in For You To Do for details). Masses of 1 and 2 kg are recommended, but certainly not required. Instead, one laboratory cart could collide with another identical laboratory cart to provide a 1:1 mass ratio, and a 2:1 (and 1:2) ratio could be obtained by loading one cart to double its mass. Similarly, air track gliders could be rigged to provide 1:1 and 2:1 mass ratios.

The colliding objects need to stick together upon colliding to move as a single object after the collision. Stick-on Velcro patches are very convenient for this purpose, but double-stick tape or modeling clay also works.

Whatever objects are used, have the students measure and use their masses, in kilograms, to call attention to and engage students in using the unit of momentum, (kg)m/s. If the masses are not to be 1 and 2 kg, have students change the values listed in the data table in For You To Do to the values to be used in your class.

The speed, in meters per second, must be measured before and after the collision. It is necessary to measure the speed of the incoming mass before the collision and the speed of the combined masses after the collision. One way to accomplish this would be to use a sonic ranging device to monitor the speed of Object 1 (the mass moving before the collision) before, during, and after the collision. Other possibilities for measuring the speeds include stop action video, strobe photography, a ticker-tape timer or a spark timer. The particular method to be used depends on the equipment available at your school. Whatever method of measuring speeds is used, students should record speeds in m/s.

For better data if friction is involved (as when using laboratory carts) it would be best to use the speed values which occur just before and just after the collision; this would help to avoid changes in speed (deceleration) as a source of error.

You may wish to provide additional mass ratios for students to use for staging additional collisions. A 3:1 mass ratio run "both ways," 3:1 and 1:3, gives particularly interesting results; if organized on bifilar supports to collide head-on as pendulums, hard wooden or metal spheres having a 3:1 mass ratio provide a cyclically repeating sequence of collisions.

"Nonsticky" collisions present problems for measuring speed because both masses move simultaneously at different speeds. This is difficult, but not impossible to overcome with ordinary equipment and may present an interesting challenge to interested students.

Teaching Notes

Monitor the class as students begin to stage collisions. They may need assistance choosing appropriate push-off speeds for the incoming mass and getting equipment to function.

Sample data is not provided because it cannot be anticipated what masses and speeds will be used in your class. However, conservation of momentum allows you to predict with ease what the velocity of the combined masses after the collision should be compared to the velocity of Object 1 before the collision:

Conservation of momentum applied to this kind of collision:

$$m_1{}^{v}Before = (m_1 + m_2){}^{v}After$$

Solving the above equation for ${}^{v}After$:

$${}^{v}After = {}^{v}Before\ m_1\ /(m_1 + m_2)$$

The final equation above will allow you to predict before class begins what the relative speeds before and after each collision should be. Another way of saying the same thing is that ${}^{v}After\ /\ {}^{v}Before = m_1\ /(m_1 + m_2)$.

Allowing for errors of measurement, students should find that the momenta before and after each collision are equal (in the Analysis table, the momentum of Object 2 will be zero, unless an additional collision is staged which is not included in the written procedure; this is pointed out only to avoid confusion).

It is important to discuss the collision analyzed as an example in Physics Talk because it is of a general kind where both objects are moving before and after a head-on collision. The same method of analysis is needed to solve some of the problems in Physics To Go.

Assigning positive and negative values to directions of velocities is very useful, if not necessary, for solving many collision problems.

2

NOTES

ACTIVITY EIGHT: Conservation of Momentum

Activity Eight
Conservation of Momentum

WHAT DO YOU THINK?

The outcome of a collision between two objects is predictable.

• **What determines the momentum of an object?**

• **What does it mean to "conserve" something?**

Record your ideas about these questions in your *Active Physics log*. Be prepared to discuss your responses with your small group and the class.

FOR YOU TO DO

1. From the objects provided arrange to have a head-on collision between two objects of equal mass. Before the collision, have one object moving and the other object at rest. Arrange for the objects to stick together to move as a single object after the collision. Stage a head-on, sticky collision between equal masses. Measure the velocity, in meters per second, of the moving mass before the collision and the velocity of the combined masses after the collision.

SPORTS

Activity Overview

In this activity students stage collisions between objects to investigate the conservation of momentum.

Student Objectives
Students will:

• understand and apply the Law of Conservation of Momentum.

• measure the momentum before and after a moving mass strikes a stationary mass in a head-on, inelastic collision.

ANSWERS FOR THE TEACHER ONLY

What Do You Think?

An object's mass and velocity determine its momentum, *mv*.

To conserve means to keep the same amount.

2

ANSWERS

For You To Do

Sample data is not provided because it cannot be anticipated what masses and speeds will be used in your class. However, conservation of momentum allows you to predict with ease what the velocity of the combined masses after the collision should be compared to the velocity of Object 1 before the collision. (See Teaching Notes.)

a) Prepare a data table in your log similar to the one shown below. Provide enough horizontal rows in the table to enter data for at least four collisions.

Sticky Head-on Collisions: One Object Moving before Collision

Mass of Object 1 (kg)	Mass of Object 2 (kg)	Velocity of Object 1 before Collision (m/s)	Velocity of Object 2 before Collision (m/s)	Mass of Combined Objects after Collision (kg)	Velocity of Combined Objects after Collision (m/s)
1.0	1.0		0.0	2.0	
2.0	1.0		0.0	3.0	
1.0	2.0		0.0	3.0	
			0.0		

b) Record the measured values of the velocities in the first row of the data table.

2. Stage other sticky, head-on collisions using the masses listed in the second and third rows of the data table. Then stage one or more additional collision using other masses. Measure the velocities before and after each collision.

a) Enter the measured values in the data table.

3. Organize a table for recording the momentum of each object before and after each of the above collisions.

a) Prepare a table similar to the following example in your log.

Momentum of Object before and after Collisions
Momentum = Mass × Velocity

Before the Collision		After the Collision
Momentum of Object 1 kg(m/s)	Momentum of Object 2 kg(m/s)	Momentum of Combined Objects 1 and 2 kg(m/s)

ANSWERS

For You To Do (continued)

Sample data is not provided because it cannot be anticipated what masses and speeds will be used in your class. However, conservation of momentum allows you to predict with ease what the velocity of the combined masses after the collision should be compared to the velocity of Object 1 before the collision. (See Teaching Notes.)

§ b) Calculate the momentum of each object before and after each of the above collisions and enter each momentum value in the table.

§ c) Calculate and compare the total momentum before each collision to the total momentum after each collision. Allowing for minor variations due to errors of measurement, write in your log a general conclusion about how the momentum before a collision compares to the momentum afterward.

PHYSICS TALK

The Law of Conservation of Momentum

Physicists have compared the momentum before and after a collision between pairs of objects ranging from railroad cars slamming together to subatomic particles impacting one another at near the speed of light. Never has an exception been found to the statement, "The total momentum before a collision is equal to the total momentum after the collision." This statement, known as the Law of Conservation of Momentum, can be written using symbols as:

$$m_1 v_{1\ Before} + m_2 v_{2\ Before} = m_1 v_{1\ After} + m_2 v_{2\ After}$$

where m_1 and m_2 represent the masses of two colliding objects, $v_{1\ Before}$ and $v_{1\ After}$ represent the velocity of one object before and after the collision, and $v_{2\ Before}$ and $v_{2\ After}$ represent the velocity of the second object before and after the collision.

You have seen that the Law of Conservation of Momentum applies when objects "stick together" upon colliding, but the law applies to other kinds of collisions, too. For an example of an application of the law to another kind of collision, the event on the following page involves objects which do not stick together upon impact.

For You To Do *(continued)*

Sample data is not provided because it cannot be anticipated what masses and speeds will be used in your class. However, conservation of momentum allows you to predict with ease what the velocity of the combined masses after the collision should be compared to the velocity of Object 1 before the collision. (See Teaching Notes.)

PHYSICS IN ACTION

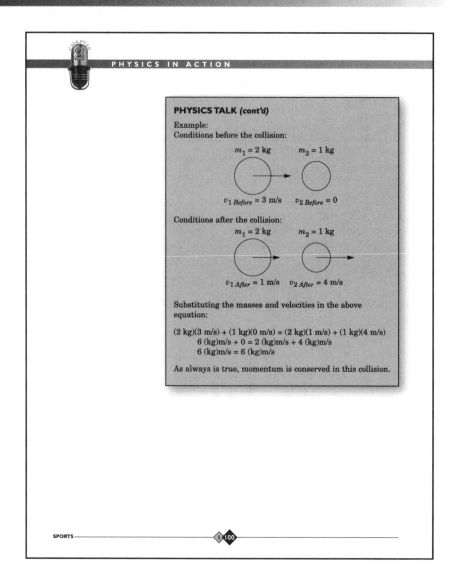

PHYSICS TALK *(cont'd)*

Example:
Conditions before the collision:

$m_1 = 2$ kg　　　$m_2 = 1$ kg

v_1 *Before* = 3 m/s　　v_2 *Before* = 0

Conditions after the collision:

$m_1 = 2$ kg　　　$m_2 = 1$ kg

v_1 *After* = 1 m/s　　v_2 *After* = 4 m/s

Substituting the masses and velocities in the above equation:

$(2$ kg$)(3$ m/s$) + (1$ kg$)(0$ m/s$) = (2$ kg$)(1$ m/s$) + (1$ kg$)(4$ m/s$)$
　　6 (kg)m/s $+ 0 = 2$ (kg)m/s $+ 4$ (kg)m/s
　　6 (kg)m/s $= 6$ (kg)m/s

As always is true, momentum is conserved in this collision.

ACTIVITY EIGHT: Conservation of Momentum

REFLECTING ON THE ACTIVITY AND THE CHALLENGE

The Law of Conservation of Momentum is a very powerful tool for explaining collisions in sports and other areas. The law works even when, as often happens in sports, one of the objects involved in a collision "bounces back," reversing the direction of its velocity and, therefore, its momentum, as a result of a collision. When describing a collision between a bat and ball, or a collision between two people, you can describe how the total momentum is conserved.

PHYSICS TO GO

1. A railroad car of 2000 kg coasting at 3.0 m/s overtakes and locks together with an identical car coasting on the same track in the same direction at 2.0 m/s. What is the speed of the cars after they lock together?

2. In a hockey game, an 80.0-kg player skating at 10.0 m/s overtakes and bumps from behind a 100-kg player who is moving in the same direction at 8.00 m/s. As a result of being bumped from behind, the 100-kg player's speed increases to 9.78 m/s. What is the 80-kg player's velocity (speed and direction) after the bump?

3. A 3-kg hard steel ball collides head-on with a 1-kg hard steel ball. The balls are moving at 2 m/s in opposite directions before they collide. Upon colliding, the 3-kg ball stops. What is the velocity of the 1-kg object after the collision? (Hint: Assign velocities in one direction as positive; then, any velocities in the opposite direction are negative.)

4. A 45-kilogram female figure skater and her 75-kg male skating partner begin their ice dancing performance standing at rest in face-to-face position with the palms of their hands touching. Cued by the start of their dance music, both skaters "push off" with their hands to move backward. If the female skater moves at 2 m/s relative to the ice, what is the velocity of the male skater? (Hint: The momentum before the skaters push off is zero.)

SPORTS

ANSWERS

Physics To Go

1. $m(3.0 \text{ m/s}) + m(2.0 \text{ m/s}) = (2m)v$

   $(3.0 \text{ m/s} + 2.0 \text{ m/s}) = 2mv$

   $v = m(5.0 \text{ m/s}) / 2m$
   $= (5.0 \text{ m/s})/2 = 2.5 \text{ m/s}$

2. $(80.0 \text{ kg})(10.0 \text{ m/s}) + (100 \text{ kg})(8.00 \text{ m/s}) = (80 \text{ kg})v + (100 \text{ kg})(9.78 \text{ m/s})$

   $800 \text{ (kg)m/s} + 800 \text{ (kg)m/s} = (80 \text{ kg})v + 978 \text{ (kg)m/s}$

   $1{,}600 \text{ (kg)m/s} = (80 \text{ kg})v + 978 \text{ (kg)m/s}$

   $622 \text{ (kg)m/s} = (80 \text{ kg})v$

   $v = [622 \text{ (kg)m/s}] / 80 \text{ kg} = 7.78 \text{ m/s}$

---

3. The direction of travel of the 3-kg ball before the collision is assigned as positive:

   $(3 \text{ kg})(2 \text{ m/s}) + (1 \text{ kg})(-2 \text{ m/s}) = (1 \text{ kg})v$

   $6 \text{ (kg)m/s} - 2 \text{ (kg)m/s} = (1 \text{ kg})v$

   $4 \text{ (kg)m/s} = (1 \text{ kg})v$

   $v = [4 \text{ (kg)m/s} / (1 \text{ kg})] = 4 \text{ m/s}$

   The 1-kg ball bounces back after the collision at twice the speed it had coming into the collision.

4. The direction of the female skater after pushoff is assigned as positive:

   $0 = (45 \text{ kg})(2 \text{ m/s}) + (75 \text{ kg})v$

   $0 = 90 \text{ (kg)m/s} + (75 \text{ kg})v$

   $-90 \text{ (kg)m/s} = (75 \text{ kg})v$

   $v = [-90 \text{ (kg)m/s}] /(75 \text{ kg}) = -1.2 \text{ m/s}$

   The male skater moves at 1.2 m/s in the direction opposite the female skater.

PHYSICS IN ACTION

5. A 0.35-kg tennis racquet moving to the right at 20 m/s hits a 0.060-kg tennis ball which is moving to the left at 30 m/s. The racquet continues moving to the right after the collision, but at a reduced speed of 10 m/s. What is the velocity (speed and direction) of the tennis ball after it is hit by the racquet?

6. A stationary 3-kg hard steel ball is hit head-on by a 1-kg hard steel ball moving to the right at 4 m/s. After the collision, the 3-kg ball moves to the right at 2 m/s. What is the velocity (speed and direction) of the 1-kg ball after the collision? (Hint: Direction is important.)

7. Write a 15- to 30-second voice-over which highlights the conservation of momentum in a sport of your choosing.

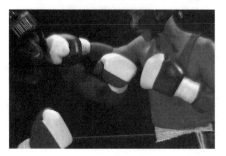

S 102

ANSWERS

Physics To Go *(continued)*

5. To the right is assigned as the positive direction:

$$(0.35 \text{ kg})(20 \text{ m/s}) + (0.060\text{kg})(30 \text{ m/s}) = (0.35 \text{ kg})(10 \text{ m/s}) + (0.060 \text{ kg})v$$

$$7.0 \text{ (kg)m/s} + 1.8 \text{ (kg)m/s} = 3.5 \text{ (kg)m/s} + (0.060 \text{ kg})v$$

$$5.3 \text{ (kg)m/s} = (0.060 \text{ kg})v$$

$$v = [5.3 \text{ (kg)m/s}] /(0.060 \text{ kg}) = 88 \text{ m/s to the right}$$

6. To the right is assigned as the positive direction:

$$0 + (1 \text{ kg})(4 \text{ m/s}) = (3 \text{ kg})(2 \text{ m/s}) + (1 \text{ kg})v$$

$$-2 \text{ (kg)m/s} = (1 \text{ kg})v$$

$$v = -2 \text{ m/s}$$

The 1-kg ball rebounds from the collision, moving to the left at half the speed it had coming into the collision; its speed after the collision also is observed to be the equal and opposite of the 3-kg ball's speed after the collision. If after this collision each ball hit a bumper and the balls came back to collide again, this would take us back to the beginning of Problem #3 above, and if they rebounded after that collision, we'd be back to this problem again, and so on...

7. Answers will vary. Encourage students to provide as much entertainment value as possible in their descriptions, as well as making sure that the principle of momentum is correctly described.

NOTES

2

ACTIVITY NINE
Circular Motion

Background Information

It already has been established that if an object is moving along a straight line at constant velocity (constant speed, always in the same direction, along the line of motion) all of the forces on the object are balanced; in other words, the net force on the object is zero. If a sudden, momentary force is applied to the same object in a direction exactly sideways, the speed of the object neither increases nor decreases because the sideways force has no effectiveness in either the same or opposite direction as the object's motion. The result of applying a sudden, momentary, sideways force to the object is to cause the object to turn in the direction of the applied force; after the force has been removed, the object would be found moving at the same constant speed as it was before application of the force, but it would be moving along a different direction line, its direction of motion having been changed by the force. Such a force is called a "deflecting force," and the result of a deflecting force—applied exactly sideways to the motion of an object—is to cause the object to change, or deflect, to a new direction of motion.

If instead of being applied suddenly and momentarily, a deflecting force is applied continuously, always in the same amount, and continuously adjusted in a direction to act exactly sideways to the object's motion, the object moves in a circular path at constant speed. In this case the force is called a "centripetal force" which is defined as "the force required to keep a mass moving in a circular path at constant speed." It can be experimentally determined that the relationship between the centripetal force, the object's mass and speed, and the radius of the object's circular path is:

$$F_c = mv^2/r$$

where F_c is the centripetal force in newtons, m is the object's mass in kilograms, v is the objects speed in meters/second, and r is the radius of the circular path in meters. A convenient alternate equation for centripetal force can be used when the period, T, of the objects circular motion (T is the time, in seconds, for the object to travel once around the circle) is known but the speed is not known:

$$F_c = m4\pi^2 R/T^2$$

According to Newton's Second Law of Motion, when there is an unbalanced force acting on an object, the object must be accelerating. This certainly must apply to an object moving in a circular path under the influence of an unbalanced, centripetal force. However, it is puzzling to contemplate that an object moving at constant speed, even on a circular path, somehow can be construed to be accelerating, because it is neither gaining nor losing speed. Consider the below "strobe" diagram showing an object moving at constant speed along a circular path at four instants equally separated in time:

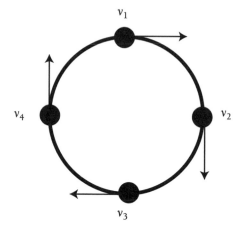

Diagram of object moving in a circular path.

The definition of acceleration,
$a = \Delta v/\Delta t = (v\Delta - v_1)/\Delta t$ will be applied to the above diagram to attempt to discover any basis for existence of acceleration. Since the instantaneous velocities v_1 and v_2 in the diagram are vectors which do not share a common direction, the quantity Δv in the defining equation for acceleration must be found by treating v_1 and v_2 as vectors during the subtraction process; that is, a vector subtraction method must be applied to find the difference $v_2 - v_1 = \Delta v$. To do so, the negative of vector v_1 (a vector having the same length but opposite direction as v_1) will be added to vector v_2 using the tip-to-tail method of adding vectors:

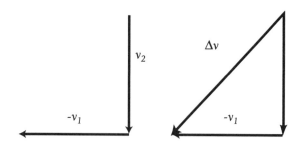

Clearly in the above diagram a change in velocity, Δv, occurred during the time interval, Δt, that the object moved at constant speed 1/4 of the distance around its circular path. If a numerical values were assigned to time and to the object's speed, a value for Δv could be determined by comparing the length of the Δv vector to the velocity vectors, and the average acceleration for the time interval could be calculated from $a = \Delta v/\Delta t$. Therefore, there is an acceleration associated with a centripetal force, but the nature of the acceleration is that it does not alter the object's speed, but does alter the direction of the object's velocity. It is informative to move the vector Δv in the above diagram into the original diagram to see Δv in relationship to the motion along the circular path. This is done in the below diagram. Notice that both the length and direction of the vector Δv are preserved and that Δv is positioned at the middle of the time interval from which it was derived:

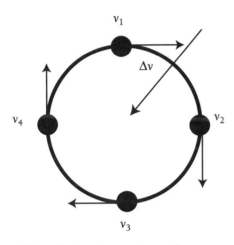

Object in circular motion with vector

Interestingly and not accidentally, Δv points toward the center of the circular path. Since the acceleration calculated from Δv using the defining equation $a = \Delta v/\Delta t$ would have the same direction as Δv, the acceleration also points toward the center of the circle. Indeed, it is a centripetal acceleration, caused, as one may suspect by the centripetal force, which also points toward the center of the circular path. Now it can be reasoned that the equations presented above for centripetal force are nothing more that Newton's Second Law, $f = ma$, applied to the special case of circular motion. The terms on the right-hand side of the equation other than the mass, m, are the centripetal acceleration, a_c:

$$F_c = ma_c = m(v^2/r) = m(4\pi^2 r/t^2)$$

In summary, the velocity, acceleration, and force vectors of an object moving at constant speed on a circular path constantly change in direction and remain constant in amount. Each vector rotates once during each trip of the object around the circular path.

Active-ating the Physics InfoMall

When teaching about circular motion, you will almost certainly have to dispel ideas about centrifugal forces. Perhaps you noticed, while performing some of the searches mentioned in the previous activities, the article "Centrifugal force: fact or fiction?," in *Physics Education*, vol. 24, issue 3, 1989. If not, you may want to check it out now.

Don't forget the references and searches done for acceleration in Chapter 1. Some of these may be useful to you again, especially regarding accelerometers.

Search the InfoMall using keywords "circular motion" AND "misconcept*" for a list of articles discussing known problems students have with this common concept. You may also find useful demonstrations by searching the Demo & Lab Shop using the keywords "circular motion."

Planning for the Activity

Time Requirements
• One class period.

Materials Needed
For each group:

• cork accelerometer
• rotating stool or chair (optional)
• force meter (plastic strip used in Activity Two
• bowling ball
• soccer ball

Advance Preparation and Setup

A ball will be needed to serve as the subject of each group's inquiry. The ball needs to be sufficiently massive so that, when rolling at low speed across the floor, students can use a force meter (the force meter used in Chapter 2, Activity Two) to push sideways to the ball's motion to cause it to turn on a curve of observable, reasonable radius. The best choice of ball may be the bowling balls used for Chapter 2, Activity Seven; if not available another kind of ball must be substituted. Try this yourself in advance of class to be sure that the ball's mass, the ball's speed, and the amount of force used, combine to produce a nice, observable curve of the ball's path.

If you have not done *Sports*, Chapter 1 yet, you will need to take the time to construct a cork accelerometer. Instructions for construction are provided in this Teacher's Edition, p. 60.

Teaching Notes

The cork accelerometer will indicate an acceleration toward the center of the circular path in which the accelerometer moves.

SAFETY PRECAUTION: If a rotating stool or chair is used to place a student holding an accelerometer in a state of circular motion, provide close supervision and maintain safe conditions; the effect can be observed without a rotating stool or chair if students simply twirl around while holding an accelerometer in the hands.

If students have not already completed *Sports*, Chapter 1, some time may be required to familiarize the students with the use of the accelerometer.

Students can be expected to need practice at keeping alongside the ball while applying a constant force always sideways (at a right angle to) the ball's motion; in fact, to do so extremely well perhaps is nearly impossible, but "close" will do well enough for students to observe the tendency for the ball to move in a circular path.

Some students may raise the question, "How can there be an acceleration when the object moves at constant speed?" You may wish to see the explanation in the Background Information for the Teacher for this activity to decide how you will deal with that question.

Students also may ask about, or bring up, "centrifugal force." This is addressed in the Background Information for the Teacher. A first response to a question about centrifugal force would be to refer students to Physics To Go, question 2.

Students may have an inclination to "run together," or treat as the same phenomenon, circular motion with spinning motion. They are separate, but related, phenomena. This activity applies to the former, an object whose center of mass is moving along a circular path. When a figure skater does an in-place spin, the skater's center of mass does not move; it is a different phenomenon. It is possible, however, to treat part of the skater, such as an extended foot at the end of the spin, as an object in circular motion to which the ideas in this activity could be applied.

Activity Overview

In this activity students use an accelerometer to identify the direction of centripetal acceleration. They then use a force meter to provide the centripetal force needed to deflect an rolling ball moving in a straight line into a curved path.

Student Objectives

Students will:

• understand that a centripetal force is required to keep a mass moving in a circular path at constant speed.

• understand that a centripetal acceleration accompanies a centripetal force, and that, at any instant, both the acceleration and force are directed toward the center of the circular path.

• apply the equation $F_c = ma_c = m(v^2/r)$ to calculations involving circular motion.

• understand that centrifugal force is the reaction to centripetal force.

ANSWERS FOR THE TEACHER ONLY

What Do You Think?

The forces acting on a race car include gravity downward, the force of the road up and centrifugal force.

ACTIVITY NINE: Circular Motion

Activity Nine
Circular Motion

WHAT DO YOU THINK?

Race cars can make turns at 150 mph.

• **What forces act on a race car when it moves along a circular path at constant speed on a flat, horizontal surface?**

Record your ideas about this question in your *Active Physics log*. Be prepared to discuss your responses with your small group and the class.

FOR YOU TO DO

1. Hold an accelerometer in your hands and observe it as you either sit on a rotating stool or spin around while standing. What is the direction of the acceleration indicated by the accelerometer? (You can find out how the cork indicates acceleration by holding it and noting its behavior as you accelerate forward.)

⚠ To avoid becoming too dizzy, limit your spins while standing to about four.

S 103   SPORTS

ANSWERS

For You To Do

1. a) Students provide a sketch similar to the one shown.

ANSWERS

For You To Do
(continued)

2. a) Students should indicate that the greater the force applied to a mass, the greater the acceleration. The acceleration occurs in the direction of the unbalanced force.

3. a) Students provide a sketch similar to the one shown.

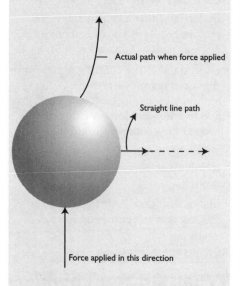

Actual path when force applied

Straight line path

Force applied in this direction

b) The speed of the ball does not change.

c) A constant force sideways needs to be applied to the ball.

d) If you stop pushing on the ball, the ball will follow a straight path. Students should provide a sketch similar to the one shown at right.

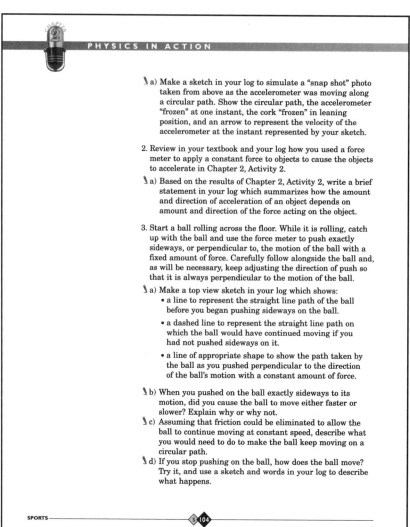

PHYSICS IN ACTION

a) Make a sketch in your log to simulate a "snap shot" photo taken from above as the accelerometer was moving along a circular path. Show the circular path, the accelerometer "frozen" at one instant, the cork "frozen" in leaning position, and an arrow to represent the velocity of the accelerometer at the instant represented by your sketch.

2. Review in your textbook and your log how you used a force meter to apply a constant force to objects to cause the objects to accelerate in Chapter 2, Activity 2.

a) Based on the results of Chapter 2, Activity 2, write a brief statement in your log which summarizes how the amount and direction of acceleration of an object depends on amount and direction of the force acting on the object.

3. Start a ball rolling across the floor. While it is rolling, catch up with the ball and use the force meter to push exactly sideways, or perpendicular to, the motion of the ball with a fixed amount of force. Carefully follow alongside the ball and, as will be necessary, keep adjusting the direction of push so that it is always perpendicular to the motion of the ball.

a) Make a top view sketch in your log which shows:
   • a line to represent the straight line path of the ball before you began pushing sideways on the ball.
   • a dashed line to represent the straight line path on which the ball would have continued moving if you had not pushed sideways on it.
   • a line of appropriate shape to show the path taken by the ball as you pushed perpendicular to the direction of the ball's motion with a constant amount of force.

b) When you pushed on the ball exactly sideways to its motion, did you cause the ball to move either faster or slower? Explain why or why not.

c) Assuming that friction could be eliminated to allow the ball to continue moving at constant speed, describe what you would need to do to make the ball keep moving on a circular path.

d) If you stop pushing on the ball, how does the ball move? Try it, and use a sketch and words in your log to describe what happens.

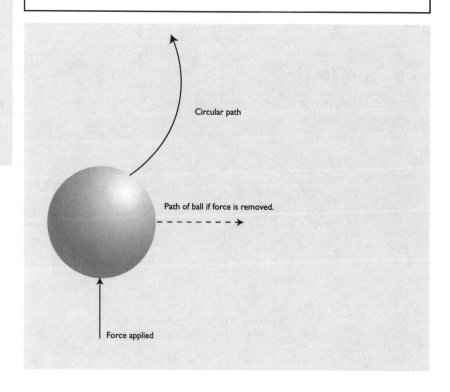

Circular path

Path of ball if force is removed.

Force applied

4. Review each of the items listed below in this book at the location given for each item. Copy each item into your log and write a statement to discuss how each item is related to an object moving along a circular path. If an item does not apply to circular motion, explain why.

a) Galileo's Principle of Inertia, Chapter 2, Activity 1, For You to Read, page S58.

b) Newton's First Law of Motion, Chapter 2, Activity 1, Physics Talk, page S59.

c) Newton's Second Law of Motion, Chapter 2, Activity 2, Physics Talk, page S64.

FOR YOU TO READ

The Unbalanced Force Required for Circular Motion

During the above activities you saw two things which are related by Newton's Second Law of Motion, $f = ma$. First, the accelerometer showed that when an object moves in a circular path there is an acceleration which at any instant is toward the center of the circle. This acceleration has a special name, "centripetal acceleration." The word "centripetal" means "toward-the-center;" therefore, centripetal acceleration refers to acceleration toward the center of the circle when an object moves in a circular path.

You also saw that a "centripetal force," a toward-the-center force, causes circular motion. When a centripetal force is applied to a moving object, the object's path curves; without the centripetal force, the object follows the tendency to move in a straight line. Therefore a centripetal force, when applied, is an unbalanced force, meaning that it is not "balanced off" by another force.

Newton's Second Law seems to apply to circular motion just as well as it applies to accelerated motion along a straight line, but with a strange "twist." It is a clearly correct application of $f = ma$ to say that a centripetal force, f, causes a mass, m, to experience an acceleration, a. However, the strange part is that when an object moves along a circular path at constant speed, acceleration is happening with no change in the object's speed. The force changes the direction of the velocity.

Velocity describes both the amount of speed and the direction of motion of an object. Thinking about the velocity of an object moving with constant speed on a circular path, it is true that the velocity is changing from one instant to the next not in the amount of the velocity, but with respect to the direction of the velocity. The diagram shows an object moving at constant speed on a circular path. Arrows are used to represent the velocity of the object at several instants during one trip around the circle.

For You To Do
(continued)

4. a) Inertia is the natural tendency of an object to remain at rest or to remain moving with constant speed in a straight line.

   When there is no centripetal force acting on an object, it will move in a straight line.

   b) In the absence of an unbalanced force, an object at rest remains at rest and an object already in motion remains in motion with constant speed on a straight line path.

   In the absence of a sideways, towards-the-center force (centripetal force), the object going in a circle continues to move straight (at a tangent to the circle).

   c) The acceleration of an object is directly proportional to the unbalanced force acting on it and is inversely proportional to the object's mass. The direction of the acceleration is the same as the direction of the unbalanced force.

   When a centripetal force is applied to an object, it moves in a circular path. The acceleration is toward the center of the circle. It is definitely an application of $F = ma$ because the force acts (toward the center of the circle) on the mass giving it an acceleration (toward the center of the circle).

2

PHYSICS IN ACTION

Physicists have shown that a special form of Newton's Second Law governs circular motion:

$$fc = mac = \frac{mv^2}{R}$$

where fc is the centripetal force in newtons, m is the mass of the object moving on the circular path in kilograms, ac is the centripetal acceleration in m/s², v is the velocity in m/s, and R is the radius of the circular path in meters.

Example:
Find the centripetal force required to cause a 1,000-kg automobile travelling at 27 m/s (60 miles/hour) to turn on an unbanked curve having a radius of 100 m.

$$fc = \frac{mv^2}{R}$$

$$= \frac{(1,000 \text{ kg})(27 \text{ m/s})^2}{100 \text{ m}}$$

$$= \frac{(1,000 \text{ kg} \times 730 \text{ m}^2/\text{s}^2)}{100 \text{ m}}$$

$$= 7,300 \text{ N}$$

If the force of friction is less than the above amount, the car will not follow the curve and will skid in the direction in which it is travelling at the instant the tires "break loose."

REFLECTING ON THE ACTIVITY AND THE CHALLENGE

Both circular motion and motion along curved paths which are not parts of perfect circles are involved in many sports. For example, both the discus and hammer throw events in track and field involve rapid circular motion before launching a projectile. Track, speed skating, and automobile races are done on curved paths. Whenever an object or athlete is observed to move along a curved path, you can be sure that a force is acting to cause the change in direction. Now you are prepared to provide voice-over explanations of examples of motion along curved paths in sports, and in many cases you perhaps can estimate the amount of force involved.

PHYSICS TO GO

1. For the car used as the example in the For You To Read, what is the minimum value of the coefficient of sliding friction between the car tires and the road surface which will allow the car to go around the curve without skidding? (Hint: First calculate the weight of the car, in newtons)

2. If you twirl an object on the end of a string, you, of course, must maintain an inward, centripetal force to keep the object moving in a circular path. You feel a force which seems to be pulling outward along the string toward the object. But the outward force which you detect, called the "centrifugal force," is only the reaction to the centripetal force which you are applying to the string. Contrary to what many people believe, there is no outward force acting on an object moving in a circular path. Explain why this must be true in terms of what happens if the string breaks while you are twirling an object.

3. A 50-kg jet pilot in level flight at a constant speed of 270 m/s (600 miles per hour) feels the seat of the airplane pushing up on her with a force equal to her normal weight, 50 kg × 10 m/s² = 500 N. If she rolls the airplane on its side and executes a tight circular turn which has a radius of 1,000 m, with how much force will the seat of the airplane push on her? How many "*g's*" (how many times her normal weight) will she experience?

ANSWERS

Physics To Go

1. μ = (frictional force)/(weight)
   = 7,300 N / mg
   = 7,300 N / (1,000 kg x 10 m/s²)

   = 7,300 N / 10,000 N = 0.73

2. If there were an outward force acting on the object, it would be expected to fly radially outward when the string breaks; when the string breaks, the object flies tangent to the circular path, indicating that there is no outward force.

3. $Fc = m(v^2/r)$
   = (50 kg) (270 m/s)²/(1,000 m)
   = (50 kg)(73,000 m²/s²)/(1,000 m)
   = 3,600 N

   The pilot's normal weight is
   mg = 50 kg x 10 m/s² = 500 N

   Therefore, the pilot "pulls" 3,600 N / 500 N = 7.2 *g's* during the turn; that is, she feels 7.2 times her normal weight. Assuming an inside turn with the top of the pilots head facing the center of the circle, the blood in her brain would tend to keep going straight ahead, tangent to the circle, draining from her brain and causing her to lose consciousness. However, her automatic pressure suit would inflate to squeeze against her legs, pushing blood upward to her brain to keep her from "blacking out."

ANSWERS

Physics To Go
(continued)

4. For the discus event in track & field, assume mass of disc = 1 kg, radius of twirling action before throw = 1 m, and speed during rotation before throw = 10 m/s:

$$Fc = m(v^2/r)$$
$$= (1 \text{ kg}) (10 \text{ m/s})^2/(1 \text{ m})$$
$$= (1 \text{ kg})(100 \text{ m}^2/\text{s}^2)/(1 \text{ m}) = 100 \text{ N}$$

The athlete must "hold on" to the disc, using the throwing hand to provide an inward force of 100 N while twirling prior to release of the disc.

5. Viewed from a helicopter above the event, the passenger would be observed to keep going in a straight line as the car turns. The seat would slide under the passenger, and the door on the passenger side would hit the right shoulder of the passenger, providing the centripetal force thereafter to push the passenger into the same curve as the car.

6. The tilt-a-whirl ride is fun (but sometimes sickening) because two circular motions are "superimposed" on the body. One motion carries the body around in a large circular path corresponding to the circular path around which the chair moves; the second, superimposed, motion is provided as the chair spins as it revolves on the circular path, causing the body simultaneously to move in a circle of small radius. Sometimes the two centripetal forces add together, and sometimes they are in different directions. All of the effects on the body are too numerous to understand or explain, but one thing is for sure, the hot dogs in the stomach don't know which way to go!

7. a) They need a frictional force to turn. On a wet field, without friction, they continue moving in the same direction.

   b) Friction supplies the centripetal force. On a wet field, without friction, the players continue moving in the same direction, obeying Newton's First Law.

   c) Student work.

---

4. Imagine a video segment of an athlete or an item of sporting equipment moving on a circular path in a sporting event. Estimate the mass, speed, and radius of the circle. Use the estimated values to calculate centripetal force and identify the source of the force.

5. Below are alternate explanations of the same event given by a person who was not wearing a seat belt when a car went around a sharp curve:

   a) "I was sitting near the middle of the front seat when the car turned sharply to the left. The centrifugal force made my body slide across the seat toward the right, outward from the center of the curve, and then my right shoulder slammed against the door on the passenger side of the car."

   b) "I was sitting near the middle of the front seat when the car turned sharply to the left. My body kept going in a straight line while, at the same time due to insufficient friction, the seat slid to the left beneath me until the door on the passenger side of the car had moved far enough to the left to exert a centripetal force against my right shoulder."

   Are both explanations correct, or is one right and one wrong? Explain your answer in terms of both explanations.

6. People seem to be fascinated with having their bodies put in a state of circular motion. Describe an amusement park ride based on circular motion which you think is fun and describe what happens to your body during the ride.

7. a) Explain why football players fall on a wet field while changing directions during a play.
   b) Include the concepts centripetal force and Newton's Laws in a revised explanation.
   c) In a new revision, make the explanation exciting enough to include in your sports video voice-over.

SPORTS

PHYSICS AT WORK

Dean Bell

**TELEVISION PRODUCER
USES SPORTS TO TEACH
MATH AND PHYSICS**

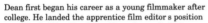

Dean Bell, an award-winning filmmaker and television writer, director, and producer, is beginning his third season of *Sports Figures*, a highly acclaimed ESPN educational television series designed to teach the principles of physics and mathematics through sports. His approach has been to tell a story, pose a problem, and then follow through with its mathematical and scientific explanation. But always, he says, you must make it fun. It has to be both educational and entertaining.

Dean first began his career as a young filmmaker after college. He landed the apprentice film editor s position on a Woody Allen film. From there, he worked his way up in the field, from assistant editor, to editor and finally writer, director, and producer.

I always been a fan of educational TV, he states, although I never thought that was where my career would take me. It s one of life s little ironies that I ve ended up producing this type of show. You see, my father worked in scientific optics and was very science oriented. He was always delighted in finding out how things worked, and was even on the Mr. Wizard TV show a few times.

Dean writes the script for each segment, working together with top educational science consultants. We spend a day coming up with ideas and then researching each subject thoroughly. Our job is to illustrate the relationship between a sports situation and the related mathematical or physics principles.

At end of day, says Dean, it really is nice to be working on a show that means something and that is so worthwhile. I m still getting ahead in my career as a film and TV producer, but now I m also an educator.

SPORTS

PHYSICS IN ACTION

Chapter 2 Assessment

Your big day has arrived. You will be meeting with the local television station to audition for a job as a "physics of sports" commentator. Whether you will get the job will be decided on the quality of your voice-over.

With what you learned in this chapter, you are ready to do your science commentary on a short sports video. Choose a video tape from a sports event, either a school event or a professional event. Each of you will be responsible for producing your own commentary whether or not you worked in cooperative groups during the activities. You are not expected to give a play-by-play description, but rather describe the rules of nature that govern the event. Your viewers should come away with a different perspective of both sports and physics. You may produce

- **a written script, or**
- **a live narrative, or**
- **a video soundtrack, or**
- **an audio cassette.**

Review the criteria by which your voice-over dialogue or script will be evaluated. Your voice-over should:

- **use physics principles and terms correctly.**
- **have entertainment value.**

After reviewing the criteria, decide as a class the point value you will give to each criterion.

- **How important is the physics content? How many physics terms and principles should be illustrated to get the minimum credit? The maximum credit?**
- **What value would you place of the entertainment aspect? How do you fairly assess the excitement and interest of the broadcast?**

Physics You Learned

Galileo's Principle of Inertia

Newton's First Law

Newton's Second Law

Newton's Third Law of Motion

Weight

Center of mass

Friction between different surfaces

Momentum

Law of Conservation of Momentum

Centripetal acceleration

Centripetal force

Alternative Chapter Assessment

Multiple Choice: Select the letter of the choice that best answers the question or best completes the statement.

1. Which of the following best illustrates Newton's First Law of Motion?

   a) A collision between a running back and a linebacker in football.

   b) An ice hockey puck sliding along the ice after being hit by a players stick.

   c) A bowling pin being struck by a bowling ball.

   d) A volleyball being "spiked" across the net.

2. A small ball rests on a circular turntable, rotating clockwise at a constant speed as illustrated in the below diagram. Which of the below pair of arrows best describes the direction of the acceleration and the net force acting on the ball at the point indicated in the diagram?

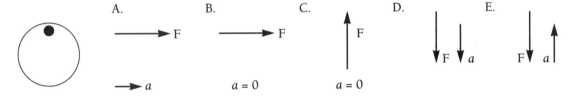

3. A person sliding into second base continues to slide past the base due to:

   a) inertia

   b) friction

   c) weight

   d) gravity

4. Newton's First Law of Motion states that an object at rest stays at rest unless acted upon by a:

   a) balanced force

   b) net force

   c) weak force

   d) strong force

5. In the absence of air, a penny and a feather dropped from the same height:

   a) fall at different rates

   b) float

   c) fall at equal rates

   d) do not have momentum

6. An object rolling across a level floor without any horizontal net force acting on it will:

a) slow down

b) speed up

c) keep moving forever

7. An object falling to Earth in the absence of air resistance:

   a) falls with a constant speed of 9.8 m/s

   b) falls with constant acceleration of 9.8m/s2

   c) slows down

8. A constant net force acting on an object causes the object to move with constant:

   a) speed

   b) velocity

   c) acceleration

   d) momentum

9. Which of the following is NOT one of Newton's Laws of Motion?

   a) An object in motion stays in motion unless acted upon by an unbalanced force.

   b) A constant net force acting on an object produces a change in the object's motion.

   c) For every action, there is an equal and opposite reaction.

   d) Energy is neither created not destroyed; it simply changes form.

10. Newton's First Law is known as the law of

   a) impetus

   b) inertia

   c) acceleration

   d) resistance

True or Replace False: Determine whether the word in bold print makes each statement true or false. If a statement is true, write "true" in the answer space. If a statement is false, write in the answer space a replacement for the word in bold print which would make the statement become true.

11. If a net force acts on an object, the object will change speed, direction or **neither**.

   Answer: _____

12. **Gravity** is the tendency of an object to resist any change.

   Answer: _____

13. A net force acting on an object causes the object to move with constant **velocity**.

    Answer: _____

14. **Forces** that are equal in amount and opposite in direction are balanced forces.

    Answer: _____

15. A bowling ball has more **inertia** than a tennis ball.

    Answer: _____

Short Answer: Write a brief response to each item. Show your work for responses which require calculations.

16. Describe how Galileo experiments with balls and ramps led to Newton's First Law of Motion.

17. Using Newton's First and Second Laws of Motion, explain why a ball thrown into the air follows a parabolic path.

18. On the diagrams below, draw arrows to show the vertical forces acting on a person jumping into the air. Use diagram A to show the forces acting during the push-off, and use diagram B to show the forces acting as the person lands on the ground. Write a paragraph comparing the forces in each situation, including a discussion on the relative sizes of the forces.

A.   B.

19. Describe the collision between a bat and a ball in terms of conservation of momentum.

20. Imagine a tug-of-war between two people. Draw a sketch indicating all of the forces acting in the tug-of-war. Using Newton's Laws and your picture, explain how and why one side will win.

21. For each of the following forces, identify the "reaction" force from Newton's Third Law:

    a) Volleyball hitting the floor.

    Answer: _____

    b) Softball bat hitting a softball.

    Answer: _____

c) Punter kicking a football.

Answer: _____

22. Explain why sprinters prefer to use longer spikes on their shoes than long distance runners, even though both run on same track surface.

23. A 75-kg ice hockey forward moving at 5.0 m/s collides with a stationary 85-kg defenseman and they become entangled. With what velocity will the pair move across the ice?

24. Explain how a soccer player can cause a ball to spin as a result of kicking it. What can the player do to change the direction of spin on the ball?

25. Two objects that have the same mass are dropped from the top of a 20-meter high building. One object is larger and flatter than the other object. Which hits the ground first? Use the terms gravity, acceleration, and air resistance correctly in your discussion.

Alternative Chapter Assessment Answers

Multiple Choice

1. B

2. D

3. A

4. B

5. C

6. C

7. B

8. C

9. D

10. B

True/Replace False Word

11.   both

12.   inertia

13.   acceleration

14.   true

15.   true

Short Answer

16. Galileo used two ramps, initially in a V-shape and demonstrated that a ball rolled from a specific height would roll up the opposite side until it reached the original height. If the angle of the second ramp was decreased, the ball still rolled to the same height but this ball traveled further on the ramp. He reasoned that the ball will roll until it reaches the height from which it was released, therefore, if the second ramp were horizontal the ball would continue to roll in the horizontal direction since it would not be able to reach this height. This supports the idea that the "natural" state of an object's motion is not necessarily to be at rest.

17. When a ball is thrown into the air the primary force acting on it is its weight. Since weight is a force that acts "down", from Newton's Second Law the acceleration of the ball will be down, therefore the vertical velocity of the ball will be changing. If we assume no air resistance, horizontally the ball will continue in its original state of motion according to Newton's First Law, minimal air resistance would at least show very little change in horizontal motion. During equally spaced time intervals the horizontal displacement will be constant and the vertical displacement will be changing, resulting in a parabolic path for the ball.

18. In each diagram the vertical forces that need to be shown are the person's weight (acting down) and the normal force from the ground (acting up). In both cases the normal force will exceed the weight since both require an acceleration in the upward direction.

19. When a bat collides with a ball the total momentum of the system will remain the same as long as no other forces acted. The bat will slow down upon collision, thus decreasing the momentum of the bat. The momentum that the bat lost will be gained by the ball thus changing the momentum of the ball, since the bat was originally traveling in the opposite direction from the ball, the ball's motion will change such that its direction will change.

20. Consider the entire tug-of-war as one object. The forces acting are as follows:
"outside forces" — the weight of each person (down) and the normal force from the ground (up). Each side will have a friction force acting (horizontally) as a reaction to each person pushing sideways on the ground.
"internal forces" — Each side applies a force to the rope, the rope applies a force back on each side.
"Outside" forces change the motion of objects, the internal forces have no effect on the total motion of the object. The vertical forces (weights and normal forces) will add to zero, the horizontal friction forces will add to show a net force in one direction that will cause the entire tug-of-war to accelerate in that direction.

21. a) Floor applies a force on volleyball.

    b) Softball applies a force on bat.

    c) Football applies force on punter's foot.

22. Sprinters require larger accelerations since they wish to reach maximum velocity in a short period of time. The spikes allow for the sprinter to push harder on the ground so the ground can push pack with a larger force.

23. $(75*5.0) + (85*0) = (75+85)v$

         $v=2.3$ m/s

24. To cause a ball to spin a force must be applied off-center. To change the direction of the spin the player need apply the force at different points on the ball, e.g., to have the ball spin to the left, the force needs to be applied to the right of center.

25. The larger and flatter object will be influenced by the air as it falls. The force of air resistance will oppose the force of gravity acting on the object causing the net downward force to be less. Therefore, the downward acceleration of the object will be less than 9.8 m/s/s. It will take longer to fall the 20 m.

SPORTS ON THE MOON

CHAPTER 3

3

Sports Chapter 3-Sports on the Moon
National Science Education Standards

Chapter Summary

Imagine when astronauts and colonists live on the moon for long periods of time. Students are challenged to write a proposal to NASA that identifies, adapts, or invents a sport that people living in such a colony on the moon would find interesting, exciting, and entertaining.

To meet this challenge, students engage in collaborative activities that explore how differences in the atmosphere and gravity between the Earth and on the moon would affect the play of different sports. These experiences provide the opportunity to develop understanding of the following contents identified in the *National Science Education Standards*.

Content Standards

Unifying Concepts

- Systems, order & organization
- Evidence, models and explanations
- Constancy, change, and measurement
- Science as Inquiry
- Identify questions and concepts that guide scientific investigations
- Use technology and mathematics to improve investigations
- Formulate & revise scientific explanations & models using logic and evidence
- Communicate and defend a scientific argument
- Physical Science
- Motions and Forces
- Conservation of energy & increase in disorder
- History and Nature of Science
- Nature of scientific knowledge
- Historical perspectives

Key Physics Concepts and Skills

Activity Summaries	Physics Principles

Activity One: What Is a Sport?

Students apply their knowledge of sports to identify attributes that define an activity as a sport. From this they begin to consider how differences between the Earth and the moon can affect sports.

- **Physical properties of matter on Earth and in space**
- **Effect of forces on motion**

Activity Two: Free Fall on the Moon

Students compare free fall of different objects, then calculate acceleration due to gravity on the moon using measurements obtained from a slow-motion video of an astronaut in space dropping objects.

- **Acceleration due to gravity**
- **Relationship of gravity to free fall**

Activity Three: Mass, Weight, and Gravity

Using a simulation that allows comparison of mass, students investigate the ratio of gravity on the Earth to that on the moon and determine force necessary to move objects on the moon.

- **Gravity and mass on the Earth and moon**
- **Inertial and gravitational mass**
- **Newton's Laws of Motion**

Activity Four: Projectile Motion on the Moon

Beginning with scale drawings, students calculate distances projected objects would travel on the moon. They then watch a video in slow motion and measure hang time and analyze vertical jumps of basketball players.

- **Gravity and mass on the Earth and moon**
- **Effect of gravity on the trajectory of projectiles**

Activity Five: Jumping on the Moon

Students measure horizontal and vertical distances of different types of jumping then analyze the force and motion involved in each. Applying what they know about gravity on the moon, they predict distances they could jump on the moon.

- **Gravity and mass on the Earth and moon**
- **Effect of force and gravity on horizontal and vertical motion**

Activity Six: Golf on the Moon

Using a variety of balls, students measure the height each bounces when dropped and when projected by a collision. They use this data to infer the speed of a golf ball when hit on Earth and on the moon.

- **Collisions**
- **Coefficient of restitution**
- **Momentum**
- **Projectile motion**

Activity Seven: Friction on the Moon

Students investigate the force necessary to overcome friction between objects and the surface on which they are moving. They then relate this to gravity and predict force needed to overcome friction against sliding motions on the moon.

- **Frictional force**
- **Gravity and mass on the Earth and moon**
- **Effect of gravity on friction**

Activity Eight: Bounding on the Moon

Using cylinders of different lengths and weights, students explore pendulum motion. They then compare the motion of the pendulums to the swinging motion of human legs when walking.

- **Period of pendulum motion**
- **Effect of gravity on pendulum motion**

Activity Nine: "Airy" Indoor Sports on the Moon

Badminton and Wiffle® balls are used by students to investigate how air resistance affects motion. They then apply what they know about the ratio of gravity on Earth to that on the moon to predict air resistance on the moon.

- **Air Resistance**
- **Free fall**
- **Projectile motion**

3

Equipment List For Chapter Three

QTY	TO SERVE	ACTIVITY	ITEM	COMMENT
1	Class	2	*Active Physics Sports* Content Video	Segment: Astronaut dropping hammer & feather on moon.
1	Class	6	*Active Physics Sports* Content Video	Segment: Astronaut hitting golf ball on moon.
1	Class	8	*Active Physics Sports* Content Video	Segment: Bounding on the moon.
1	Class	6	Adhesive for attaching string to balls	Some balls may be drilled instead.
1	Class	9	Badminton shuttlecock & racquet, volunteer	For demonstration.
1	Group	6	Ball, golf	
1	Group	6	Balls, small assorted	Diameters of balls approximately same as golf ball.
1	Group	2	Book	Any book which may be dropped.
2	Group	3	Bottle labeled "1 kg, Earth," actual mass 1 kg	See Teacher's Edition for preparation.
1	Group	3	Bottle labeled "1 kg, Moon," actual mass 0.165 kg	See Teacher's Edition for preparation.
1	Group	3	Bottle labeled "1 kg, Moon," actual mass 1 kg	See Teacher's Edition for preparation.
1	Group	7	Box, capacity to hold 1 kg sand	Serves as subject for measuring sliding friction.
1	Individual	All	Calculator, basic	One per student best; one per group minimum.
21	Group	9	Coffee filters, paper basket-type	Any size, all same size within set.
1	Individual	2	Drawing of hammer dropped on moon & Earth	Master copy provided in Teacher's Edition.
1	Group	2	Feather	
1	Class	9	Golf club, Whiffle® practice golf ball, volunteer	For demonstration
1	Group	2	Hammer	Substitute: Similar high density object.
1	Group	8	Hangible cylinders, lengths 0.1, 0.4, 0.9, 1.6 m	Option: Vary kind of material across groups.
1	Individual	4,8	Meter stick	
1	Group	8	Nail suspension	For suspending cylinders.
1	Group	5	Paper strip, vertically mounted on wall	Approximately 4' long.
1	Individual	4,5	Pencil or pen	
2	Group	2	Pencils or pens	Students' pencils or pens may be used.
1	Individual	4	Protractor	Substitute: Cardboard template for 30 degree angle.
1	Individual	2	Ruler, mm markings	
1	Group	7	Sand, at least 1 kg, in container	Substitute: Any other material to serve as ballast.
1	Group	2	Sheet paper	Any sheet of paper.
1	Individual	4	Sheet paper, 8-1/2" x 11"	
1	Individual	4	Sheet paper, approx 18" x 36"	Substitute: 8-1/2" x 11" sheets taped together.
1	Group	2	Short length of string or thread	For tying pencils together.
2	Group	3	Short lengths of string	For connecting spring scale to bottles.
1	Group	3,7	Spring scale, 0-10 newton range	To convert gram calibration, 100 g = 1.0 N (approx.).
1	Group	6	Stand for hanging pairs of balls to be collided	Ringstands, clamped crossarms.
1	Group	8	Stopwatch	Substitute: Wristwatch having stopwatch function.
1	Group	7	String	For suspending & pulling box.
1	Group	3	Table accessible from opposite sides	To serve as "Mass Station".
1	Group	7	Table-like surface	On which to slide box.
1	Individual	4	Tape	To join pieces of paper.
1	Group	6	Two-meter stick	Substitute: Two 1-meter sticks held end-to-end.
1	Class	2,6,8	VCR & TV Monitor	

Organizer for Materials Available in Teacher's Edition

Activity in Student Text	Additional Material	Alternative / Optional Activities
ACTIVITY ONE: What is a Sport?, p. S114	Space Suit Facts, p. 247	
ACTIVITY TWO: Free Fall on the Moon, p. S117	Assessment Rubrics, pgs. 257-259 Hammer Falling on Earth and on the Moon, p.260	
ACTIVITY THREE: Mass, Weight, and Gravity, p. S123	Assessment Rubric for Mass, Weight, and Gravity, p. 271	
ACTIVITY FOUR: Projectile Motion on the Moon, p. S129	Sample Scale Drawing of the Trajectory of a Projectile, p. 281	
ACTIVITY FIVE: Jumping on the Moon, p. S134		
ACTIVITY SIX: Golf on the Moon, p. S140	Taming the Golf Club, p. 299	
ACTIVITY SEVEN: Friction on the Moon, p. S145	Assessment Rubric for Graphing Skills, p. 307	
ACTIVITY EIGHT: Bounding on the Moon, p. S150		
ACTIVITY NINE: "Airy" Indoor Sports on the Moon, p. S156		

3

CHAPTER **3**

SPORTS ON THE MOON

Scenario

One day, a colony will be set up on the moon and families will live there for extended periods. Plans will have to be made for exercise and entertainment while people live on the moon. Since sports on Earth satisfy both of these needs—exercise and entertainment—it is reasonable to assume that people on the moon colony will also wish to participate in sports. It may even be possible that moon sporting events could be television entertainment for the people back home on Earth.

Challenge

Your challenge is to identify, adapt, or invent a sport that people on the moon will find interesting, exciting, and entertaining.

Write a proposal to NASA (National Aeronautics and Space Administration) which includes the following:

- a description of your sport and its rules and how it meets the basic requirements for a sport,

- a comparison of factors affecting sports on Earth and on the moon in general,

- a comparison of the play of your sport on the Earth and on the moon including any changes to the size of the field, alterations to the equipment, or changes in the rules,

- a newspaper article for the sports section of your local paper back home describing a "championship" match of your moon sport.

s 112

©1999 American Association of Physics Teachers

Chapter and Challenge Overview

Humans have been to the moon, but we have never really lived there. The adventure of the trip and the thrill of being in a new and strange place for a short period of time have kept all of those who have gone to the moon very busy while they were on the lunar surface. However, new projects are being designed to establish a human colony on the moon. Such a colony will permit extensive exploration and investigation of the moon. For the first time, people will be on the moon, not for hours or days, but for many months. Provisions must be made to sustain life for such long periods of time. To maintain peak mental and physical efficiency, there must be some activities other than work. We want to have the moon residents do more than just survive on the moon. We want them to thrive there.

Will movies, card, board, or video games be the extent of the leisure time possibilities, or is it possible to have some active recreation and sports on the moon? This question is at the heart of this chapter. Which, if any, of our familiar Earth sports can be played on the moon? What will playing conditions be like on the moon? What adjustments or modifications will be needed for those sports? Will we be in for major surprises, or do you think with our knowledge of physics we can anticipate how the sports will "work" on the moon? Will the moon be such an exciting place for some sports that literally "the whole world will be watching?"

NASA, the National Aeronautics and Space Administration, is the agency in our country that is responsible for planning a moon base. They have become expert at all sorts of specialized areas related to the technology of transporting people to the moon and providing them with a suitable environment and supplies on the moon. Nearly everything that is used or needed in the early colonies on the moon will have to be sent from Earth. NASA will have to get it there. But NASA might need advice related to human life support, social interactions, and human institutions.

It is not clear which of our familiar sports, if any, can be played in a recognizable way on the moon. NASA is in need of proposals that identify appropriate sports and recreation for the moon colony.

In this chapter students investigate the factors on Earth and the moon. They will learn about acceleration due to gravity, inertial mass and weight, projectile motion, momentum, and the effect of gravity on friction. After they have completed the activities, the students will be asked to write a proposal for submission to NASA. The proposal will explain what sort of facilities should be set up on the moon, and/or why certain facilities and sports would not be advisable. They may evaluate the suitability of one or more specific activities for use on the moon, or invent their own sport making use of the unique conditions that exist on the moon.

3

Criteria

The NASA proposal will be graded on the quality, creativity, and scientific accuracy of your invented sport as well as the description of your sport, the factors affecting sports on the Earth and on the moon, the comparison of play of your sport on the Earth and on the moon, and the newspaper article. NASA proposals which include a mathematical analysis of the sport will be considered superior to those that describe the sport qualitatively (without numbers). In your pursuit of finding the "best" sport for the moon, you may investigate sports that would not be suitable for the moon. Descriptions of these rejected sports and the reasons that they were rejected would raise the quality of your proposal.

For each subject of the final proposal, your class should decide on what should be included and what point value each part should have. How many points should be allocated for creativity and how many should be allocated for mathematical analysis? How many points should be allocated for the comparison of the play on the Earth and on the moon, and how many points should be allocated for the newspaper article? When writing the newspaper article, should points be provided for the quality of the writing, for sketches or drawings which illustrate the article, and for reader interest? What are the attributes that make a superior newspaper article?

If a group is going to hand in one proposal, how will the individuals in the group get graded? How will the grading ensure that all members of the group complete their responsibilities as well as help the other members of the group? The grading criteria should satisfy every person's need for fairness and reward.

In February, 1970, Alan Shepard was the first person to hit a golf ball on the moon.

Assessment Rubric: Training Manual

Meets the standard of excellence. **5**	• The basic requirements of a sport are clearly presented, and related with specific details to the chosen sport. • Details of the rules of the sport are specific and consistently effective. The rules are very easy to follow with no ambivalence. • A comparison of the factors affecting sports on Earth and on the moon is complete and clearly presented. All the physics concepts within the chapter are integrated in appropriate places. • A comparison of the play of the sport on Earth and on the moon includes all the details of any changes to the size of the field, alterations to the equipment, or changes in the rules. A new sport invented for the moon specifies the exact size of the field and a description and diagram of the equipment required. • Additional research, beyond basic concepts presented in the chapter, is evident. • The writing of the newspaper article holds the reader's interest.
Approaches the standard of excellence. **4**	• The basic requirements of a sport are clearly presented, and related to the chosen sport. • Significant information is most often presented in an appropriate manner. • Details of the rules of the sports are specific, effective, and generally easy to follow. • A comparison of the factors affecting sports on Earth and on the moon is complete and clearly presented. Most of the physics concepts within the chapter are integrated in appropriate places. • A comparison of the play of the sport on Earth and on the moon includes any changes to the size of the field, alterations to the equipment, or changes in the rules. A new sport invented for the moon specifies the approximate size of the field and a description or diagram of the equipment required. • The writing of the newspaper article holds the reader's interest.
Meets an acceptable standard. **3**	• The basic requirements of a sport are presented, and are partially related to the chosen sport. • Sufficient information is presented in an appropriate manner. • Details of the rules of the sport are general but effective. • A comparison of the factors affecting sports on Earth and on the moon is clearly presented. Some of the physics concepts within the chapter are integrated in appropriate places. • A comparison of the play of the sport on Earth and on the moon considers two of the following: changes to the size of the field, or alterations to the equipment, or changes in the rules. • The writing of the newspaper article generally holds the reader's interest.
Below acceptable standard and requires remedial help. **2**	• The basic requirements of a sport are presented. • Details of the rules of the sports are limited and difficult to follow. • A comparison of the factors affecting sports on Earth and on the moon is presented. Few of the physics concepts within the chapter are integrated in appropriate places. • A comparison of the play of the sport on Earth and on the moon considers one of the following: changes to the size of the field, or alterations to the equipment, or changes in the rules. • The writing of the newspaper article generally does not hold the reader's interest.
Basic level that requires remedial help or demonstrates a lack of effort. **1**	• The chosen sport is identified but does not relate to any basic requirements of a sport. • Details of the rules of the sports are limited and impossible to follow. • A comparison of the factors affecting sports on Earth and on the moon is briefly presented. Few of the physics concepts within the chapter are integrated in appropriate places. Essential information is missing. • A comparison of the play of the sport on Earth and on the moon fails to consider changes to the size of the field, or alterations to the equipment, or changes in the rules. • The writing of the newspaper article is very difficult to follow. • Much of the NASA proposal remains incomplete.

For use with *Sports*, Chapter 3

©1998 American Association of Physics Teachers

Assessment Rubric: Sports on the Moon

Meets the standard of excellence. **5**	• Scientific vocabulary is used consistently and precisely. • Sentence structure is consistently controlled. • Spelling, punctuation, and grammar are consistently used in an effective manner. • Scientific symbols for units of measurement are used appropriately in all cases. • Where appropriate, data is organized into tables or presented by graphs.
Approaches the standard of excellence. **4**	• Scientific vocabulary is used appropriately in most situations. • Sentence structure is usually consistently controlled. • Spelling, punctuation, and grammar are generally used in an effective manner. • Scientific symbols for units of measurement are used appropriately in most cases. • Where appropriate, most of the data is organized into tables or presented by graphs.
Meets an acceptable standard. **3**	• Some evidence that the student has used scientific vocabulary, although usage is not consistent or precise. • Sentence structure is generally controlled. • Spelling, punctuation, and grammar do not impede the meaning. • Some scientific symbols for units of measurement are used. Generally, the usage is appropriate. • Limited presentation of data by tables or graphs.
Below acceptable standard and requires remedial help. **2**	• Limited evidence that the student has used scientific vocabulary. Generally, the usage is not consistent or precise. • Sentence structure is poorly controlled. • Spelling, punctuation, and grammar impedes the meaning. • Some scientific symbols for units of measurement are used, but most often, the usage is inappropriate. • No presentation of data by tables or graphs.
Basic level that requires remedial help or demonstrates a lack of effort. **1**	• Limited evidence that the student has used scientific vocabulary and usage is not consistent or precise. • Sentence structure is poorly controlled. • Spelling, punctuation, and grammar impedes the meaning. • No attention to using scientific symbols for units of measurement. • No presentation of data by tables or graphs.

Maximum = 10 points

For use with *Sports*, Chapter 3

©1998 American Association of Physics Teachers

        CHAPTER 3

What is in the Physics InfoMall for Chapter 3?

This Chapter of *Active Physics* is all about how things change on the moon. The InfoMall contains plenty of information on the moon, as well as other parts of the solar system. Before even performing a search on the InfoMall, you might want to simply browse through stores looking at titles. You might find such things as "Astronomical Society Of the Pacific - Selected Readings From Mercury," in the Pamphlet Parlor, and "Moon Dance -- A Model of the Earth-Moon-Sun System," in Teachers Treasures. These were found simply by reading a few titles. Look for yourself - you may find more. A look in the Utility Closet uncovers "Physics Phact Potpourri" containing *Many Magnitudes: A Collection of Useful and Useless Numbers*. In this collection, you can find that the density of the moon is 3.342 x 104 kg/m³, as well as the mass and radius of the moon, and much more. These physical factors may affect sports on the moon. Plus, this list contains many interesting items that you may want to use simply for fun.

When considering sports, whether on the moon or elsewhere, you should also consider the ideas in the Physics Olympics Handbook, found in the Pamphlet Parlor. Some of these activities could be discussed for use on the moon.

The obvious search to do first is with keywords "moon" AND "sports". Although a reasonably long list of hits results, only a few are worth considering (often, the moon is used in one paragraph of a textbook or article, and the word "sports" appears soon after, usually as a "sports car"). However, the very first hit is great! It is from Chapter 2 of *Physics: The Excitement of Discovery* in the Textbook Trove. Many parts of this section pertain to individual activities in this Chapter, so we will examine some of them as we go through the activities. You may wish to look at it sooner.

And while doing a preliminary search, search the entire CD-ROM using just one keyword: "sport*". The list of hits contains many articles and discussions you may find helpful.

3

NOTES

ACTIVITY ONE
What is a Sport?

Background Information

Need to develop a working definition of the term "sport" is introduced in this activity.

Open-mindedness in the sense of need to respect the thoughts of others is central to the process outlined for students to arrive at a definition of "sport" in this activity. It can be expected that some students will have a rather narrow sense of what is or is not a sport, and those students also may have low tolerance for different meanings of the term held by others. The most likely point of contention will probably focus on the desire for some to limit the meaning of the term to athletic contests and for others to attach broader meaning to include activities such as fishing as sports.

The "right" definition is, of course, the definition on which the group or the entire class agrees and which would seem reasonable to include in a proposal to NASA for sports on the moon.

The main rule of the brainstorming process to be used for the activity is that, initially, no idea should be evaluated or thrown out. An unwritten extension of that rule is that tolerance and kindness to others should be exhibited when brainstormed lists of ideas are reduced and refined through discussion.

Active-ating the Physics InfoMall

When reading the WDYT introduction, you find that dancing is used as a way to provoke thought on what defines a sport. In case you wonder what physics is involved, and if dance is a sport, try a search that is almost guaranteed to give results; use keywords "dance*" and "sports". (The asterisk after "dance" searches for "dance" as well as "dancing" and "dancer", for example.) The first three hits are articles: "The physics of dance," *Physics Today*, vol. 38, issue 2, 1985; "Resource letter PS-1: Physics of sports," *American Journal of Physics* , vol. 54, issue 7, 1986; and "The physics of some dance movements," *Physics Education*, vol. 24, issue 3, 1989. These articles should be consulted before dance is accepted or rejected as a sport.

The same articles, especially the resource letter from *AJP*, may also give ideas as to what conditions must be met to designate something a sport. Looking through this letter indicates that you can find information on the physics of several sports: archery, baseball, basketball, bowling, cycling, golf, gymnastics, martial arts, racket sports, track and field, winter sports, and others.

Planning for the Activity

Time Requirements

• One class period.

Materials Needed
For each group:
• paper and pen/pencil.

Advance Preparation and Setup

• None needed.

Teaching Notes

Review NASA's purpose for inviting proposals for sports on the moon as to provide activities which moon colonists will find "interesting, exciting and entertaining" forms of "exercise."

Suggest to students that their proposals to NASA would be strengthened by including a definition of the term "sport" and the relation of the definition to their chosen sport for the moon.

Review the brainstorming procedures to be used in the activity, emphasizing tolerance of each other's ideas and opinions

The importance of demonstrating to NASA in the proposal that thought has been given to the meaning of the term "sport" is emphasized. Also, the lists of sports generated during the activity is a good starting place for groups to begin considering possibilities for sports which could be played on the moon.

3

Activity Overview

In this activity students brainstorm to arrive at a definition of a sport.

Student Objectives

Students will:

- apply brainstorming as a process for generating ideas.

- develop a working definition of the term "sport."

ANSWERS FOR THE TEACHER ONLY

What Do You Think?

A key word found in dictionary definitions of "sport" is the word "diversion;" sports provide enjoyable diversion from other, less enjoyable forms of human activity such as household tasks and career activities.

Usually, "diversions of the field" such as hunting and fishing are included with athletic games under the umbrella term "sport;" competition usually is not involved in the former and usually is involved in the latter sports.

SPORTS ON THE MOON

Activity One
What is a Sport?

WHAT DO YOU THINK?

Ballroom dancing recently was approved as an Olympic competition.

- **What separates sports from other kinds of human activities?**
- **Do all sports result in "winners" and "losers?" Should they?**

Record your ideas about these questions in your *Active Physics log*. Be prepared to discuss your responses with your small group and the class.

FOR YOU TO DO

1. In your group, brainstorm a list of at least ten words or phrases which identify attributes, or characteristics, of activities known as sports (example: team involved). All ideas should be accepted and no idea should be evaluated or thrown out until brainstorming has been completed; during brainstorming, it is "legal" to ask for clarification of an idea, but no discussion of an idea should occur until later. Continue brainstorming until ten or more attributes of sports have been identified and the group runs out of ideas.

SPORTS S 114

ANSWERS

For You To Do

Answers to questions will vary from group to group. See Background Information for suggestions on how to encourage a variety of opinions.

a) A member of your group should volunteer to record the list of attributes of sports as they are identified by all members of the group. Everyone, including the person serving as recorder, should participate in identifying attributes.

b) Discuss each attribute and arrive at consensus within your group upon a final list of attributes which apply to many, but not necessarily all, sports. Each member of the group should copy the list in his or her log under the heading, "Attributes of Many Sports."

2. In your group, brainstorm a list of names of at least 25 but not more than 50 sports (example: rock climbing). All sports named should be accepted without discussion or evaluation. Continue identifying sports until the process either "slows down" after 25 sports have been named or when the length of the list of sports reaches 50.

a) One member of your group should volunteer to record the list of sports. Everyone should participate.

3. Decide which items on the list "Attributes of Many Sports" apply to all of the sports identified by the group. In your group, consider the attributes one-at-a-time and, by consensus, answer the question, "Does this attribute apply to all of the sports on the list, or to only some of the sports?"

a) On the list of attributes of sports in your log, mark with a "*" only those attributes which apply to all sports identified by the group.

b) In your log, include the above marked items in a new list under the heading, "Attributes which Apply to All Sports."

c) Discuss within your group whether any attributes which apply to all sports seem to have been left out; if the group agrees, attributes which seem appropriate or necessary may be added to the list.

4. Define the term "sport."

a) In your group, use the list "Attributes which Apply to All Sports" to construct a written definition of the term "sport." Test drafts of definitions against the list of sports generated by the group—and other sports which may come to mind—until your group agrees upon a definition which seems to apply to all sports.

b) Write your group's definition of "sport" in your log.

 SPORTS

ANSWERS

For You To Do (continued)

Answers to questions will vary from group to group. See Background Information for suggestions on how to encourage a variety of opinions.

ANSWERS

Physics To Go

1. The student's response should recognize the need to define terms because, lacking definition of important terms, different people often attach different meanings to terms.

2. *Webster's New Collegiate Dictionary*, first two meanings: 1. That which diverts, and makes mirth; pastime; diversion. 2. A diversion of the field, as hunting, fishing, racing, games, especially athletic games, etcetera; also, any of various similar games usually played under cover, as bowling, rackets, basketball, etc.

3. Among the more obvious are reduced gravity and lack of an atmosphere.

4. In addition to food, air and water, humans need protection from the vacuum on the moon and extremes of temperature.

5. Conceivably, both indoor and outdoor sports could be considered, but each would require attention to survival needs.

6. The response depends on the student's favorite sports; it is unlikely that students will have very much knowledge of how sports would be affected on the moon, including, for example, inability to walk or run normally on the moon.

SPORTS ON THE MOON

REFLECTING ON THE ACTIVITY AND THE CHALLENGE

The first item that you must address in your proposal to NASA is how your chosen sport for moon dwellers meets the basic requirements for a sport. In order for you to convince NASA that you know what the requirements for a sport are, it seems necessary for you to include a fundamental definition of "sport" as a basis of your proposal. While you may wish to refine your definition later, you have a good start. The list of sports generated by your group during this activity probably is a good starting place for considering which sports would be good candidates for being adapted to the moon. However, you probably need more information about differences between the Earth and moon before you can make a good decision about the particular sport to include in your proposal.

PHYSICS TO GO

1. You probably learned from this activity that terms such as "sport" mean different things to different people. Write a brief paragraph to give an example of an occasion when you felt it necessary to ask someone for a definition of the meaning of a term used in a conversation with you. (Example: A parent saying, "Don't get home too late.")

2. Look up the definition of "sport" in a dictionary. Do you agree with the definition? Why or why not?

3. Do some research on physical properties of the moon. What properties seem to have great implications for sports on the moon? What properties do not?

4. What do humans need for survival on the moon, and how might the requirements for survival affect participation in sports?

5. Could both indoor and outdoor sports be considered for the moon? Why or why not?

6. Based only on what you know about the moon so far—you will know much more very soon—how would conditions on the moon affect:

   a) the sport in which you most enjoy participating?
   b) the sport which you most enjoy as a spectator?

S 116

Space Suit Facts

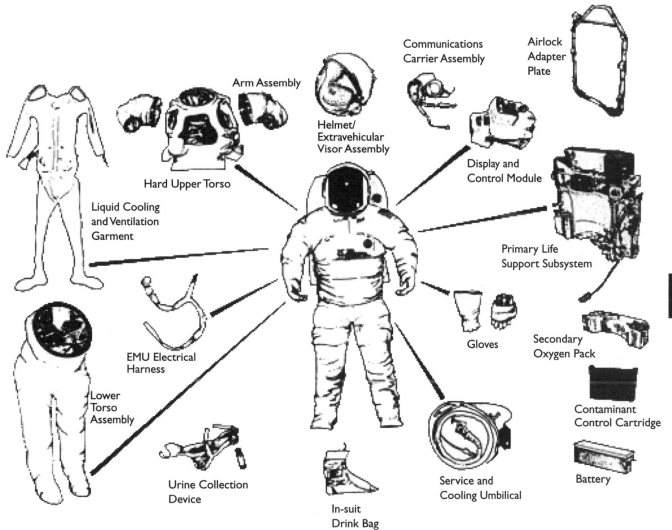

Arm Assembly

Communications Carrier Assembly

Airlock Adapter Plate

Helmet/ Extravehicular Visor Assembly

Hard Upper Torso

Display and Control Module

Liquid Cooling and Ventilation Garment

Primary Life Support Subsystem

EMU Electrical Harness

Gloves

Secondary Oxygen Pack

Lower Torso Assembly

Contaminant Control Cartridge

Urine Collection Device

Service and Cooling Umbilical

Battery

In-suit Drink Bag

3

For use with *Sports*, Chapter 3, ACTIVITY ONE: What is a Sport?

©1999 American Association of Physics Teachers

ACTIVITY TWO
Free Fall on the Moon

Background Information

Comparison of the acceleration due to gravity on Earth and the moon is the main topic addressed in this activity. The comparison is based on the comparison of the distances that an identical object falls from a "rest start" during equal time intervals on Earth and the moon.

It is recommended that you read For You To Read on page S120 of the student text before proceeding in this section. Digestion of the derivation presented in the student text will be assumed in the below presentation of background information.

The equation $d = \frac{1}{2}at^2$ derived in the student text is a special case which applies only to situations where the initial speed of an object is zero when it begins to undergo uniform acceleration. The general case which applies to situations where the initial speed of an object is not zero, resulting in the equation $d = v_i t + \frac{1}{2}at^2$ is derived below only for your information as the teacher.

Acceleration is defined as: $a = \Delta v/\Delta t = (v_f - v_i)/t$

where a is the uniform acceleration, v_i is the object's (non-zero) initial speed, v_f is the object's final speed, and t is the amount of time elapsed since the object began accelerating.

Solving for v_f, $v_f = at + v_i$

The object's average speed during the period of acceleration is:

$v_{Average} = (v_f + v_i)/2$

The distance travelled during the period of acceleration is:

$d = v_{Average}\, t = [(v_f + v_i)/2]t = [v_i + (at + v_i)]t\, /2 = [2v_i t + at^2]2$

$d = v_i t + \frac{1}{2}at^2$

This equation applies to cases where an object already is moving when acceleration begins, such as a car already traveling down the highway accelerating to pass another car. The equation also serves very well to calculate how far an object thrown straight upward with an initial speed will be from its starting point at any time, t; in such a case, it is

necessary to assign a negative value to the acceleration due to gravity if distances upward from the starting point are assigned positive values.

The special case of $v_i = 0$, or starting from rest, applies to this student activity, in which case the equation reduces to $d = \frac{1}{2}at^2$.

The acceleration due to gravity on the moon is, for practical purposes with your students, 1/6 of the acceleration due to gravity on Earth. You have a choice here. If you have been using 10/m/s^2 as the "rounded-off" value of g on Earth, then

$10 \div 6 = 1.7$ m/s^2 would be the value of g on the moon rounded to two significant figures. If you choose at this point to "shift" to using the more refined, widely used value of g, 9.8 m/s^2, then, to two significant figures, the value on the moon would be $9.8 \div 6$ m/s$^2 = 1.6$ m/s^2

Active-ating the Physics InfoMall

The classic experiment in which astronaut David R. Scott (on the Apollo 15 moon flight) dropped the feather and hammer is discussed on the InfoMall, including a graphic of a ball and feather being dropped in a vacuum. Just do a search with "hammer" AND "feather" and you will find appropriate references. Further down the list, you can find another graphic, but with a feather and silver coin. It seems no one wants to drop hammers in their vacuum chambers.

In Physics To Go, questions 7 and 8 refer to specific sports. Notice that you can find articles or references to articles on basketball and gymnastics on the InfoMall. As mentioned above, "Resource letter PS-1: Physics of sports," *American Journal of Physics*, vol. 54, issue 7, 1986, discusses both. In addition, with a search using keyword "sport*" you can find "Physics of basketball," in the *American Journal of Physics*, vol. 49, issue 4, 1981. This article discusses such things as the effect of air resistance. A search for gymnastics is not so fruitful, however. Other than the mention of gymnastics in the resource letter, the InfoMall only makes reference to "mental gymnastics" and biographical references to women in science.

Planning for the Activity

Time Requirements

• One class period.

Materials Needed

For the class:

• *Active Physics Sports* content video (sequence: astronaut dropping hammer and feather on the moon)

• VCR & TV monitor

For each group:

items to be dropped:

• 2 pencils

• short length of string or thread

• 1 book

• 1 sheet paper

• 1 hammer

• 1 feather

For each student:

• ruler, millimeter markings

• reproduced side-by-side drawings of astronaut dropping a hammer on the moon and on Earth

Advance Preparation and Setup

Reproduce for each student the side-by-side drawings of the astronaut dropping a hammer on the moon and on Earth provided at the end of this activity in the Teacher's Edition.

Reserve a VCR and TV monitor for showing the sequence of the astronaut dropping a hammer and feather on the moon from the *Active Physics Sports* content video.

Obtain feathers and hammers (or similar high density substitutes for a hammer) to compare as they are dropped; the feather/hammer pair could be done as a demonstration for the class if a supply of feathers for each group is not available. However, feathers are readily available at craft and hobby stores.

Teaching Notes

Clearly, sports involving free fall will have "slow-motion" effects due to the reduced effect of gravity on the moon. Warn students that while it may be tempting to assume that everything in sports on the moon would be different by a factor of six, that is not necessarily true as will be seen in future activities. Make them aware of what effects they are observing in this activity.

3

NOTES

Activity Two
Free Fall on the Moon

WHAT DO YOU THINK?

The diameter of the moon is only one-fourth the diameter of the Earth.

• **Describe the motion of an object falling on the moon.**

• **Compare this motion with that of an object falling on the Earth.**

Record your ideas about these questions in your *Active Physics log*. Be prepared to discuss your responses with your small group and the class.

FOR YOU TO DO

1. Compare how pairs of objects fall on Earth when released from equal heights. For each of the following pairs of objects, hold one object in each hand and release both objects at the same instant from equal heights.

   • a single pencil / two pencils tied together with thread
   • a closed book / an open sheet of paper
   • a closed book / a tightly crumpled sheet of paper
   • a hammer / a feather

    a) Record which, if either object, hits the ground first or if the objects strike the ground at the same instant. Try to explain each case in terms of what you know about gravity and air resistance.

──────SPORTS

Activity Overview

In this activity students compare the acceleration due to gravity on the moon and on Earth by observing actual pairs of objects falling on Earth and a video of objects falling on the moon. They will take measurements from a diagram and calculate the ratio of acceleration due to gravity on the moon to the Earth.

Student Objectives

Students will:

• understand what is meant by deriving an equation.

• understand and apply the derived equation $d = 1/2at^2$ to compare the acceleration due to gravity on Earth and the moon.

ANSWERS FOR THE TEACHER ONLY

What Do You Think?

In the same way as on Earth, all objects fall with the same acceleration on the moon, but with less acceleration than on Earth. The acceleration due to gravity on the moon is 1/6 that of Earth. Objects dropped on the moon will not have as great an acceleration as those on Earth.

3

ANSWERS

For You To Do

1. a) Only the sheet of paper and the feather should have their falls significantly affected by air resistance.

ANSWERS

For You To Do

(continued)

2. a) The hammer and feather land together on the moon because there is no atmosphere.

   b) The moon has no atmosphere because the molecular velocity of air on the moon would be greater than the velocity required to escape from the moon's gravitational pull.

   c) The acceleration due to gravity on Earth is greater than on the moon. Accurate measurements will confirm this observation.

SPORTS ON THE MOON

2. Observe a video sequence of an astronaut dropping a hammer and a feather while standing on the surface of the moon. Record answers to the following questions in your log.

   a) How did the times for the hammer and the feather to fall equal distances to the surface of the moon compare? What do you think is the reason for the result?

   b) Do you think the moon has a gaseous atmosphere similar to Earth's air? Why or why not?

   c) Do you think the acceleration of the falling hammer which you dropped on Earth was greater, about the same, or less than the acceleration of the hammer dropped by the astronaut on the moon? What information would you need to make a careful comparison of the acceleration in the two cases of the falling hammer?

3. Examine the two "double exposure" diagrams shown. On the left is an astronaut dropping a hammer while standing on the moon. Two images of the hammer are visible. The first image was made at the instant the astronaut released the hammer to allow it to fall. The second image was made at an instant 0.50 second after the hammer began to fall. On the right is shown the same astronaut dropping the same hammer while standing on Earth. Again, one image of the hammer was made at the instant of release and another image was made after the hammer had fallen for 0.50 second.

Hammer being dropped on the moon   **Hammer being dropped on earth**

SPORTS

S 118

ACTIVITY TWO: Free Fall on the Moon

a) Place a ruler on the page and measure the distance between corresponding points on the two images of the hammer in each diagram. Record each distance, measured to the nearest millimeter, in your log.

b) Again place the ruler on the page and measure the height of the image of the astronaut on Earth to the nearest millimeter. Record the measurement.

c) The astronaut shown in the diagrams is known to have a real height of 1.9 m. See if you agree that the "shrink factor" of the diagram is approximately 3 cm/m. Show your work in your log.

d) Divide each of the fall distances in above step (a) by the shrink factor of the diagram to convert the distance the hammer falls in 0.50 second on the moon and on Earth to real world distances.

e) In your log, substitute the values of the distance, in meters, that the hammer fell during 0.50 seconds on the moon and on Earth in the below equation:

$$\frac{d_{moon}}{d_{earth}} = \frac{a_{moon}}{a_{earth}}$$

f) Perform the division $\frac{d_{moon}}{d_{earth}}$ and enter the answer in your your log as equal to the ratio $\frac{a_{moon}}{a_{earth}}$.

g) Compare your calculated value of the ratio $\frac{a_{moon}}{a_{earth}}$ to the fraction $\frac{1}{6}$. (To do so, you may wish to convert $\frac{1}{6}$ into decimal form by dividing 1 by 6.) Do your calculations show that the acceleration of gravity on the moon is about $\frac{1}{6}$ of the value on Earth? Comment on the comparison in your log.

h) The acceleration due to gravity on Earth is 9.8 m/s² (meters per second every second). From your results for this activity, what should be the value of the acceleration due to gravity, in m/s², on the moon? Show how you arrive at your answer in your log.

 119

SPORTS

ANSWERS

For You To Do

(continued)

3. The values expected for measurements and calculations of the drawings of the astronaut dropping the hammer on Earth and on the moon are:

a) Diagram measurement of hammer fall distance, moon = 7.0 mm

Diagram measurement of hammer fall distance, Earth = 42.0 mm

b) Diagram measurement of astronaut's height = 63.0 mm

c) Shrink factor
= (Real height)/(Height in diagram)
= 1.9 m/63 mm
= 0.030 m/mm

d) Real distance hammer falls on moon in 0.50 s
= 7.0 mm x 0.030 m/mm
= 0.21m

Real distance hammer falls on Earth in 0.50 s

= 42 mm x 0.030 m/mm
= 1.3 m

---

e) In Chapter 1, Activity Nine, it was calculated that an object falling from rest for 0.50 s on Earth would travel 1.25 m (1.3 m to two significant figures); within error of measurement, the distance the hammer dropped on Earth by the astronaut fell during 0.50 seconds, 1.3 m, is the same, equal distance.

f) $d_{Moon}/d_{Earth} = a_{Moon}/a_{Earth}$ = 0.21 m /1.3 m

g) a_{Moon}/a_{Earth} = 0.16

h) The fraction 1/6 in decimal form is, to two significant figures, 0.17; the experimental value of a_{Moon}/a_{Earth} , 0.16, is 94% of the decimal value of the fraction 1/6. Therefore, it is reasonable to use 1/6 of Earth's acceleration due to gravity as the acceleration due to gravity on the moon.

If g_{Earth} = 10 m/s², then g_{Moon} = 10m/s² ÷ 6 = 1.7 m/s²

FOR YOU TO READ

You are able to use what you already know to reason toward an answer to the question, "Do the distances that objects fall from rest during equal time intervals on Earth and the moon compare in the same way as the acceleration due to gravity on Earth and the moon?"

- **Acceleration is defined as:**

  $a = \Delta v / \Delta t$

- **Change in speed at the end of time interval Δt:**

  $\Delta v = a(\Delta t)$

Since the object starts from rest, both its speed and time of fall are zero at the instant it is dropped.

- **Final speed after falling for an amount of time, t:**

  $v = at$

- **Average speed while falling for time t:**

  $v_{ave} = \left(\dfrac{1}{2}\right)v$

  $= \left(\dfrac{1}{2}\right)at$

The distance an object falls during a time of fall, t, is simply the object's average speed while falling, v_{ave}, multiplied by the time of fall:

- **Fall distance during time of fall, t:**

  $d = v_{ave}t$

  $= \left(\dfrac{1}{2}\right)at \times t$

  $= \left(\dfrac{1}{2}\right)at^2$

The equation, $d = \left(\dfrac{1}{2}\right)at^2$, was developed by combining equations already known. Physicists call this process "deriving" an equation. This new equation is very useful because it allows calculation of the distance an object falls when only the object's acceleration and time of fall are known. In addition, the equation works for falling objects on either Earth or the moon:

- **Fall distance during time of fall, t, on moon:**

  $d_{moon} = \left(\dfrac{1}{2}\right)a_{moon}t^2$

- **Fall distance during time of fall, t, on Earth:**

  $d_{earth} = \left(\dfrac{1}{2}\right)a_{earth}t^2$

Dividing the above equation for d_{moon} by the above equation for d_{earth}, with the condition that the times of fall, t, for objects on the moon and Earth are equal:

$$\frac{d_{moon}}{d_{earth}} = \frac{\left(\dfrac{1}{2}\right)a_{moon}t^2}{\left(\dfrac{1}{2}\right)a_{earth}t^2} = \frac{a_{moon}}{a_{earth}}$$

The above equation was simplified by cancelling the equal "$\dfrac{1}{2}$" and "t^2" terms which appear in both the numerator and the denominator.

The equation $\dfrac{d_{moon}}{d_{earth}} = \dfrac{a_{moon}}{a_{earth}}$ provides the answer, "Yes, the distances that objects fall from rest during equal time intervals on Earth and the moon compare in the same way as the acceleration due to gravity on Earth and the moon." Therefore, it is valid to compare the acceleration due to gravity on the moon and on Earth by comparing the distances that the hammer dropped by the astronaut fell during equal time intervals on the moon and Earth.

REFLECTING ON THE ACTIVITY
AND THE CHALLENGE

Objects take much longer to fall on the moon than on Earth.
* A ball that requires only 2 sec to fall to the ground on Earth
would take 12 sec to fall on the moon. This is due to the moon's
gravity being only $\frac{1}{6}$ the gravity on the Earth.

Now that you are equipped with a specific value for the
acceleration due to gravity on the moon, it is possible for you to
do calculations to show exactly how anything in a sport that
involves free fall would be affected if the sport were played on
the moon. This would include not only simple "up" and "down"
cases of free fall—such as vertical jumps— but also all cases of
projectile motion—such as the shot put—in sports on the moon.

When developing a sport for the moon, you will have to take
into account how long an object is in the air. A sport can get
very boring if most of the time is spent waiting for gravity to
cause a ball to drop.

PHYSICS TO GO

1. Prove that the equation $d = \left(\frac{1}{2}\right)at^2$ can be rewritten as $a = \frac{2d}{t^2}$.

2. When exploring a planet, it was found that a rock dropped
from 2.0 m above the planet's surface took 0.50 second to fall
to the surface. What is the acceleration due to gravity on that
planet? (Hint: Use the information in the above problem.)

3. Prove that the equation $d = \left(\frac{1}{2}\right)at^2$ can be rewritten as
$t = \sqrt{\frac{2d}{a}}$.

4. If a rock were dropped from 2.0 m above the surface of the
moon, how much time would it take to fall to the surface?
(Hint: Use the information in the above problem.)

5. A baseball player on the moon hits a pop-up fly straight
upward at an initial speed of 32 m/s.

   a) How much time would it take the ball to reach the highest
   point in its flight?
   b) How much time would the catcher have to prepare to
   catch the ball when it comes back down?
   c) What would be the maximum height of the ball in its
   flight? Compare your answer to the length of a football
   field; 100 yards between goal lines equals about 91 m.

* Referring to line two, please be sure
to note that it is a "kicked" ball,
not just any ball.

3

Physics To Go

1. $d = 1/2at^2$. Multiply both sides of the equation by 2 and divide both sides of the equation by t^2.
Then $a = 2d/t^2$.

2. $a = 2d/t^2 = 2(2.0 \text{ m}) / (0.50 \text{ s})^2 = 1.0 \text{ m/s}^2$

3. $d = 1/2at^2$. Multiply both sides of the equation by 2, divide both sides of the equation by a,
and then take the square root of both sides of the equation. Then $t = \sqrt{2d/a}$.

4. $t = \sqrt{2d/a} = \sqrt{(2 \times 2.0 \text{ m})/(1.7 \text{ m/s}^2)} = \sqrt{2.6 \text{ s}^2} = 1.6 \text{ s}$

5. a) Losing 1.7 m/s of its speed every second, it will take the ball $(32 \text{ m/s})/1.7 \text{ m/s}^2 = 19 \text{ s}$
to reach the peak of its upward flight.

   b) It will also take 19 s to fall back down; therefore, the catcher has 38 s.

   c) The average speed on the way up will be 16 m/s; maintained for 19 s, the ball will travel
   16 m/s x 19 s = 300 m, which is over three football fields of distance.

ANSWERS

Physics To Go

(continued)

6. Using the more accurate value of
gMoon = (9.8 m/s²) ÷ 6
= 1.6 m/s², students should enter
the values for the empty cells
shown below in bold. (Students
who use 1.7 m/s² as the acceleration due to gravity on the moon
will arrive at slightly different
answers than those shown; such
answers should be accepted.)

7. - 9. These refer to specific sports.
Students should make
mention that flight time on
the moon would greatly
increase. Sports, such as
gymnastics, might benefit
greatly. However, other
sports, such as basketball,
would suffer. Refer to
Reflecting on the Activity
and the Challenge on
page S121.

SPORTS ON THE MOON

6. A group of physics students plans to adapt the Soap Box Derby
to the moon. The contestants' cars will start from rest and coast
down a 160-m mountainside on the moon which has a straight
slope. The slope is great enough that a derby car which has low
friction will accelerate at about one-half of the acceleration due
to gravity on the moon. Before each car's run, the race sponsors
will place a high-tech instrument package on the car which will
allow the driver to read the elapsed time, acceleration, speed,
and distance travelled throughout the run. Copy and complete
the table below to show the highest possible readings which the
accelerometer, speedometer, and odometer could show at the
end of each 2 s during an ideal, friction-free run. Be sure to fill
in each empty cell in the table.

Clock reading (seconds)	Accelerometer reading (meters/second²)	Speedometer reading (meters/second)	Odometer reading (meters)
0	0.8	0	0
2.0	0.8	1.6	
4.0	0.8		
6.0			
8.0			
10			
12			
14			
16			
18			
20	0.8		160

7. How would the difference in time for the flight of the ball
affect the Earth game of basketball if played on the moon
with no modifications?

8. How would the difference in time for the flight of the
gymnast in the air affect Earth gymnastics if done on the
moon with no modifications?

9. How would the difference in time for the flight of a projectile
affect the throw of the javelin on the moon?

S 122

Clock reading (seconds)	Accelerometer reading (meters/second²)	Speedometer reading (meters/second)	Odometer reading (meters)
0	0.8	0	0
2	0.8	1.6	0.6
4	0.8	3.2	6.4
6	0.8	4.8	14.4
8	0.8	6.4	25.6
10	0.8	8.0	40
12	0.8	9.6	57.6
14	0.8	11.2	78.4
16	0.8	12.8	102.4
18	0.8	14.4	129.6
20	0.8	16	160

Assessment Rubrics for Activity Two: Free Fall on the Moon

Place a check mark (√) in the appropriate box.

Student name: _____

Date: _____

1. Compare how pairs of objects fall on Earth when released from equal heights. For each of the below pairs of objects, hold one object in each hand and release both objects at the same instant from equal heights.

   • a single pencil/two pencils tied together with thread
   • a closed book/an open sheet of paper
   • a closed book/a tightly crumpled sheet of paper
   • a hammer and a feather

Laboratory skills

Observation/Recording Data	Excels	Acceptable	Remedial
1. Student observes and records time needed for different objects to fall.	Detailed and accurate observations.	Little detail provided but mostly accurate observations.	Observations lack detail and accuracy.
Conclusions/Deductive Thinking	**Excels**	**Acceptable**	**Remedial**
2. Uses concepts of gravity and air resistance to provide an explanation. Explains that objects with greater surface area have greater air resistance.	Explains that the force of gravity acts on all objects in the same manner.	Is able to explain one of the concepts correctly and relate it to the dropping of the objects.	Believes that heavy objects fall faster than light ones regardless of their shape.

3

For use with *Sports*, Chapter 3, ACTIVITY TWO: Free Fall on the Moon

Student name: _____

Date: _____

2. Observe a video sequence of an astronaut dropping a hammer and a feather while standing on the surface of the moon. Record answers to the questions below in your log.

Laboratory skills

Observation/Recording Data	Excels	Acceptable	Remedial
1. Student observes the falling of the feather and the hammer on the surface of the moon.	Quantitative data recorded indicates that both objects fall at the same rate. Identifies need to record distance and time to determine acceleration of falling objects.	Correct observation is provided but no quantitative data or measurement is presented.	Observations lack detail and accuracy.
Conclusions/Deductive Thinking	**Excels**	**Acceptable**	**Remedial**
2. Uses concepts of gravity and air resistance to provide an explanation.	Concludes that objects fall to the surface of the moon more slowly than Earth. Identifies differences in gravitational force of moon and Earth.	Is able to explain one of the concepts correctly and relate it to the dropping of the objects.	Is not aware that objects fall more slowly on the moon because of reduced gravitational force.

Student name: _____

Date: _____

3. Examine the two "double exposure" diagrams shown. On the left is an
   astronaut dropping a hammer while standing on the moon. Two images
   of the hammer are visible. The first image was made at the instant the
   astronaut released the hammer to allow it to fall. The second image was
   made at an instant 0.50 second after the hammer began to fall.

Laboratory skills

Observation/Recording Data	Excels	Acceptable	Remedial
1. Student measures the distance between corresponding points on the two images. Measurement of the height of the astronaut is recorded.	Both measurements are accurate and units for measurement are provided to nearest 1 mm.	Both measurements are accurate (no measurement is off more than 2 mm) and units for measurement are provided.	Measurement inaccurate or no units for measurement provided.
Manipulating Data	**Excels**	**Acceptable**	**Remedial**
2. Uses a ratio to determine fall distances on Earth and moon.	Accurately calculates the "shrink factor" to determine actual height of the fall distance. Correctly substitutes values for distance of Earth and moon. Correctly calculates ratio as 1/6.	Accurately calculates the "shrink factor" to determine actual height of the fall distance. Correctly substitutes values for distance of Earth and moon. Requires assistance to calculate the ratio as 1/6.	Requires assistance to perform two or more calculations.
Manipulating Data	**Excels**	**Acceptable**	**Remedial**
3. Uses a ratio to determine the acceleration due to gravity on the moon.	Works independently to calculate the moon's acceleration due to gravity, by using Earth's value (9.8 m/s/s) and the ratio 1/6.	Requires assistance to calculate the moon's acceleration due to gravity from a ratio.	Is unable to use a ratio to calculate the acceleration due to gravity on the moon.

For use with *Sports*, Chapter 3, ACTIVITY TWO: Free Fall on the Moon

Hammer Falling on Earth and on the Moon.

Hammer being dropped on the moon   **Hammer being dropped on Earth**

For use with *Sports*, Chapter 3, ACTIVITY TWO: Free Fall on the Moon

©1999 American Association of Physics Teachers

Hammer Falling on Earth and on the Moon.

Hammer being dropped on the moon   **Hammer being dropped on Earth**

For use with *Sports*, Chapter 3, ACTIVITY TWO: Free Fall on the Moon

©1999 American Association of Physics Teachers

NOTES

3

ACTIVITY THREE
Mass, Weight, and Gravity

Background Information

It is suggested that you read the section "For You To Read: Gravity on the Planets and the Moons" in the student text for this activity before proceeding in this section. For a review of the meanings of mass and weight on Earth, you also may wish to review the Background Information for Chapter 2, Activities One and Two before reading the following background information.

This activity has students compare the mass and weight of a 1-kg object on Earth and on the moon.

Recall from the Background Information from Chapter 2, Activity Two that an object's gravitational mass (determined by using an equal arm balance to compare the object to a standard mass) is equal to an object's inertial mass (determined by comparing the object's acceleration in response to an equal force applied to a standard mass); in this activity, the inertial masses of objects are compared as students estimate the amount of force required to produce equal accelerations of a 1-kg object on Earth and the moon. The inertia, or inertial mass, of an object does not vary with location; an object should respond to an applied force in the same way on Earth and the moon.

A simplified version of Newton's Universal Law of Gravitation is used for students to reason toward an explanation of the reduced effect of gravity on the moon and to predict, in general, what would be the local acceleration due to gravity on any planet (or another moon) in terms of the planet's mass and radius. For your information as the teacher, and not intended to be shared with students unless you deem them capable of understanding it, the basis in Newton's Universal Law of Gravitation for the proportionality derived in "For You To Read: Gravity on Planets and Moons" is presented below.

Assume that it is desired to calculate the gravitational force of attraction between a small object whose mass is m_{small} and a huge spherical object, such as a planet, whose mass is m_{huge}. Also assume that the distance between the centers of masses of the huge and small objects is R—this can be any distance, large or small, but the smallest possible value occurs when the surfaces of the objects are in contact. According to Newton's Universal Law of Gravitation, each object experiences a force toward the center of mass of the other object equal to:

$$F = (Gm_{small}\, m_{huge})/R^2$$

Where F is the force experienced by each object (an equal and opposite pair of forces, consistent with Newton's Third Law) and G is the universal gravitational constant equal to 6.67×10^{-11} $m^3/(kg)s^2$. According to Newton's Second Law each object will accelerate toward the other object with an amount of acceleration $a = F/m$. The acceleration of small should be equal to F/m_{small}, which can be found by rearranging the above equation as:

$$a_{small} = F/m_{small} = (G\, m_{huge})/R^2$$

If the small object is located at or very near the surface of the huge spherical object, then, for practical purposes, $R = R_{huge}$, and the above equation becomes:

$$a_{small} = F/m_{small} = (Gm_{huge})/R_{huge}^2$$

The above equation gives the acceleration which will be experienced by any small object located near the surface of a huge object, such as a planet or a moon, in terms of properties of only the huge object; this acceleration is called the acceleration due to gravity, or g, of that planet or moon. Therefore:

$$g_{huge} = (Gm_{huge})/R_{huge}^2$$

You may wish to substitute values for G, the mass of Earth, and Earth's radius in the equation; if you do, the answer obtained for a_{small} surely will be $g = 9.8$ m/s^2. Doing the same for the moon will result in 1.6 m/s^2.

Only the proportionalities between g, m and R in the above equation are used to reason toward comparison of the accelerations due to gravity at the surface of Earth and the moon in the student activity.

Active-ating the Physics InfoMall

A reasonable search for this activity is with keywords "moon" AND "acceleration" AND "gravity". However, you get "too many hits" with this search. There are the normal suggestions to overcome this: limit your search to only a few stores at a time, add other keywords, or require that the keywords must appear in the same paragraph (one of the choices in the compound search window). You can also adjust the order of the keywords, as well as how close they must be to one another. If you choose "within paragraph" and "exact order" you will get a list with many nice entries. (Change the search parameters to find different lists. This example is only one of many different lists you can get.) For example, one hit is from Chapter 2 of *Physics: The Excitement of Discovery* in the Textbook Trove. There is a section of this chapter devoted to the effects on the astronauts due to the moon's smaller gravitational acceleration.

One of the concepts explored in this activity is inertial mass. To stay with the moon theme, perform a search using keywords "moon" AND "inertial mass". (Don't forget to change to parameters of your search to "within article" if you changed them above.) The very first result is pretty good, including some amusing (and informative) graphics. This hit is from the Textbook Trove, *Physics for the Inquiring Mind,* Chapter 7. For the purpose at hand, check out Figure 7.19. This shows how mass and weight can be different. (For an amusing comparison, see Figure 18.) Check out the rest of the hits; many should be quite useful. Many of the hits on this list also discuss the effect of the moon's weaker gravity on objects in flight.

Planning for the Activity

Time Requirements

- One class period.

Materials Needed

For each group:

A Mass Station consisting of:

- 1 table accessible from opposite sides
- 1 plastic bottle labeled "1 kilogram, Earth" (actual mass = 1 kg)
- 1 plastic bottle labeled "1 kilogram, Moon" (actual mass = 1 kg)

A Weight Station consisting of:

- 1 floor surface
- 1 spring scale calibrated in newtons, 10 N capacity
- 1 plastic bottle fitted with a loop of thread around the neck for lifting and labeled "1 kilogram, Earth" (actual mass = 1 kg)
- 1 plastic bottle fitted with a loop of thread around the neck for lifting and labeled "1 kilogram, Moon" (actual mass = 0.165 kg)

Advance Preparation and Setup

Each group will visit two stations during the For You To Do activity, a Mass Station and a Weight Station. Although the procedure indicates visiting the Mass Station first and the Weight Station last, the order does not matter. The number of stations needed for the class is halved if half of the small student groups begin working at the Mass Station while the remaining half of the groups begin working at the Weight Station, and then the groups switch stations. For example, for 6 small groups, 3 Mass Stations and 3 Weight Stations would be needed.

In advance of the activity you will need to begin collecting 1-quart plastic bottles fitted with caps. Sports drink bottles in 1-quart size work very well. The desired shape is that which is most nearly cylindrical; a design having a sharply tapered neck is preferred over a design having a gentler taper of the neck, the reason for the preference being that the more cylindrical shape will tend to roll straight when, during the activity, the bottle is laid on its side and pushed. It will do no harm if the design of the bottles includes, as many do, fluted indentations on the sides. Two bottles are needed for each Weight Station and Mass Station, a total of 12 bottles for the example scheme given in the above paragraph.

All of the bottles, usually made of clear plastic, should be painted to make the contents invisible. This can be done using paint from an aerosol spray can or by dipping in a can of leftover paint — be sure that the paint used will not dissolve the plastic bottle!

Once painted and dry, three-fourths of the bottles (a total of 9 bottles for the example used) should be filled with water until the total mass of the bottle, cap and water equals one kilogram. Each bottle and cap should be placed on a balance having a capacity of at least one kilogram, and water should be added to bring the total mass to one kilogram. You probably will find that the mass of the painted bottle and cap almost makes up for the amount by which the mass of one quart of water is less than one kilogram; the bottle probably will be almost full. Once the masses of the bottles have been "trimmed" by adding water to bring the mass of each to one kilogram (if you are a few grams off, it will do no harm), you may wish to permanently seal each bottle by using silicone cement to glue the tops to the bottles. Label 2/3 of the bottles "1 kilogram, Earth" and label 1/3 of the bottles "1 kilogram, Moon."

3

To one-fourth of the bottles add sand to bring the total mass of each bottle, cap and sand to 0.165 kg, or 165 grams (sand is used to prevent "sloshing" of the filler within the partially filled bottle, making the simulation of lifting the bottle in a moon environment more realistic). Since the mass of a 1-quart plastic bottle typically is 50 g, only a small amount, 115 or so grams, of sand will be needed. Each of these bottles should be sealed and labeled "1 kilogram, Moon"—this label, of course, will not indicate the true mass of each of these bottles, but is "faked" for the simulation of weighing 1 kilogram in a moon environment. The choice of the number 0.165 is based on the acceleration due to gravity on the moon being, to three significant figures, 0.165 times the acceleration due to gravity on Earth.

Teaching Notes

Preview the rules of the simulation of the moon condition at each station with students, pointing out that, in each case, they will be experiencing what would happen if the object was manipulated the same way on the moon as manipulated in the activity. If any students persist in suggesting that something must be faked, point to the final question which invites students to explain how the simulation was "rigged."

For You To Read section explains why, in terms of Newton's Universal Law of Gravitation, the ratio of the accelerations due to gravity on Earth and the moon is 6:1, and not some other ratio.

Point out that this section reveals that inertial effects in sports would be unchanged on the moon.

Activity Three
Mass, Weight, and Gravity

WHAT DO YOU THINK?

Newton's Laws of Motion apply on the moon as well as on Earth.

- If an object has a mass of 1 kilogram on Earth, what would be its mass on the moon?
- If a 1-kilogram object weighs about 10 newtons on Earth, what would be its weight on the moon?

Record your ideas about these questions in your *Active Physics log*. Be prepared to discuss your responses with your small group and the class.

FOR YOU TO DO

1. Your teacher has prepared a simulation that will allow you to compare the mass of an object on Earth to the mass that the same object would have on the moon. At the Mass Station you will find two bottles lying on a table, one labelled "1 kilogram, Earth" and another labelled "1 kilogram, moon." To keep the simulation accurate and realistic, follow the rules below:
   - Leave the bottles lying on their sides on the table; do not stand the bottles upright.

⚠ Use only plastic bottles.
   - You may move the bottles only by rolling them; do not lift the bottles.

2. Select the bottle labelled "1 kilogram, Earth" and use a push of your hand to start the bottle rolling freely across the table to a partner from your group. Your partner should use a push to stop the bottle and then push to send it rolling back to you. Play rolling-the-bottle-catch until you and your partner have the "feel" of the pushing force needed to accelerate and decelerate the "1 kilogram, Earth" bottle.

Activity Overview

In this activity students investigate the difference between inertial mass and weight.

Student Objectives

Students will:

- use semi-quantitative comparison of the responses of objects to applied forces as a way of comparing the masses of objects.

- use a spring balance to measure the weights of objects.

- understand and apply Newton's Universal Law of Gravitation to comparing the acceleration due to gravity on Earth and the moon.

ANSWERS FOR THE TEACHER ONLY

What Do You Think?

The mass on the moon would be unchanged, 1 kg.

The weight on the moon would be about 1/6 x 10 N = 1.6 N.

3

ANSWERS

For You To Do

1. - 2. Student activities.

3. Pretend that you and your partner are on the surface of the moon, standing across a table from one another. Select the bottle labelled "1 kilogram, moon." Play rolling catch to compare the amount of pushing force needed to accelerate and decelerate the "1 kilogram, moon" bottle to the amount needed to accelerate 1 kilogram when you were back on Earth.

   a) Based on your observations, how does the amount of force needed to accelerate a 1-kilogram mass on Earth compare to the amount of force needed to accelerate a 1-kilogram object by the same amount while standing on the moon? Is the amount of force needed to produce equal accelerations in the two locations significantly different? About the same? Write your observations in your log.

   b) Isaac Newton explained that an object's mass is a measure of its inertia, or, in other words, its natural tendency to resist being accelerated. Do you think Newton should have allowed that an object's tendency to resist being accelerated might depend on whether the object is on the Earth or the moon? Explain why or why not.

   c) Keeping in mind Newton's Second Law, $f = ma$, if equal forces applied to two objects produce equal accelerations of the objects, what else must be equal about the objects? Explain your answer.

4. Your teacher has prepared another simulation that will allow you to compare the weight of an object on Earth to the weight that the same object would have on the moon. At the Weight Station you will find two bottles resting upright on the floor, one labelled "1 kilogram, Earth" and another labelled "1 kilogram, moon." To keep the simulation accurate and realistic, do only what is listed below.

5. Grasp the string attached to the bottle labelled "1 kilogram, Earth" and lift the bottle vertically. Get the "feel" of the downward gravitational pull of the Earth on the bottle and then carefully lower it back to the floor to rest in upright position. Attach a spring scale calibrated in newtons to the string, lift the bottle and measure its weight. Lower the bottle to the floor.

   a) Record the weight of the "1 kilogram, Earth" bottle in your log.

SPORTS S 124

ANSWERS

For You To Do (continued)

3. a) Students should not be able to detect any differences in the two forces.

   b) At the Mass Station, students should find no difference in the inertial effects of the two bottles, one on Earth and one "on the moon" because there would be no difference in the inertial properties of a 1-kg object between the two locations.

   c) If equal forces are applied, and equal acceleration takes place, then according to Newton's Second Law, the masses must also be the same.

5. a) At the Weight Station, depending on the accuracy of the spring scale, students should find the weight of the "1-kg Earth" to be approximately 10 N.

6. Pretend you have been transported to the moon.
Repeat step 5 for the bottle labelled "1 kilogram, moon."

 a) Record the weight, in newtons, of the "1 kilogram, moon" bottle in your log.

 b) Divide the weight of 1 kilogram on Earth by the weight of 1 kilogram on the moon and round off the answer to the nearest integer. Show your work and record your answer in your log.

 c) If 2-kilogram masses instead of 1-kilogram masses had been used, what do you think would have been the individual weights on the Earth and moon? The ratio of the weights? Write your answers in your log.

 d) Why do you think the weights of equal masses, one on Earth and the other on the moon, are different? Be as specific as you can.

 e) To satisfy the above two simulations, it may have been necessary for your teacher to "fake" some of the bottles. Which, if any bottles, do you think were faked? Why and how? Explain your answer.

3

ANSWERS

For You To Do (continued)

6. a) The weight of the "1-kg Moon" bottle should be approximately 1.6 N.

 b) The ratio of the weights to the nearest integer should be 6:1.

 c) If 2-kg masses had been used, the individual weights would have doubled, but the ratio of the weights would have remained the same.

 d) The weights of equal masses on Earth and the moon are different because the attractive forces of Earth and the moon are different.

 e) Students should realize that the 1-kg Moon weight would have to be "rigged."

See Advance Preparation and Setup for further details.

SPORTS ON THE MOON

FOR YOU TO READ

Gravity on the Planets and the Moons

In Activity 2 you saw that the acceleration due to gravity on the moon is $\frac{1}{6}$ of the acceleration due to gravity on Earth. Therefore, you probably were not surprised when the simulation in this activity showed that the weight of an object on the moon is $\frac{1}{6}$ of the weight of the same object on Earth. Since, according to $f = ma$, the amount of acceleration of an object depends directly on the amount of applied force, the reduced rate of free fall acceleration on the moon must be caused by a gravitational pull which is smaller on the moon than on Earth. On the moon, both the free fall acceleration and the force causing the acceleration are $\frac{1}{6}$ of the amounts on Earth, regardless of what object is compared at both locations. But why $\frac{1}{6}$, and not some other number? Isaac Newton answered that question, too.

Newton reasoned that any massive object— a star, planet, moon, comet, or even a speck of dust—pulls other objects toward it with a force called gravity. Newton explained that the free fall acceleration of a small object—such as a hammer—dropped near the surface of a huge object—such as a planet or moon—depends on two factors: the mass of the planet and the square of the radius of the planet.

Newton further explained that the mathematical ways in which the two factors affect the acceleration due to gravity near the surface of the huge object are:

- g_{huge} is directly proportional to m_{huge}, or, in symbols, $g_{huge} \propto m_{huge}$

- g_{huge} is inversely proportional to the square of R_{huge}, or, in symbols, $g_{huge} \propto \dfrac{1}{(R_{huge})^2}$

Combining the above statements of proportionality into one statement:

$$g_{huge} \propto \frac{m_{huge}}{(R_{huge})^2}$$

Since the mass of the moon is known to be only about $\frac{1}{100}$ of Earth's mass, g_{moon} should be expected to be $\frac{1}{100}$ of g_{Earth}. However, the fact that the moon's radius is only about $\frac{1}{4}$ of Earth's radius also must be considered; the moon's smaller radius suggests that g_{moon} should be $\dfrac{1}{\left(\frac{1}{4}\right)^2} = \dfrac{1}{\left(\frac{1}{16}\right)} = 16$ times greater than g_{Earth}.

Combining the effects of differences in mass and radius:

$$g_{moon} = g_{Earth} \times \left(\frac{16}{100}\right) = \frac{g_{Earth}}{6}$$

It works! Isaac Newton's explanation, made over 300 years before anyone went to the moon, also relates to the moon and can be verfied. When an astronaut standing on the moon drops a hammer, the free fall acceleration is, according to Newton's prediction, $\frac{1}{6}$ of the free fall acceleration due to gravity on Earth.

On the moon, an athlete would find it easier to lift a shot ball from his equipment bag because it would weigh only $\frac{1}{6}$ as much as on Earth. When twirling and extending his throwing hand to accelerate the ball prior to launch, the athlete would find that the same amount of force is needed as "back home" on Earth; it's the same mass, and the force needed to accelerate it by the same amount as on Earth hasn't changed. However, the athlete would be thrilled at the result. For an amount of muscular work done by the athlete equal to a shot put effort on Earth, the shot would fly six times further.

ACTIVITY THREE: Mass, Weight, and Gravity

REFLECTING ON THE ACTIVITY AND THE CHALLENGE

The fact that the weight of objects is different on the moon, but that the mass and inertial properties of objects remain unchanged on the moon has great implications for sports on the moon. Many sports involve lifting objects against the force of gravity and placing objects in a condition of free fall; these aspects of sports will be different in the "$\frac{1}{6}g$" condition on the moon. Many sports involve applying forces to accelerate objects; these aspects of sports will be no different on the moon than on Earth. In fact, many sports involve combinations of actions, some of which may be different on the moon than on Earth, and some of which may be the same. It will be necessary for you to consider what parts of the sport you choose to play on the moon will be affected by reduced gravity and what parts will not be affected. Remember, lifting is six times easier on the moon; pushing is just as difficult as it is on Earth.

PHYSICS TO GO

1. How would the sport of weight lifting be affected on the moon? If a person can "press" 220 pounds on Earth, what weight could that person press on the moon?

2. How would a baseball player's ability to swing (accelerate) a bat be affected on the moon? Assume that a space suit does not inhibit the batter's movement.

3. For equal swings of the bat and equal speeds of the ball on Earth and the moon, how would the speed of a baseball at the instant it loses contact with the bat compare on Earth and the moon?

4. Imagine an outdoor game of baseball at "Lunar Stadium."
   a) If, typically, the center field wall is 400 feet from home plate at a baseball park on Earth, how far from home plate should the wall be located at Lunar Stadium? Why?
   b) Should the pitcher's mound and the bases at Lunar Stadium be the located at the same or different distances from home plate as on Earth? How fast would a major league pitcher capable of a 100 mph "Earth pitch" be able to throw the ball on the moon? How fast would players be able to run the bases on the moon? Why?

SPORTS

Physics To Go

1. 220 pounds x 6 = 1,320 pounds.

2. A bat's inertia, or resistance to being accelerated would be unchanged on the moon; therefore, a batter's ability to swing a bat would be unchanged on the moon.

3. A baseball's inertia would be equal on the moon and Earth; therefore, a ball's response to being hit by a bat would be the same on the moon as on Earth. An equally good explanation is that the Law of Conservation of Momentum applies equally on the moon and Earth.

4. a) 400 ft x 6 = 2,400 ft

   b) A baseball pitcher would be able to throw a ball at the same speed on the moon and on Earth because the "launch speed" depends on the inertia of the ball which is the same in both locations; without air resistance on the moon, the ball would maintain the horizontal component of its speed and a "curve ball," which depends on the effect of air on a spinning ball, would not be possible on the moon. A significant change in the distance from the pitcher's mound to home plate would not be called for. Players would not be able to run as usual on the moon (as will be explained in detail in Activity Eight later in this chapter); therefore, research is needed to find out how fast players would be able to move from base to base in order to establish how to adjust the distance between bases.

3

c) Fans in the center field bleachers would be nearly 1/2 mile
from the infield — one mile is
5,280 feet.
A fly ball would rise 6 times
higher on the moon than on
Earth; a high fly ball would
rise nearly out of sight.
Outdoor baseball would
require space helmets for spectators; eating a hot dog would
be difficult.
Because there is no air on the
moon, sound is not transmit-
ted; astronauts on the moon's
surface used radio communication.

5. On the moon, 1 liter of water
would have a mass of 1 kg and a
weight of 1.6 N.

6. a) The amount of force needed to
overcome friction between the
disc and the court would be
1/6 the amount on Earth, but
the force required to accelerate
the disc to the same speed as
on Earth would not change.

b) If the speed of the disc upon
leaving the cue stick were the
same on the moon as on
Earth, the disc would decelerate at 1/6 the rate as on Earth
because the force of friction,
assuming the same surfaces in
both locations, would be 1/6
the amount on the moon.

c) In the same way as on Earth,
the moving disc would stop
and the stationary disc would
move away from the collision
with a speed equal to the
incoming speed of the
moving disc.

7. The same amount of time.

SPORTS ON THE MOON

c) What problems might fans in the center field bleachers have:
• seeing a player slide into second base or home plate?
• watching a high fly ball?
• eating a hot dog?
• shouting at the umpire?

5. Water would be very expensive on the moon because there isn't any there. If you purchased a precious 1-L bottle of "Genuine Earth Water" at the Lunar Mall, what would be its mass, in kilograms? What would be its weight, in newtons? (Hint: On Earth, one liter of water has, by definition, a mass of one kilogram.)

6. How would the game of shuffleboard be different on the moon compared to on Earth? Be sure to compare stages of shuffleboard when:
• a player is pushing a cue stick to accelerate a shuffleboard disc to send the disc sliding down the court.
• a disc is sliding on the surface of the shuffleboard court before entering the scoring area.
• a disc slides into the scoring area and collides head-on with a stationary disc to knock it out of the scoring area.

7. If you were to buy one kilogram of potatoes on the moon, how long would it last compared to one kilogram of potatoes purchased on Earth?

S 128

Assessment Rubric for Mass, Weight, and Gravity

Formative report:

Student name: _____

Date:_____

Place a check mark (√) in the appropriate box.

Descriptor	Concept attained	Concept requires further work
1. Student is able to differentiate between weight and mass.		
2. Student explains that objects that are thrown on the moon will travel 6x further.		
3. Student explains that objects that are thrown on the moon will travel at the same speed if the same force is applied.		
4. Student explains that although lifting is 6x easier on the moon, pushing an object requires the same effort.		
5. Student explains that the smaller mass of the moon accounts for the reduced gravitational force.		

3

For use with *Sports*, Chapter 3, ACTIVITY THREE: Mass, Weight, and Gravity

©1999 American Association of Physics Teachers

ACTIVITY FOUR
Projectile Motion on the Moon

Background Information

In this activity, the mathematical and physical models of projectile motion developed by students in Chapter 1, Activity Nine are adapted to the 1/6 g environment of the moon. You may wish to review the student text and the Background Information for Chapter 1, Activity Nine before proceeding in this section.

All of the equations for projectile motion presented in the Background Information for Chapter 1, Activity Nine may be applied to comparing quantities on the moon to corresponding quantities on Earth by substituting $g/6$ for g in the equations as shown below. In each equation below, g represents the acceleration due to gravity on Earth.

For a projectile launched horizontally at speed v from height h on the moon:

the time of flight is $t = \sqrt{(2h/(g/6))} = \sqrt{(6)2h/g}$ $= (\sqrt{6})\sqrt{2h/g} = (2.45)\sqrt{2h/g}$

the range (horizontal distance) is $R = vt = v(\sqrt{6})\sqrt{2h/g} = v(2.45)\sqrt{2h/g}$

Compared to on Earth, both the time of flight and the range of a projectile launched horizontally are increased by a factor of $\sqrt{6}=2.45$ on the moon.

For the general case of a projectile launched from ground level at speed v at an angle q above the horizontal, and traveling over flat ground on the moon (again, in each equation g represents the acceleration due to gravity on Earth):

the speed in the horizontal direction is $v_x = v \cos\theta$ (remains constant)

the initial speed in the vertical direction is $v_{yo} = v \sin\theta$

the speed in the vertical direction at time t is $v_y = (v \sin\theta) - (g/6)t$

the horizontal position at time t is $x = (v \cos\theta)\, t$

the vertical position at time t is $y = (v \sin\theta)t - (1/2)(g/6)t^2$

the time to reach maximum height is $t_{max} = (v \sin\theta)/(g/6) = 6\,(v \sin\theta)/g$

the total time of flight is: $t = 2t_{max}$ (see t_{max}, above)

the range (horizontal distance) is: $R = (v^2 \sin2\theta)/(g/6) = 6\,(v^2 \sin2\theta)/g$

Compared to on Earth, both the range and time of flight of a projectile launched at an angle θ are increased by a factor of 6 on the moon. Since the time for which the projectile rises is increased by a factor of 6, it follows that the maximum height of the projectile also is increased by a factor of 6.

Active-ating the Physics InfoMall

You may wish to search the InfoMall for information on projectile motion in general first, before restricting it to the moon. This is another of the big concepts in physics, and you can either perform a search, or simply browse the textbooks. However you choose, there is abundant information on the CD-ROM discussing projectiles.

In addition to the information found in the search at the end of Activity Three, you may also wish to check one of the hits from an earlier search. When we searched for "sport*", one of the hits was from *Physics in the Real World* (in the Textbook Trove). Chapter 5 is on Some Sporting Controversies. (Note that the title of this book suggests it might be a great reference for many things!) One of the controversies discussed is the "magic angle" of 45 degrees. This angle is the best angle only under certain circumstances, and one must be careful to understand that. For example, air resistance makes a difference. But perhaps one of the most important things to know is that this equation angle is best only when the launch height and landing height are identical. Since many sports involve launching something from shoulder height, and the object lands on the ground, 45 degrees is not the best angle. In addition, people do not throw with the same force at different angles. Check this text for more information. Check any of the textbooks for information on how the "range equation" is found, and what assumptions must be made.

Planning for the Activity

Time Requirements

- One class period.

Materials Needed

For each group:

- protractor
- calculator

For each student:

- sheet paper, 8-1/2" x 11"
- sheet paper, 18" x 36"
- meter stick
- tape

Advance Preparation and Setup

A roll of wrapping paper 18" wide will be needed to provide each student a piece 18" x 36". A ready source is freezer paper in 18"-wide rolls available at supermarkets. If necessary, students can tape 6 pieces of 8-1/2" x 11" sheets of paper into a 2 x 3 array as a substitute.

Teaching Notes

The point intended to be made in this activity is that the great ranges which projectiles in sports would have on the moon would be a serious limiting factor for what sports could be transferred without adjustment to the moon.

3

NOTES

Activity Four
Projectile Motion on the Moon

WHAT DO YOU THINK?

A baseball has $\frac{1}{6}$ the weight on the moon as on Earth, but a baseball's mass is equal on Earth and the moon.

• **Can a batter hit or a player throw a baseball <u>faster</u> on the moon than on Earth?**

• **Can a batter hit or a player throw a baseball <u>farther</u> on the moon than on Earth?**

Record your ideas about these questions in your *Active Physics log*. Be prepared to discuss your responses with your small group and the class.

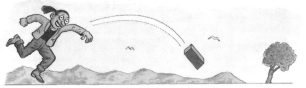

S 129   ———— SPORTS

Activity Overview

In this activity students use a model to investigate the differences and similarities between projectile motion on the moon and on Earth.

Student Objectives

Students will:

• understand and apply the acceleration due to gravity on the moon to projectile motion on the moon.

• create a mathematical model and a physical model of the trajectory of a projectile on the moon.

ANSWERS FOR THE TEACHER ONLY

What Do You Think?

No, since the inertial properties of a bat and ball are the same on the moon as on Earth, a player cannot throw or hit a ball faster on the moon than on Earth.

Yes, since g on the moon is 1/6 g on Earth, a ball can be thrown 6 times farther on the moon than on Earth.

3

ANSWERS

For You To Do

1. d) The "Earth" fall distances, in meters, for the portable physical model developed in Chapter 1, Activity Nine are listed below with the "Moon" fall distances, in each case 1/6 of the Earth fall distance, shown in parenthesis.

Earth fall distance (moon fall distance), in meters: 0.050 (0.0083), 0.20 (0.033), 0.45 (0.075), 0.80 (0.13), 1.25 (0.21), 1.80 (0.30).

For the 1/10 scale drawing produced in this activity, each of the above fall distances must be divided by 10. Dividing each distance by 10 and changing the unit of distance from meters to millimeters (multiplying by 1,000), the fall distances on the scale drawing, in millimeters should be: 5.0 (0.83), 20 (3.3), 45 (7.5), 80 (13), 125 (21), 180 (30).

SPORTS ON THE MOON

FOR YOU TO DO

1. Use the instructions listed below to produce a $\frac{1}{10}$ scale drawing—that is, a drawing $\frac{1}{10}$ of the actual size—of a trajectory model of a projectile (the path an object you throw will take) launched at a speed of 4.0 m/s. Work with members of your group.

a) On an $8\frac{1}{2}" \times 11"$ sheet of paper in "landscape" orientation as shown reduced in size below, mark a starting point 2 cm above and 2 cm to the right of the bottom-left corner of the paper. From the starting point, draw two straight lines entirely across the sheet, one horizontally and another inclined at an angle of 30°. Add the title shown in the sketch.

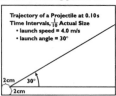

Trajectory of a Projectile at 0.10 s
Time Intervals, $\frac{1}{16}$ Actual Size
• launch speed = 4.0 m/s
• launch angle = 30°

2cm   30°
    2cm

b) The horizontal line represents ground level, and the inclined line represents the path that a projectile launched from the starting point at a 30° angle would follow if there were no gravity. Measuring from the starting point, mark points at 4.0-cm intervals on the inclined line (4.0 cm is $\frac{1}{10}$ of the actual distance, 40 cm, that the projectile would travel in 0.10 s). The marked points represent the position of the projectile every 0.10 s for a zero gravity condition. Beginning by labeling the starting point as 0.00 s, label successive points on the inclined line as 0.10 s, 0.20 s, 0.30 s and so on.

c) Also mark points at 4.0-cm intervals on the horizontal line. Beginning by also labeling the starting point as 0.00 m, mark successive points on the horizontal line as 0.40 m, 0.80 m, 1.20 m and so on. These points represent distance along the ground, reduced, of course, by a factor of 10 from real world distances.

d) Use the equation $d = \frac{1}{2}at^2$ (where a = 980 cm/s²) to calculate the total distance an object on the Earth would fall, starting from rest, in 0.10 s, 0.20 s...0.60 s. To fit the $\frac{1}{10}$ scale of the drawing, first divide each fall distance by 10.

SPORTS

 S 130

ACTIVITY FOUR: Projectile Motion on the Moon

Next, draw a line vertically downward from each marked point on the inclined line to show the projectile's position at that time. For example, the line at point labelled 0.10 s should extend 5 cm ÷ 10 = 0.50 cm (or 5.0 mm) downward from the inclined line.

e) The bottom ends of the vertical fall lines represent the projectile's position at 0.10 s intervals during its flight. Connect the bottom ends of the lines with a smooth curve to show the shape of the trajectory and label the curve "Trajectory on Earth."

f) Use the distance scale established on the horizontal line to measure, to the nearest 0.10 m, the projectile's "real world" maximum height above ground level and the horiontal range of the projectile before striking the ground. Record the maximum height and range on the drawing.

g) Use the time scale established on the inclined line to measure, to the nearest 0.010 s, the projectile's time of flight. Record the time of flight on the drawing.

2. You will now draw the trajectory that would result if the projectile were launched at the same speed and in the same direction on the moon.

a) Use the equation, $d = \frac{1}{2}at^2$, to calculate the total distance an object on the moon would fall, starting from rest, in 0.10 s, 0.20 s, 0.30 s, and so on. The value of acceleration to use in the equation is the acceleration due to gravity on the moon, 1.6 m/s^2, or 160 cm/s^2. Prepare a table in your log to show the calculated value of the total distance of fall at the end of each 0.10 s of flight on the moon. Notice that this projectile launched on the moon will not even have reached maximum height when the projectile launched with the same velocity on Earth hits the ground. Therefore, it will be necessary to extend your calculations for the projectile launched on the moon.

b) Divide each distance of fall by 10, and draw a vertical line downward from each marked point on the inclined line to show the projectile's position at that time on the moon. For example, the line at point labelled 0.30 s should extend 0.72 cm, or 7.2 mm, downward from the inclined line. This line, and others, will need to be drawn on top of the lines drawn earlier for fall distances on Earth.

 131

SPORTS

Projectile Motion on the Moon

Time of fall (sec)	Distance (m)	Scaled distance (cm)
0.1	0.01	0.1
0.2	0.03	0.3
0.3	0.08	0.8
0.4	0.14	1.4
0.5	0.21	2.1
0.6	0.3	3
0.7	0.41	4.1
0.8	0.54	5.4
0.9	0.67	6.7
1.0	0.84	8.4
1.1	1.01	10.1
1.2	1.2	12
1.3	1.42	14.2
1.4	1.64	16.4
1.5	1.68	18.8
1.6	2.14	21.4
1.7	2.41	24.1
1.8	2.72	27.2
1.9	3.02	30.2
2.0	3.34	33.4
2.1	3.7	37
2.2	4.05	40.5
2.3	4.42	44.2
2.4	4.82	46.2

ANSWERS

For You To Do

(continued)

f) The scale drawing should show that the projectile on Earth would have a range (cross the horizontal line at the level of launch) at a scale (on paper) distance of 13.8 cm (about 14 cm), corresponding to a real world distance of about 1.4 m. The maximum height would be a scale distance of about 2.5 cm, about 0.25 m in the real world.

g) The time of flight would be about 0.40 s.
See sample scale drawing following this activity in the Teacher's Edition.
The scale drawing should show that the projectile on the moon would have a range at a scale distance of about 84 cm, corresponding to a real-world distance of about 8.4 m. The maximum height would have a scale distance of about 150 cm, about 1.5 m in the real world. The time of flight would be about 2.4 s.
See the table for Projectile Motion on the Moon.

3

ANSWERS

For You To Do

(continued)

3. a-c) Students should find that all the ratios are 6:1.

d) The moon's gravity increases the range, maximum height, and time of flight by a factor of 6 over that on Earth.

SPORTS ON THE MOON

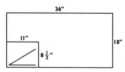

Extend the size of the paper to be able to show the entire trajectory. As shown in the sketch, tape the sheet of paper containing your drawing to the lower left-hand corner of a sheet of wrapping paper approximately 18 inches high and 36 inches wide.

✎ c) The bottom ends of the vertical fall lines represent the projectile's position at 0.10 s intervals during its flight on the moon. Connect the bottom ends of the lines with a smooth curve to show the shape of the trajectory and label the curve "Trajectory on the moon."

✎ d) Use the distance and time scales on the drawing to measure the projectile's maximum height, range, and time of flight on the moon. Record the values on the drawing. Fold and save your drawing.

3. Use the above measurements of the maximum heights, ranges, and times of flight of a projectile launched with equal initial velocities on Earth and the moon to complete the calculations below. Show your work in your log.

✎ a) $\dfrac{\text{(Max. height of projectile on moon)}}{\text{(Max. height of projectile on Earth)}}$

✎ b) $\dfrac{\text{(Range of projectile on moon)}}{\text{(Range of projectile on Earth)}}$

✎ c) $\dfrac{\text{(Time of flight of projectile on moon)}}{\text{(Time of flight of projectile on Earth)}}$

✎ d) Write a summary in your log of the effects of the moon's "$\frac{1}{6} g$" on the maximum height, range, and time of flight of a projectile launched on the moon compared to a projectile launched at the same speed and angle of elevation on Earth.

REFLECTING ON THE ACTIVITY AND THE CHALLENGE

This activity clearly demonstrates that some sports may be, quite literally, "out of sight" on the moon. For example, a 300-yard golf drive on Earth should translate into an 1,800-yard drive on the moon; that's over a mile (1,760 yards)! Could a golf ball be found after such a drive? Probably not.

Adapting "Earth sports" to the moon isn't as simple as it seems at first glance. A proposal to play golf on the moon with no adjustments would, without doubt, "not fly" with NASA. In another example, consider a baseball hit to the outfield in a moon stadium. It may take so long for the ball to fall that everyone would be bored in the middle of the play. Any sport involving projectile motion will need careful analysis to see if it is feasible to be used on the moon.

PHYSICS TO GO

1. What adjustments due to the increased time and distance of travel of a projectile may be needed to play each of the following sports on the moon?

   a) football   b) gymnastics
   c) trapeze   d) baseball

2. A typical sports arena on Earth has a playing field 120 m long and 100 m wide surrounded by tiered seats for spectators. World class shot put athletes throw the steel shot 23 m. Would spectators be safe if a shot put event were held in a stadium of similar size on the moon? Explain why or why not.

3. The maximum range of a projectile occurs when the launch angle is 45°. Physicists have shown that the range of a projectile launched at 45° is given by the equation $R = \frac{v^2}{g}$, where R is the range, v is the launch speed and g is the acceleration due to gravity on the planet or moon where the projectile is launched. How would this equation be useful for estimating the size of facilities needed for sports on the moon?

4. If a golf ball were hit at a speed of 40 m/s at a launch angle of 45° on the moon, what would be its range? (Hint: Use the equation from the previous question.)

5. Since the moon's gravity is weak and since projectiles near the moon do not experience air resistance, does it seem possible that an object could be thrown straight upward from the surface of the moon and "escape" the moon's gravity, never to fall back down to the moon? Write a brief statement about your thoughts on this possibility.

S 133

Physics To Go

1. Students should somehow recognize the need to limit the range of projectiles involved in sports in order to transfer sports involving projectiles to the moon.

2. No. Many spectators would be within range of, or be able to be hit by, a shot put ball having a range of 23 m x 6 = 138 m in a stadium on the moon.

3. Adapted to the moon by multiplying the maximum range by six, the equation would be very useful because it could be used to predict the maximum range of projectiles used in sports on the moon; the facilities would need to be able to fit the maximum range.

4. $R = (6) (40 \text{ m/s})2/(10 \text{ m/s}^2) = 960$ m

5. Students should be able to reason that an "escape speed" exists for any planet or moon. For your information as the teacher, each planet or moon has an "escape speed" equal to the square root of the quantity [2 x (acceleration due to gravity) x (radius of planet or moon)]. For Earth, the escape velocity is about 11 km/s (25,000 mi./h); for the moon the escape velocity is less by a factor of about 3.3 km/s.

3

NOTES

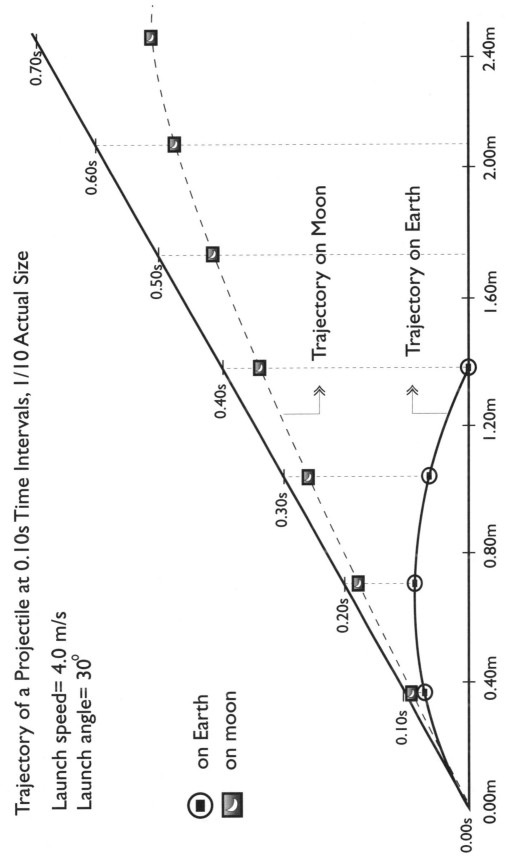

Trajectory of a Projectile at 0.10s Time Intervals, 1/10 Actual Size

Launch speed= 4.0 m/s
Launch angle= 30°

on Earth
on moon

Trajectory on Moon

Trajectory on Earth

0.00s   0.10s   0.20s   0.30s   0.40s   0.50s   0.60s   0.70s

0.00m   0.40m   0.80m   1.20m   1.60m   2.00m   2.40m

For use with *Sports*, Chapter 3, ACTIVITY FOUR: Projectile Motion on the Moon

©1999 American Association of Physics Teachers

ACTIVITY FIVE
Jumping on the Moon

Background Information

In this activity, the analysis of a vertical jump from Chapter 2, Activity Four is applied to a vertical jump on the moon with the surprising result that humans would jump more than six times higher on the moon than on Earth.

You may find it surprising that the predicted height of a vertical jump by a person standing on the moon would not simply be, due to the 1/6 g moon environment, six times greater than a jump executed in the same way by the same person on Earth. It is true that a ball thrown straight upward with equal launch speeds on Earth and the moon would fly six times higher on the moon than on Earth, so why would a human body also not fly six times higher on the moon than on Earth? The answer to the dilemma is that human body would fly six times higher on the moon than on Earth if the launch speeds were equal at both locations, but that would not be the case if equal jumping efforts were made on Earth and on the moon.

As will be shown below, the Law of Conservation of Energy shows that if a person were to replicate the same pre-launch technique used for a vertical jump on Earth to a jump from the surface of the moon—that is, at both locations the person applies the same average muscular force to push the feet against the ground while rising the same distance from crouched to launch positions—the person's speed at the instant of launch would be significantly higher on the moon than on Earth. The person's mass also is assumed to be equal at both locations; for direct comparison of gravitational effects on jumping height, it is assumed for this discussion that the person jumping on the moon is inside an air-filled facility which does not require wearing a 100-kg space suit/life-support system which complicates the comparison by introducing mass and limitations on movement as variables. The enhanced launch speed of a jumper on the moon would cause the person's body to fly upward significantly higher than the "six times Earth" height for equal launch speeds on the moon and Earth. Conservation of energy can be used to determine launch speeds for vertical jumps on Earth and the moon:

(KE at launch) = (Work done during push phase) − (Gain in PE of C of M during push phase)

$$mv^2/2 = Fh - mgh = h(F - mg)$$

Rearranging the above equation yields the predicted launch speed on Earth:

$$v_{Earth} = \sqrt{(2h/m)(F - mg)}$$

The launch speed on the moon is predicted by replacing the jumpers weight in the above equation, mg, by $(mg/6)$:

$$v_{Moon} = \sqrt{(2h/m)[F - (mg/6)]}$$

Comparing the above two equations, v_{Moon} for a particular jumper is greater than v_{Earth} because the term $[F - (mg/6)]$ in the equation for v_{Moon} is greater than the term $(F - mg)$ in the equation for v_{Earth}.

Conservation of energy is used below to arrive at a general equation to predict a person's vertical jump height (the vertical distance that the person's center of mass rises from the instant of launch to the peak of flight), H, in terms of the average force, F, exerted during the push phase, the vertical distance, h, that the person's center of mass is lifted during the push phase, the person's mass, m, and the acceleration due to gravity, g. The person's weight is, of course, represented as mg. Other symbols used are C of M to represent center of mass, KE to represent kinetic energy, and PE to represent gravitational potential energy.

(Work done during push phase) = (Gain in PE of C of M during push phase) + (Gain in PE of C of M during flight phase)

$$Fh = mgh + mgH = mg(h + H)$$
$$Fh/mg = H + h$$
$$H = (Fh/mg) - h = h\,[(F/mg) - 1]$$

Substituting data from the Example Analysis in Chapter 2, Activity Four of the student text as a consistency check of the above equation:

$$H = h\,[(F/mg) - 1]$$
$$= 0.35\ m\ [(1{,}200\ N)/(44\ kg)(10m/s^2) -1]$$
$$= 0.60\ m$$

The equation provides the same result as the Example Analysis.

As presented in the student text for this activity, the equation predicts a jump height 9.3 times higher when data for the same person used in the Example Analysis is used in the 1/6 moon environment:

$$H = h\,[(F/mg) - 1] = 0.35\ m\ [(1{,}200\ N)/(44\ kg)(1.6\ m/s^2) -1] = 5.6\ m$$

Is this a "universal" answer? Would everyone jump 9.3 times higher on the moon than on Earth? No, the height to which a person would be able to jump on the moon compared to on Earth depends not only on the different effects of gravity, but, in addition, three factors which may vary among individuals. As may be seen in the equation $H = h [(F/mg) - 1]$, the jump height, H, depends not only on the "local" value of g ("g" on Earth, "g/6" on the moon), but also on the vertical distance that the person's center of mass is lifted during the push phase, h, the average force exerted during the push phase, F, and the person's weight, mg.

The equation for jump height is somewhat simpler if the quantity (F/mg) is simplified. Since F represents the average force exerted during the push phase and since mg represents the person's weight, it makes sense to define $R = (F/mg)$ as a symbol to represent the "force-to-weight ratio" for an individual jumper. R, a jumper's force-to-weight ratio, is a meaningful attribute because it describes how much pushing force a jumper can generate compared to his or her body weight; the higher a jumper's R-value, the better the jumper. Substituting R for (F/mg), the equation becomes:

$$H = h(R - 1)$$

Training directed toward muscles of the lower extremities no doubt can improve a jumper's R-value. An individual jumper's "h" seems essentially fixed because it depends on the lengths and angles of leg parts involved as the jumper rises from crouch to takeoff positions, although it may be possible for an individual to train to crouch lower before pushing off while still maintaining high force and acceleration during push-off. But how does this explain lack of a "universal multiplier" for jumps on the moon compared to on Earth? For that, more probing at the above equation is needed. First, keeping in mind that R in the above equation represents the ratio of the average force that the jumper exerts during the push phase to the jumper's weight on Earth, let us specify that the below equation is appropriate for predicting the height of a jump on Earth, H_{Earth}:

$$H_{Earth} = h(R - 1)$$

To adapt the above equation to predict the height of a jump on the moon, H_{Moon}, it is necessary to substitute $g/6$ for g in the quantity (F/mg) which R has been defined to represent:

$$F/[m(g/6)] = 6F/mg = 6R$$

Therefore:
$$H_{Moon} = h(6R - 1)$$

In the above equation, R has the same value for an individual jumper as on Earth, and the equation is adjusted for the $1/6$ g moon environment.

Now it is possible to see why the enhancement of jump height on the moon would vary among individuals. The difference between a person's jump heights on the moon and Earth would be:

$$\Delta H = H_{Moon} - H_{Earth} = [h(6R - 1)] - [h(R - 1)]$$
$$= h[(6R - 1) - (R - 1)]$$
$$= h(6R - 1 - R + 1)$$
$$\Delta H = 5hR$$

By the above equation, it is clear that the best jumpers on Earth, those who have the highest product of their h and R values, would add the most to their jump heights on the moon. Naturally, they, too, would be the champion jumpers on the moon. But, compared to on Earth, would the champion jumpers on the moon be as much better in all ways than average jumpers on the moon? What about the ratio of jump height on the moon to jump height on Earth?

$$\frac{H_{Moon}}{H_{Earth}} = \frac{h(6R - 1)}{h(R - 1)} = \frac{(6R - 1)}{(R - 1)} = \frac{6 - (1/R)}{1 - (1/R)}$$

The above ratio shows that anyone who can jump at all on Earth (the minimum requirement for being able to jump on Earth is to possess an R-value greater than 1) would be able to jump more than six times higher on the moon. If R were to approach a very large number—not likely, remembering what R represents—the limiting value of the ratio would approach six. Therefore, any jumper would fly more than six times higher on the moon than on Earth. Furthermore, average jumpers on the moon at least would be able to tell the champions, "Well, you can outjump me both here and on Earth, but when I jumped on the moon, I multiplied my Earth jump height more than you did." To see that this is true, show that an excellent jumper having an R-value of 3 would have an H_{Moon}/H_{Earth} ratio equal to 8.5 compared to a poorer jumper's H_{Moon}/H_{Earth} ratio equal to 11 based on an R-value of 2.

Active-ating the Physics InfoMall

As surprising as it might seem, you can find articles on jumping on the InfoMall. For example, the search we have examined several times already (using "sport*") contains the article "Notes on Jumping," from the *American Journal of Physics*, vol.

25, issue 9, 1957. This article is not the only one. Perform a search using "jumping" and you will find "Physics and the vertical jump," in the *American Journal of Physics*, vol. 38, issue 7, 1970 among others.

Search using "jumping" AND "moon" (select "within paragraph") to find an alternative approach to the one used in *Active Physics*. Look at the *Physics Including Human Applications* textbook.

Make a slight change to "jump*" AND "moon" (select "within paragraph"), and the first selection on the hit list is also another approach to the same problem. (This hit is from *Physics: The Excitement of Discovery*.)

Note that the concept involved is Conservation of Energy (and Work), so you can always look in the textbooks for similar analysis.

Planning for the Activity

Time Requirements

- One class period.

Materials Needed

For each student:

- calculator
- paper strip, vertically mounted on wall
- pencil or pen

Advance Preparation and Setup

No significant advanced preparation is needed.

Teaching Notes

You may wish to "walk the class through" the calculations for the example given in this section which uses data from the Example Analysis in Chapter 2, Activity Four to predict the height of a vertical jump on the moon.

You may wish to adapt the Additional Materials provided for Chapter 2, Activity Four, Calculating Hang Time and Force during a Vertical Jump Worksheet, p. 170, and Activity Four A, p. 169 to this activity.

It is suggested that you make yourself available to assist students with their calculations during the activity.

As students arrive at their predicted jump heights on the moon, encourage them to check their personal ratio of moon-to-Earth jump heights to see if they are more than 6. Ask why. (You may wish to check the Background Information for your own understanding.)

To predict maximum jump heights which might be expected on the moon, perhaps suggest to students that it might be wise to find out how high the best jumpers would fly on the moon — how high would the ceiling need to be to allow a Michael Jordan to jump?

NOTES

3

Activity Overview

In this activity students analyze jumping on the Earth and use their results to predict what would happen to a person if they jumped on the moon.

Student Objectives

Students will:

- measure changes in height during a vertical jump.

- calculate changes in the gravitational potential energy during a vertical jump.

- apply conservation of energy to analysis of a vertical jump, including weight, force, height, and time of flight.

- make predictions about jumping on the moon using information gained from analyzing jumping on Earth.

ANSWERS FOR THE TEACHER ONLY

What Do You Think?

The answer varies with the individual and is not simply six times higher.

Expect students to predict that they will probably jump about six times higher on the moon. Encourage them to record their predictions in their log, so they are able to confront their misconception after their analysis of the data.

Activity Five
Jumping on the Moon

WHAT DO YOU THINK?

Michael Jordan has a hang time of two seconds on Earth.

- **What would be a typical NBA star's "hang time" on the moon?**
- **How high could you jump on the moon?**

Record your ideas about these questions in your *Active Physics log*. Be prepared to discuss your responses with your small group and the class.

FOR YOU TO DO

1. In an area free of obstructions crouch and jump straight up as high as you can. Next, crouch in the same way, as if you are ready to jump, and then rise without jumping. Discuss how to answer the following questions within your group and be prepared to share your group's responses with the class.

ACTIVITY FIVE: Jumping on the Moon

a) What is the source of the energy used to push your body upward in each of the above cases?

b) Why does your body leave the floor in one case but not in the other?

2. Stand with your shoulder near a wall on which a vertical strip of paper has been mounted. Hold a marker in your hand nearest the wall. Crouch in a deep knee bend as if you are ready to jump straight up as high as you can. While in this "ready" position, raise your arm holding the marker straight upward and make a mark on the paper strip.

a) Measure the distance from the floor to the mark on the wall to the nearest 0.10 m and record the measurement in your log as the ready distance.

3. Rise to your tiptoes as if you are ready to leave the floor in a vertical jump. While in this "launch" position, raise your arm straight up and make another mark on the paper.

a) Measure and record the distance from the floor to the mark in your log as the launch distance.

4. Crouch to the "ready" position and jump straight up as high as you can. Raise your arm straight up and make a mark when you are at the "peak," or highest, position of your jump.

a) Measure and record the distance from the floor to the mark in your log as the peak distance.

b) By subtraction, calculate and record in your log how high you can jump. Use the equation:
Jump height = (Peak distance) – (Launch distance)

5. Use the example presented in Physics Talk to analyze your vertical jump. Replace the data used in the example with your personal data. You will need to know your body mass in kilograms and your body weight in newtons for the analysis.

a) Show your calculations and answers in your log.

b) Predict how high you would be able to jump on the moon by using your personal data.

SPORTS

For You To Do

1. a) The energy is transformed by the leg muscles. The stored energy in the body cells is converted to provide the energy to push the body upwards.

b) The legs apply a downward force on the ground, and the ground provides the upward force to launch the body into the air.

2.– 4. Data for the jump will vary with each student.

5. a) Students will use the following equation to find their body weight in newtons.
Weight in newtons = Weight in pounds x 4.38 N/lb.
The body mass in kilograms may be calculated using the following equation.
Weight = mg
m (kg) = weight (kg.m/s^2) g(m/s^2)

b) Calculations will depend on the students' personal data.

SPORTS ON THE MOON

PHYSICS TALK

Analysis of Jumping on Earth

This analysis uses sample data for a 100-pound person who performed a vertical jump:

- **Body mass = 100 lb. × 1 kg/2.2 lb. = 44 kg**
- **Body weight = mg = 44 kg × 9.8 m/s² = 430 N**
- **Ready distance (from floor to ready mark) = 1.70 m**
- **Launch distance (from floor to launch mark) = 2.05 m**
- **Peak distance (from floor to peak mark) = 2.65 m**

You can apply this analysis to your vertical jump by substituting your personal data for the sample data.

The total energy of the jump is equal to the overall gain in the jumper's potential energy from the ready position to the peak position:

$$\text{Total energy} = \text{Change in potential energy}$$
$$= mg\,[(\text{Peak distance}) - (\text{Ready distance})]$$
$$= 430\ \text{N} \times (2.65\ \text{m} - 1.70\ \text{m})$$
$$= 430\ \text{N} \times 0.95\ \text{m}$$
$$= 410\ \text{J}$$

This 410 J of energy was used by the jumper during the push phase, while the feet were in contact with the ground. It is helpful to think about the energy during the push phase in two parts:

1) the gain in gravitational potential energy of the body as it is lifted against gravity before leaving the ground without accelerating the body.

2) the kinetic energy to propel the body off the ground if the body is accelerated.

Energy to lift body before leaving ground (without acceleration)
$$= mg\,[(\text{Launch distance}) - (\text{Ready distance})]$$
$$= 430\ \text{N} \times (2.05\ \text{m} - 170\ \text{m})$$
$$= 430\ \text{N} \times 0.35\ \text{m}$$
$$= 150\ \text{J}$$

The kinetic energy to propel the body off the ground must be equal to any amount of the total energy which is "leftover" after subtracting the amount of energy needed to lift the body against gravity before leaving the ground without acceleration.

Kinetic energy at launch = (Total energy) − (Energy to lift body without acceleration)
= 410 J − 150 J
= 260 J

Predictions about Jumping on the Moon

What would happen if the same person used in this example were to repeat the same jumping technique on the moon, using the same amount of energy, 410 J, while lifting the body the same distance, 0.35 m, during the push phase? What would be the person's jump height?

The person's mass, 44 kg, would remain the same on the moon, but the person's weight would be less on the moon due to the reduced acceleration due to gravity on the moon:

Weight on the moon = 44 kg × 1.6 m/s^2
= 70 N

The energy needed to lift the body against gravity without acceleration during the push phase on the moon would be:

Energy to lift body without acceleration = mg [(Launch Distance) − (Ready Distance)]
= 70 N × (2.05 m − 170 m)
= 70 N × 0.35 m
= 25 J

Therefore, the kinetic energy to propel the body off the ground on the moon is equal to the total energy minus the energy needed to lift the body against gravity without acceleration:

Kinetic energy at launch = (Total energy) − (Energy to lift body without acceleration)
= 410 J − 25 J
= 385 J

Much more of the leg energy can go into propelling the body off the ground!

3

SPORTS ON THE MOON

The jump height can be predicted by assuming that the kinetic energy at launch is transformed into the gain in gravitational potential energy from launch to the peak of the jump:

KE at launch = PE gained from launch to peak

385 J = mg (Jump height)

Therefore:

$$\text{Jump height} = \frac{395 \text{ J}}{(44 \text{ kg} \times 1.6 \text{ m/s}^2)}$$

$$= 5.47 \text{ m}$$

Notice that the jump height on the moon is
5.47 m ÷ 0.60 m = 9 times higher on the moon than on Earth.

It is tempting to jump to the conclusion that jump heights on the moon and Earth would compare in the same way—different by a factor of six—as the accelerations due to gravity on the moon and Earth. As shown by this analysis, that is not true.

The equations used for analyzing the vertical jump in the above analysis are based on the assumption that a jumper applies a constant downward force to the ground beneath him—and also that the ground pushes with an equal and opposite constant upward force on the jumper—during the pre-launch phase of jumping. In real jumps, the force on the jumper changes rapidly with time. However, the equations used here provide reasonably good estimates of jump characteristics. Research shows that the best jumpers are able to accelerate to high speed in a very short amount of time and are able to maintain a fairly constant force while rising from a crouch to launch position. Not enough is known about jumping on the moon to be sure that a jumper there would have enough time before launch to build up the muscular force assumed in this example of a moon jump. Therefore, the estimated jump height in this example may be somewhat high.

REFLECTING ON THE ACTIVITY AND THE CHALLENGE

It appears that sports involving jumping would be interesting on the moon. Shot from a circus cannon on the moon, a human body would, in the same way as any projectile launched on the moon, fly six times higher and six times farther than on Earth. Launched vertically using body muscles, a human body should fly more than six times higher on the moon than on Earth. In a sport like gymnastics, people are propelled by their leg muscles. In basketball, as well, the height and hang time are determined by the leg muscles. These sports will be very different on the moon. Adjustments in rules may have to be made to keep these sports fun, challenging, and competitive.

PHYSICS TO GO

1. How much do you think that wearing a cumbersome, heavy space suit and carrying a life support system backpack that have a combined mass of about 225 kg would reduce the height of a person's vertical jump

   a) on the moon?
   b) on Earth?

   Use sample data in your answers.

2. On Earth, the top edge of a volleyball net is placed 8 feet above the ground and a basketball hoop is 10 feet above the ground. At what heights would they need to be placed on the moon to keep the sports equivalent in difficulty?

3. Would jumping on a trampoline be different on the moon than on Earth? Why or why not? If so, how?

4. How would events in gymnastics be affected if transferred to the moon? Choose an event and describe how it would be different on the moon. Use numbers as well as descriptions in your response.

5. What do you think will be the winning height in the high jump during the first Olympiad held on the moon?

6. A student riding in a chair moving at constant speed throws a ball into the air and catches it when it comes back down. What would be the same and different if the activity were done in exactly the same way on the moon?

S 139

Physics To Go

1. Answers to this question will vary according to the assumptions made about the increase in mass of the jumper caused by wearing a space suit and backpack and about the restricting effect of the same gear on the jumper's ability to crouch deeply and to straighten the legs quickly during the push phase of jumping. Using data for the jumper used as an example in this activity and assuming that (a) the gear worn causes a 50% reduction in the total work done during the jump phase (from 420 joules to 210 joules) and (b) the gear doubles the jumper's mass from 44 kg to 88 kg:

   (Work during push phase) – (PE gain in C of M during push phase) = (PE gain during flight phase)

   $210 \text{ J} – [(88 \text{ kg})(1.6 \text{ m/s}^2)(0.35 \text{ m})]$
   $= [(88 \text{ kg})(1.6 \text{ m/s}^2)]\text{H}$
   $210 \text{ J} – 49 \text{ J} = [140 \text{ (kg)m/s}^2]\text{H}$
   $\text{H} = (210 – 49) \text{ J} / 140 \text{ (kg)m/s}^2$
   $= 1.1 \text{ m}$

   Even with such severe restrictions the jump height on the moon, 1.1 m, is nearly twice the same person's jump height on Earth, 0.60 m. The backpack would tend to shift the jumper's center of mass toward the rear of the body, perhaps causing the jumper to rotate backward while in flight.

3

2. As shown in the analysis of a vertical jump on the moon during this activity, everyone who can jump on Earth would be able to jump more than six times higher on the moon (assuming indoor conditions). Since there is considerably more individual variation in vertical jump heights on the moon than on Earth (as could be shown by sampling the jump heights calculated by individual members of your class for this activity), there is no simple, right answer to this question.

3. A trampoline works in the same way on the moon as on Earth, involving a repeating chain of energy transformations for each jump cycle: Initially, the trampolinist does work to jump; part of the work done during jumping is transformed into kinetic energy at take-off; the kinetic energy at take-off is transformed into gravitational potential energy at the peak of the jump; the gravitational potential energy at the peak of the jump is transformed into kinetic energy at landing; the kinetic energy at landing is transformed into work done to stretch the trampoline; the elastic potential energy stored in the stretched trampoline is transformed into work done to push the jumper's body upward; the work done by the trampoline is transformed into kinetic energy at take-off. If during the second and succeeding take-offs the jumper does "legwork," the legwork energy is added to the take-off kinetic energy provided by the trampoline, and the jumper "builds up" height with each jump. Due to reduced g on the moon, the additional height gained due to work done by the jumper's muscles during each cycle of legwork will be greater on the moon than on Earth, and, therefore, much greater height should be able to be attained on the moon.

4. The floor exercise could be sensational on the moon because the athletes are among the world's best jumpers. They would have more time to do spins during flight on the moon than on Earth.

5. A men's world record for the high jump set in 1993 was 2.45 m. It seems sure that a similar effort on the moon would result in a jump more than six times higher (6 x 2.45 m = 14.7 m), but, for certainty, it would be necessary to measure— perhaps from a slow-motion video of a high jump—the position of a high jumper's center of mass during push and flight phases and then apply the methods of analysis and prediction similar to the methods used for vertical jumps on Earth and the moon in Chapter 2, Activity Four and in this activity.

6. The ball would fly six times higher and would be in flight six times longer, but still would land in the student's hand.

ACTIVITY SIX
Golf on the Moon

Background Information

Two new phenomena are introduced in this activity:

- compare the "bounciness" of various balls to a golf ball.

- relativity of motion applied to a head-on collision between a golf ball and club.

It is suggested that you read The Physics of Taming the Golf Ball before proceeding in this section.

When testing the possibility of "taming" the ball to adapt golf to the moon, a golf ball and a variety of "less bouncy" balls are tested in a simulation of a golf club hitting a ball where the balls are dropped from a standard height to bounce back upward from a hard floor surface. In this simulation, the floor represents the face of the golf club hitting the ball. The assumption is made that the forces acting on the ball during the collision and the resulting velocity of the ball after impact are equal regardless of which object, the face of the golf club (simulated by the floor surface) or the ball, is moving before the collision; that is, the effect of a collision between a golf club and ball depends on the relative motion of one object with respect to the other before the collision. It is true that the velocity of a stationary ball after being hit by a club head moving at a certain speed would be equal to the velocity that would result if the same ball moving in the opposite direction at the same speed collided with a stationary golf club head. Therefore, the initial speeds of a variety of balls hit by the moving head of a golf club would compare in the same way as the initial rebound speeds of a variety of balls dropped from equal heights to strike a hard floor surface.

Conservation of energy is used to calculate the rebound speeds of balls dropped from a standard height:

(PE of ball at peak of rebound) = (Initial rebound KE)

$$mgh = mv^2/2$$

$$v = \sqrt{2gh}$$

It is desired to find a kind of ball which would have a rebound speed equal to $1/\sqrt{6}$ times the rebound speed of a golf ball. A durable ball about the same size as a golf ball which would have such a speed when hit by a golf club would be difficult, if not impossible to find. Therefore, taming the golf ball does not seem to be a good possibility.

The second possibility considered for limiting the range of golf balls on the moon considered in the activity is to tame the club. As explained in the second For You To Read section in the student text, the requirement is to find a mass ratio for a golf ball and club head which would result in reducing the speed of a golf ball hit by a club from the typical value of 60 m/s by a factor of $\sqrt{6}$, or to 60 ÷ 2.45 = 24 m/s, which is about 1/2 of the speed of the head of a golf club for a normal golfer's swing. To meet this condition, conservation of momentum demands that the mass ratio of the head of the golf club to the golf ball would be about 1:3 (a very lightweight club head compared to standard golf club heads) and, more problematic, the club head would need to bounce back from the collision at a speed equal to about 1/2 of its speed before the collision—follow-though of the golfer's swing would not be possible, and the bounce-back of the club would not be pleasant for the golfer. (For detailed analysis of the conservation of momentum of a head-on collision between objects having a 1:3 mass ratio when the less massive object is moving before the collision, see Problem # 6 in Physics To Go, Activity Eight, Chapter 2; only the velocities need to be changed to apply to this situation.) Therefore, taming the golf club also does not seem a good possibility for adapting the game of golf to the moon.

Only for your information as the teacher, the "bounciness" of balls, which of course depends on the materials from which balls are made, usually is quantified by the "Coefficient of Restitution" which is defined as the ratio of the rebound height to drop height for a ball striking a hard surface. The coefficient of restitution ranges from zero (no bounce at all) to, ideally, one. Balls made of highly "bouncy" materials such as golf balls and "superballs" have coefficients of restitution in the range 0.8 to 0.9. Ball bearings made of hardened steel have high coefficients of restitution compared to balls made of most other materials.

Active-ating the Physics InfoMall

One of the textbooks mentioned earlier, Physics in the Real World, Chapter 5, has a section on the effect of dimples on golf balls. This leads to a search using "golf" AND "dimple*". The four hits are "The Physics of the Drive in Golf", American Journal of Physics, vol. 59, issue 3, 1991; "Effect of spin and speed on the lateral deflection (curve) of a baseball; and the Magnus effect for smoother spheres," *American Journal of Physics*, vol. 27, issue 8, 1959; Teacher Treasure "A Different Kind of Physics Final for Enriched Physics;" and *Physics in the Real World,* Chapter 5. Check these out, and notice how important the atmosphere is for golf.

In addition, the collision involved between the golf club and the ball can be analyzed in terms of collisions, momentum, and the coefficient of restitution. These are discussed in almost every textbook, and you are encouraged to look for yourself for a discussion of these.

Planning for the Activity

Time Requirements

• One class period.

Materials Needed

For each class:

• *Active Physics Sports* content video (sequence: astronaut hitting golf ball on moon)

• VCR & TV monitor

For each group:

• calculator

• 2-meter stick

• golf ball

• assortment of balls each having approximately the same diameter as a golf ball

• stand for hanging pairs of balls to be collided

Advance Preparation and Setup

A set of balls containing one golf ball and a variety of balls made from other materials but having about the same diameter as a golf ball will be needed for each group. Possibilities for balls similar in size to golf balls include Wiffle® practice golf balls, small superballs, table tennis (ping pong) balls, rubber balls used for playing jacks, and rubber balls used on paddle ball sets. If you can secure other kinds, use them too.

Each group also will need a collision apparatus (see the diagram in the student text). The apparatus can be arranged using a ring stand having a horizontal

rod clamped to the upright rod. A variety of spheres need to be able to be mounted two-at-a-time from bifilar (V-shaped) suspensions to assure good alignment for head-on collisions. It is not critical in this part of the activity that the balls be of nearly equal diameters; instead, it is important to have several combinations of mass ratios available for students to test. Try to include two spheres having a mass ratio of about 1:3.

Epoxy cement or the glue product called Goop® will work when threads cannot be attached to the spheres by simpler methods such as tying. Also, consider that students will need to be able to exchange the spheres easily, so use a simple method of attaching the threads to the horizontal rod.

See the Additional Materials: Taming the Golf Club following this activity in the Teacher's Edition.

Teaching Notes

You may wish to break the class session into two separate parts, the first dealing with taming the golf ball and the second taming the golf club. Also, it probably would be a good idea to have students read and discuss the For You To Read section applying to each of the two parts before beginning the activity which applies to each part, this to assure that students know what they are seeking to find in qualities of balls and clubs which would provide the desired taming effects.

The Physics Talk section lays the theoretical groundwork for the requirements for altering the golf ball and the golf club to reduce the range of golf balls hit on the moon by a factor of six to accommodate Earth-sized golf courses on the moon..

It is pointed out that other sports which involve projectile motion also may be in need of significant adjustments to be played on the moon.

3

Activity Overview

In this activity students use collisions between swinging pendulums to predict what changes should be made to the golf club or the golf ball to adapt it to play on the moon.

Student Objectives

Students will:

- compare the bouncing qualities of balls made from a variety of materials.

- analyze required characteristics of a replacement for a standard golf ball which would limit the range of a ball hit on the moon to the typical range of a golf ball hit on Earth.

- analyze how a golf club would need to be modified to limit the range of standard golf balls hit on the moon to the typical range of a golf ball hit on Earth.

ANSWERS FOR THE TEACHER ONLY

What Do You Think?

Show the video segment of astronaut Alan Shepard hitting a golf ball on the moon.

As will be seen in this activity, taming the game of golf to be played on Earth-sized golf courses on the moon is beset with serious problems. Many sports which involve objects moving as projectiles (e.g., baseball) would need to be tamed to be adapted to the moon.

SPORTS ON THE MOON

Activity Six
Golf on the Moon

Astronaut Alan Shepard accomplished two firsts. He was the first American to ride a rocket into space (May, 1961) and he was the first person to hit a golf ball on the moon (February, 1970).

- **How could the game of golf be modified to be played on the moon?**

Record your ideas about this question in your *Active Physics log*. Be prepared to discuss your responses with your small group and the class.

FOR YOU TO DO

1. Explore the possibility of substituting a "dead" ball for a golf ball to reduce the distance the ball would travel when hit by a golf club on the moon. Obtain a standard golf ball and a collection of other balls of similar sizes and masses which may have potential to be used for "moon golf." Identify each ball in some way.

✎ a) Make a descriptive list of the balls in your log.

✎ b) With your group members decide whether or not you agree with the following statements. Confer with your teacher if anyone disagrees with the statement.

SPORTS S 140

ANSWERS

For You To Do

1. a) See Advanced Preparation and Setup for suggestions of balls to use in this activity. Students' descriptions will depend on the type of ball you select.

   b) Students must agree with each statement to proceed with the activity effectively.

i) When different kinds of balls are dropped to the floor from equal heights, all of the balls hit the floor with the same speed.

ii) Each ball rebounds with a particular speed relative to the floor.

iii) The speed of each ball after impact would be the same if the balls stood still and the floor moved upward at impact speed to hit each ball from below.

2. Position a 2-m stick (or two ordinary meter sticks clamped end-to-end) vertically with the zero-end resting on the floor. Secure the 2-m stick to a wall or the edge of a table so that it will not move. Be sure to allow space for a member of your group to observe a falling ball with the stick in the near background.

3. Drop each ball from a height of 2.00 m so that it falls along a line directly in front of, but not touching, the stick. One member of the group should be prepared to read the maximum height reached by the bottom edge of the ball when it bounces back up from the floor. Allow the person measuring the bounce height to have some practice before recording data. Decide within your group if you wish to have more than one trial for each ball and average the bounce heights.

🖎 a) Record the bounce height in your log beside the description of each ball.

🖎 b) Divide the bounce height of each ball by the bounce height of the standard golf ball. Record the answers in your log.

🖎 c) Would a ball which bounces only $\frac{1}{6}$ as high as a golf ball fly only $\frac{1}{6}$ as far when hit by a golf club?

🖎 d) Do any of the balls bounce only $\frac{1}{6}$, or 0.167, as high as a golf ball? Which, if any of the balls, comes closest to bouncing $\frac{1}{6}$ as high?

4. Since golfers on the moon will want to swing their clubs at normal speed and since problems exist for making changes in the ball, consider reducing the mass of the head of the golf club as a way of reducing the range of a golf ball on the moon. Use the apparatus shown in the diagram to simulate hitting a golf ball with the head of a golf club. Take some time to become familiar with the apparatus.

One ball, representing the head of a golf club, is pulled back and released to collide with a stationary ball representing a golf ball.

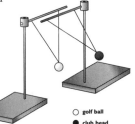

○ golf ball
● club head

For You To Do

(continued)

2. Student activity.

3. Below is sample data for balls dropped from a 2-m height to a concrete floor.

Golf ball:

Bounce height = 1.60 m

Launch speed = 5.7 m/s

Speed reduction factor = 1.00

Ball used for playing jacks:

Bounce height = 0.94 m

Launch speed = 4.8 m/s

Speed reduction factor = 1.19

Wiffle® practice golf ball:

Bounce height = 0.74 m

Launch speed = 3.8 m/s

Speed reduction factor = 1.5

None of the balls tested in the above sample data had a speed reduction factor near the desired value of 2.45; a ball having the desired speed reduction factor would, compared to the golf ball tested for the sample case, bounce to a height of 0.27 m and have a launch speed of 2.3 m/s. It seems unlikely that a ball will be found which is "dead" enough to have the desired speed reduction factor while retaining its original shape after colliding.

4. Student activity.

3

ANSWERS

For You To Do

(continued)

5. In the collisions to explore what would be needed to tame the golf club, students will find that it is necessary for the head of the club to have less mass than the ball in order to meet the requirement that the ball's speed after impact will be about 1/2 of the club head's speed before impact. In addition, students will find that the club head will reverse direction after impact, this being a serious drawback that golfers would be sure not to like.

6. Student answers will vary. See Backround information and Taming the Golf Club for additional comments.

SPORTS ON THE MOON

5. Typically on Earth, the launch speed of a golf ball is 1.5 times the speed of the club at impact. Perhaps the head of the club could be made less massive, so that the ball's launch speed would be reduced to about one half ($1.5 \div \sqrt{6} = 0.6$ or about 1/2) the speed of the club.

Try various combinations of masses representing the golf ball and the head of the golf club. Find a combination for which, as shown in the sketch, the ball moves away just after the collision at about half of the speed of the head of the club just before the collision.

Speed of ball after collision   Speed of club before collision

a) What combination of masses for the golf ball and the head of the golf club met, or nearly met, the above condition? When the condition is met, which is the more massive, the ball or the club head? What is the ratio of the masses, (mass of ball)/(mass of head)? Write your answers in your log.

b) Describe how the head of the golf club moves after it hits the ball. Do you think golfers would find this satisfactory? What problems are apparent for this method of reducing the range of a golf ball on the moon?

6. Discuss within your group whether it seems possible to use some combination of altering both the golf ball and the golf club to fit the game of golf on the moon to an Earth-size golf course. Write the reactions of your group and your personal opinions to the following questions in your journal.

a) Does playing golf on the moon seem feasible?

b) Would golfers on the moon be likely to have complaints?

c) Does it seem worth the trouble to propose golf to NASA as a sport for the moon?

SPORTS ——————— ◆ S 142 ——————

PHYSICS TALK

The Physics of Taming the Golf Ball

Is it true that a ball which bounces only $\frac{1}{6}$ as high as a golf ball when dropped from the same height would have $\frac{1}{6}$ of the range of a golf ball when hit by a golf club? Two equations help to answer this question.

The first equation is found by equating the kinetic energy of the bouncing ball at the instant it leaves the floor to the gravitational potential energy the ball has at the peak of its bounce: $\frac{1}{2}mv^2 = mgh$. Dividing both sides of this equation by m gives the simplified equation to $v^2 = 2gh$ where v is the speed of a bouncing ball at the instant it leaves the floor, g is the acceleration due to gravity and h is the bounce height. Notice that the equation shows that the bounce height of a ball is directly proportional to the ball's "takeoff" speed squared.

The second equation is $R = \frac{v^2}{g}$, where R is the maximum range, or horizontal travel distance, of a projectile launched with speed v at a 45° angle of elevation at a location where the acceleration due to gravity is g (this equation is explained in many advanced physics textbooks if you wish to see how it is derived). Notice that this equation shows that the maximum range of a ball is directly proportional to the ball's launch speed squared.

Both the bounce height of a dropped ball and the range of a ball launched like a golf ball are proportional to the ball's speed squared. If the bounce height and the range of a ball are directly proportional to one another; then it is true that a ball which bounces $\frac{1}{6}$ as high as a golf ball also would have $\frac{1}{6}$ of the maximum range of a golf ball when hit by a golf club.

The data you have gathered about the bounce heights of balls can be used to infer how the speeds of various kinds of balls, when hit, compare to the speed of a golf ball after it experiences a similar hit. Such a comparison will help you to decide if any of the balls you have tested would serve for playing golf on the moon.

SPORTS

3

ANSWERS

Physics To Go

1. Air resistance reduces the range of golf balls on Earth compared to what the range would be if no air were present. Since there is no air resistance on the moon, outdoor golf on the moon probably would result, for hits equal to those on Earth, in ranges more than six times the ranges on Earth.

2. Hooks and slices of golf balls hit on Earth are caused when a golf ball having a sideways spin interacts with the air surrounding the ball to reduce the pressure on one side of the ball. Since there is no air on the moon, golf balls would not hook or slice on the moon.

3. Hitting a golf ball resting on lunar soil would be similar to hitting a golf ball out of a sand trap on a golf course on Earth. Particles ranging in size from sand to dust would be ejected from the surface as projectiles. On Earth, air resistance causes such particles to slow down quickly, so they do not travel a great horizontal distance from the point of impact before falling back to the surface; also on Earth, tiny dust-like particles fall back to the surface slowly because air resistance causes them to reach terminal velocity soon after they begin to fall. With no air resistance on the moon, particles launched in a divot all would fly along parabolic trajectories, some traveling great distances, and all particles would accelerate downward at the moon's g, 1.6 m/s/s.

4. The space suits and backpacks worn by astronauts during Apollo missions were very bulky and limited the range of motion of arms and legs; a two-handed golf swing with complete follow-through of the swing probably would not be possible when wearing such gear.

SPORTS ON THE MOON

REFLECTING ON THE ACTIVITY AND THE CHALLENGE

What you learned in this activity would apply in similar ways to other sports in which a bat, racquet, paddle, club, or other hitting device is used to launch an object into a state of projectile motion. As you may have gathered from this analysis of ways to "tame down" the motion of a golf ball, similar problems can be expected to arise when trying to limit projectile motion in other sports.

PHYSICS TO GO

1. To keep things simple, the effect of air resistance on golf balls in flight was not considered in this activity. Should it have been considered? What difference do you think it would have made, and why?

2. Describe briefly why you would have to change the racquet or ball in a game of tennis.

3. On Earth, golfers sometimes hit "divots," chunks of grass sod, when the club hits the ground in the process of hitting the ball. On the moon, a divot would be a cloud of sand-and-dust-like lunar soil. With weak gravity and no wind due to lack of air, would "moon divots" present a problem for golfers on the moon?

4. Many golfers say that they enjoy the social part of golf as much as the game. It's a good chance to visit with golfing partners. Would golfers be able to visit as usual on the moon? Also, golfers holler "Fore" to warn a far away person who might be in the way of a drive. Would that method of warning work on the Moon? Explain your answers.

5. Make a list of three reasons:
   a) in favor of proposing golf to NASA as a sport for the moon.
   b) against proposing golf to NASA as a sport for the moon.

6. Name three sports which use bats, clubs, or racquets. Describe the changes that you would make in the ball or hitting device to insure that the sport is fun on the moon.

 S 144

5. There is no air to transmit sound waves on the moon, so communication is accomplished via radio. Communication is possible only with those who are "tuned" to the same radio frequency.

6. a) Some reasons in favor of golf on the moon:
   • Many people enjoy golf and it provides exercise.
   • There is plenty of available space on the moon's surface for golf courses.
   • The differences between trajectories of golf balls on the moon and on Earth would be interesting.

   b) Some reasons against golf on the moon:
   • It may not be possible to find a way to limit the range of golf balls to reasonable distances on the moon while at the same time preserving the nature of the game.
   • Need for space suits and survival equipment would make it time-consuming and expensive to play a round of golf on the moon.
   • It is not possible to walk in a normal way on the moon, so the nature of the exercise provided would be different from playing golf on Earth.

Taming the Golf Club

In order to sidestep the possible issue of the role of the diameter of the ball in the collisions dynamics, it is important to have the spheres used in this part of the investigation of as nearly the same diameter as possible. An actual golf ball makes a good target and adds to the realism of the activity. An ideal set of "clubs" can be made from a set of cork balls of about the same size as a golf ball. Different mass "clubs" can be made by drilling through the diameter of the cork and filling the hole partially or completely with metal cylinders cut from rod stock. Since the balls are suspended from a bifilar support, the rod can be kept perpendicular to the line of motion of the ball. The coefficient of restitution is also kept reasonably constant in this arrangement, so the observed differences in behavior are essentially due to the mass ratios.

In this system, the golf ball also has a hole drilled for the support thread. It may also be possible to use foam balls in place of the cork ones. Try a dense, resilient foam to prevent the surfaces from being crushed.

Another way of setting up the apparatus gives a better visual analog to the club and ball. Again, a golf ball plays itself. The tops of the cork balls are drilled to a slightly smaller diameter than a piece of rigid aluminum tubing which has been threaded at one end.

The upper end of the rod is attached to a round cross piece by any convenient means. A tubing T is shown. A repeatable starting position can be achieved if the support rod is in the horizontal position when it is released.

When a pure cork ball is used, a slight amount of recoil of the club may be observed (more so in the bifilar pendulum arrangement where the additional mass of the rod is not a factor). This behavior is predictably undesirable for the golfer, and it will probably lead to a whole new class of sports injury.

From this activity, students are led to the proposition that to make the game of golf playable on the moon, a relatively massive ball which produces as nearly a "dead" collision with the club as possible is going to be required. The relative masses can be adjusted by making the club very light, but that will speed up the club at the end of the downswing, and that result is counterproductive.

If we accept that golfers are a hardy lot who enjoy the challenge of adversity, we may conclude that the sport can be transplanted to the moon. But it surely won't be the "same" game. Whether it is worth the trouble is a matter of personal opinion, appropriately supported, of course.

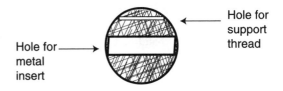

Cross section of cork ball

Side view

Cross section and side view of ball.

3

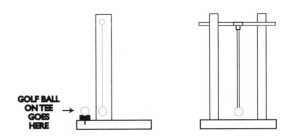

Golf ball on tee with club.

ACTIVITY SEVEN
Friction on the Moon

Background Information

It is recommended that you read Frictional Force in the student text before proceeding in this section. You may also wish to review the Background Information for Chapter 2, Activity Six where the coefficient of sliding friction was introduced.

In general, the frictional force between the surfaces of two objects on the moon is 1/6 of the amount that the frictional force between the same two objects would be in a similar situation on Earth. This is true because the frictional force is directly proportional to the normal force pushing the surfaces of the objects together; since, for particular similar situations on the moon and on Earth, the normal force on the moon is 1/6 the value of what it would be on Earth, so is the frictional force on the moon 1/6 of what it would be on Earth.

It is critical to recognize that the coefficient of sliding friction between a particular pair of surfaces does not change from Earth to the moon, this because the coefficient of sliding friction is determined by characteristics of the two surfaces in contact; changing the location of the surfaces does not affect the way in which the surfaces interact during sliding contact.

Active-ating the Physics InfoMall

FYTD asks you to think about how walking requires friction. Search the InfoMall with "walking" AND "friction", but make sure you require that they are in the same paragraph, and only about 10 words apart, or you will get too many hits with too little good information. This search results in a few hits that explain friction, and mention that walking is impossible without it.

Friction is another of those concepts that almost requires browsing though textbooks to get the information you want. However, (and this applies directly to Physics To Go, number 7), you may want to do a search with "friction" AND "turn*" within a paragraph. You should find several references on how turning can depend on friction.

Planning for the Activity

Time Requirements
• One class period.

Materials Needed
For each group:
• table-like surface on which to slide box
• box, capacity to hold 1 kg of sand
• sand, at least 1 kg in container
• spring scale, 0-10 newton range
• calculator
• string for suspending and pulling box

Advance Preparation and Setup

In advance of the activity you will need to collect flat-bottomed containers for which the coefficient of sliding friction between the container and a table top will be measured at each station. A set of shoe boxes would work very well.

A ballast material such as sand will be added to the containers.

Since the coefficient of friction depends on the kind of surface across which sliding occurs, it would be necessary for the table top surfaces for all stations to be identical if it is desired for all groups to have identical data (the floor could be used instead of table tops if it is a continuous, smooth surface material).

Teaching Notes

Sample data is not given because the data will depend on the surfaces used in your situation.

Assessment

See the template for the assessment of Graphing Skills following this activity in the Teacher's Edition.

Activity Seven
Friction on the Moon

WHAT DO YOU THINK?

The Lunar Rover proved that there is enough frictional force on the moon to operate a passenger-carrying wheeled vehicle.

• **How do frictional forces on Earth and the moon compare?**

Record your ideas about this question in your Active Physics log. Be prepared to discuss your responses with your small group and the class.

FOR YOU TO DO

1. Walk forward for a few steps and then come to a quick stop. Make the observations needed to write answers to the following questions in your log.

　a) In what direction do you push your feet to make your body go forward?

　b) Identify the force that makes your body go forward with each step.

　c) In what direction do you push your feet to make your body stop?

　d) Identify the force that makes your body stop.

　e) Explain in terms of forces why it is difficult to walk forward or to come to a quick stop on a slippery surface such as ice.

SPORTS

Activity Overview

In this activity students investigate the effect of weight on friction. They then using graphing to predict what might happen on the moon.

Student Objectives
Students will:

• use a spring balance to measure forces of sliding friction.

• understand and apply the definition of the coefficient of sliding friction to comparing frictional forces on Earth and the moon.

ANSWERS FOR THE TEACHER ONLY

What Do You Think?

Frictional forces on the moon are generally 1/6 the amount of frictional forces on Earth. However, the coefficient of friction is not affected by location.

3

ANSWERS

For You To Do

1. a) You push your feet backwards.

b) Friction makes your body go forward.

c) You push your feet forward.

d) Friction makes you stop.

e) On a slippery surface there is very little friction, therefore there is very little force to help you move forward or stop.

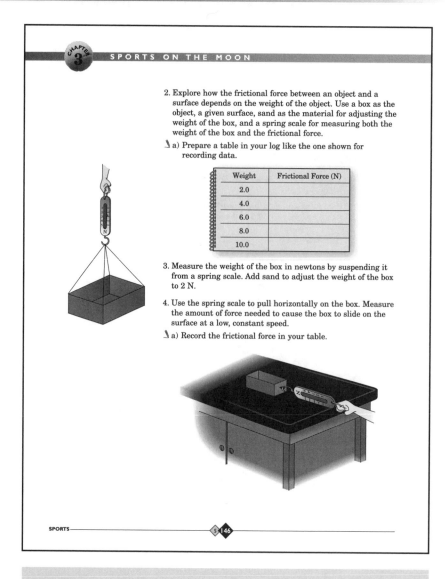

SPORTS ON THE MOON

2. Explore how the frictional force between an object and a surface depends on the weight of the object. Use a box as the object, a given surface, sand as the material for adjusting the weight of the box, and a spring scale for measuring both the weight of the box and the frictional force.

a) Prepare a table in your log like the one shown for recording data.

Weight	Frictional Force (N)
2.0	
4.0	
6.0	
8.0	
10.0	

3. Measure the weight of the box in newtons by suspending it from a spring scale. Add sand to adjust the weight of the box to 2 N.

4. Use the spring scale to pull horizontally on the box. Measure the amount of force needed to cause the box to slide on the surface at a low, constant speed.

a) Record the frictional force in your table.

SPORTS

S 146

ANSWERS

For You To Do (continued)

2. a) Students copy table into their logs.

3. Student activity,

4. - 5. The frictional force recorded by the students will vary, depending on the surface used.

---

5. Continue adding sand to increase the weight of the box to 4.0, 6.0, 8.0, and 10.0 N, measuring the frictional force for each weight.

✎ a) Record the frictional force for each weight to complete your data table.

6. Plot a graph of Frictional Force versus Weight. Plot frictional force on the vertical axis and weight on the horizontal axis.

✎ a) Plot the points from the data table and sketch a line to connect the plotted points. Carefully consider whether the sketch should be a straight or curved line.

✎ b) Based on the graph of the data, write a statement in your log which summarizes the relationship between weight and frictional force.

✎ c) You learned in Activity 3 that the weight of an object on the moon is $\frac{1}{6}$ of the object's weight on Earth. When the box used in this activity weighs 9 N on Earth, what would the same box weigh on the moon? Show your calculation in your log.

7. Use interpolation of the graph Frictional Force versus Weight to determine what the force of friction would be on the same surface on the moon if the box weighs 9 N on Earth.

✎ a) Locate the point on the horizontal weight axis which corresponds to the weight of the box on the moon and draw a vertical line from that point to touch the curve.

✎ b) From the point where the vertical line touches the curve, draw a horizontal line to touch the vertical frictional force axis. Read the value where the horizontal line touches the axis. This is the frictional force on the moon.

✎ c) Write a statement in your log which summarizes how the frictional force between an object and a surface on Earth compares to the same object and surface on the moon.

8. Apply what you have learned about friction on the moon compared to friction on Earth to walking and running on the moon compared to on Earth.

✎ a) Write a statement in your log which summarizes your thoughts about how friction may affect the ability to walk and run on the moon.

ANSWERS

For You To Do *(continued)*

6. b) The graph should be a straight line.

c) As the weight increases, so does the frictional force.

d) The weight will be 1/6 of 9 N, or 1.5 N.

7. Students will find that the frictional force between an object and a surface on the moon will be 1/6 less than that between the same object and surface on Earth.

8. Reduced friction will present problems for walking on the moon.

SPORTS ON THE MOON

PHYSICS TALK

Frictional Force

A force called friction arises when an attempt is made to slide an object on a surface. When an object resting on a horizontal surface is pushed or pulled horizontally, the amount of the force of friction between the object and the surface is equal to the amount of the horizontal force required to make the object move at constant speed. As the object moves at constant speed, the applied force causing the motion is equal in amount but opposite in direction to the frictional force.

Constant Speed →

Frictional Force ←   → Applied Force

If the amount of the applied force is less than the frictional force, the object does not slide on the surface; if the amount of the applied force is greater than the force of friction, the object accelerates as it slides across the surface.

INQUIRY INVESTIGATION

Astronauts on the moon found that the soil at the surface is powdery but firm. Do you think the kind of surface beneath an object also affects the frictional force? How could you find out? Might this also affect the ability to walk or run on the moon?

REFLECTING ON THE ACTIVITY AND THE CHALLENGE

Friction is involved somehow in most if not all sports. Any sport involving walking or running also involves friction. Sliding friction is the basis for some sports such as shuffleboard and curling. Most winter sports also are based on sliding; since there is no water, snow, or ice on the moon, are all winter sports "out," or could some winter sports equipment be adapted to slide on moon soil? One thing is certain, your proposal to NASA won't "slide through" if you don't demonstrate that you understand frictional forces on the moon.

ACTIVITY SEVEN: Friction on the Moon

PHYSICS TO GO

1. a) Based on what you have learned about friction on the moon, what problems do you see for walking and running on the moon? Why?

   b) What problems do you see for quick starts and quick stops on the moon? Why?

2. How many 10-pound bags of potatoes (that is, 10 pounds weight on Earth, or 4.5 kg of mass) would a 70-kg person need to carry on the moon to have the person's weight on the moon (body + potatoes) equal the person's weight on Earth (body only)?

3. Would carrying extra weight—perhaps a material other than potatoes—be a possibility for achieving normal frictional force for walking or running on the moon?

4. What problems might race cars or bikes encounter going around curves on the moon?

5. How would sliding into second base be different on the moon than on Earth?

6. Identify one sport which would in no way be affected by differences in frictional effects between Earth and the moon.

7. If you were to give a shuffleboard disk a push on a shuffleboard court on the moon, would it slow down just as it does on Earth, or will it take a longer or shorter distance to slow down? Give evidence to support your answer.

8. Will friction between your hand and a football or your hand and a bat be different on the moon? Why?

S 149

ANSWERS

Physics To Go

1. Reduced friction combined with loose soil on the moon could present problems for the ability to walk or run on the moon—for you as the teacher at this stage, be aware that the pendulum action of the human leg during walking, introduced in the next activity, is another problem which affects the ability to walk on the moon.

2. Weight on Earth = Desired weight on moon = mg = 70 kg x 10 m/s^2 = 700 N

   Person's weight on moon = 70 kg x 1.6 m/s^2 = 112 N

   Additional weight needed on moon = 700 N – 112 N = 588 N

   Weight of bag of potatoes on moon = 4.5 kg x 1.6 m/s^2 = 7.2 N

   Number of bags needed to carry on moon = 588 N ÷ 7.2 N/bag = 82 bags

3. Yes, carrying extra weight on the moon equal to 5 times the person's body weight on Earth would provide the normal amount of "Earth" frictional force for walking on the moon; for

---

example, a person who weighs 100 pounds on Earth carrying 500 pounds (500 Earth pounds, that is) would feel normal weight and frictional force on the moon.

4. Reduced friction on the moon would increase the tendency to "spin out" on curves.

5. A person would slide much more because of the decreased friction. The person may slide right past the base.

6. If air resistance is included as a frictional force, it is difficult to think of a sport which would not be affected by differences in frictional effects between Earth and the moon.

7. It will take much longer to slow down and would travel much further.

8. The friction would be the same. Here the friction is not dependent on gravity but on the pressure of your fingers on the ball.

NOTES

Assessment Rubric for Graphing Skills

Place a check mark (√) in the appropriate box.

Descriptor	Yes	No
1. Graph of Frictional Force and Weight is given a title.		
2. Units of measurement are found on the *x*-axis (N) and *y*-axis (N).		
3. Line graph is used to demonstrate change.		
4. Frictional force is plotted on *y*-axis and weight on the *x*-axis.		
5. Appropriate scale is used for both axes.		
6. Coordinates are plotted in correct places.		
7. Best fit curve is used to join coordinates.		
8. Uses interpolation to determine the force of friction for a box that weighs 12 N on Earth. (See 6a)		
9. Uses interpolation to identify the frictional force of the moon. (See 6b)		
10. Is able to summarize graphical data to explain comparative frictional force between objects on the moon and Earth.		
Marks Awarded	$\overline{10}$	

Each check mark (√) is awarded one mark for a maximum of 10 points.

For use with *Sports*, Chapter 3, ACTIVITY SEVEN: Friction on the Moon

©1999 American Association of Physics Teachers

ACTIVITY EIGHT
Bounding on the Moon

Background Information

This activity introduces the pendulum as a model of human legs during walking, showing that the pendulum action of legs during walking is a problem in the reduced gravitational field of the moon. This effect, in addition to the reduced frictional force and loose soil on the moon, renders walking (or running) virtually impossible on the moon. That is why astronauts on the moon developed the "bounding" technique for locomotion on the moon's surface.

It is recommended that you study the section Pendulums and Gravity in the student text before proceeding in this section.

The general equation for the period of a pendulum (the time for a pendulum to complete one complete cycle of motion such as to swing over-and-back when displaced from the equilibrium position and released) is:

$$T = 2\pi \sqrt{I/(mgd)}$$

Where T is the period in seconds, I is the moment of inertia in (kg)m^2 of the object acting as a pendulum (to be described immediately below), m is the mass of the object in kilograms, g is the acceleration due to gravity in m/s^2, and d is the distance in meters from the point of suspension of the pendulum to the object's center of mass.

Just as an object's mass is a measure of its tendency to resist translational acceleration (acceleration along a straight line path), an object's moment of inertia is a measure of its tendency to resist angular acceleration (changes in angular speed during rotation of the object about an axis). When an object rotates on an axis, its resistance to angular acceleration is determined not only by the object's mass, but also by how the object's mass is distributed in respect to the object's axis of rotation.

If m_1, m_2, m_3, etc. represent the masses of infinitely small particles of a rotating object, and if r_1, r_2, r_3, etc., represent their respective distances from the axis of rotation, the moment of inertia of the object in respect to the axis of rotation is:

$$I = [m_1r_1^2 + m_2r_2^2 + m_3r_3^2 + ...] \text{ ,or}$$

$$I = \Sigma(mr^2)$$

For the simple case of an object being twirled on the end of a string (when the object's size is negligible compared to the length of the string and the string has negligible mass) the moment of inertia is $I = mr^2$; the same equation would apply to a thin cylindrical shell, hoop, or ring (such as a bicycle wheel) because, even though the mass is spread along the circumference, all of the mass is at the same distance from the spin axis.

For more complex shapes where the mass is distributed at various distances from the axis of rotation, calculus methods are used to determine the moment of inertia.

When an object acts as a pendulum, the moment of inertia is involved because a pendulum is an object which has its rotation arrested by gravity acting somewhat as a spring; if there were no gravity, a pendulum would continue in circular motion at constant angular speed.

For this activity wherein the human leg is modeled as a solid cylinder rotated about one end of the rod, application of calculus results in the moment of inertia:

$$I = (1/3)(mL^2)$$

where m is the mass of the cylinder and L is the length of the cylinder (or leg).

Substituting the above as the moment of inertia and substituting $L/2$ for d in the general equation for the period of a pendulum:

$$
\begin{aligned}
T &= 2\pi \sqrt{I/mgd} \\
&= 2\pi \sqrt{[(1/3)mL^2]/[mg(L^2)]} \\
&= 2\pi \sqrt{2L/3g} \\
&= 2\pi \sqrt{2/3} \sqrt{L/g}
\end{aligned}
$$

The above equation dictates the period of a human leg swinging freely as a pendulum; this is what a human leg does during the forward swing of the leg during each stride. It should be noted that the $\sqrt{6}$ in the equation is not related to the factor of 6 by which thr acceleration due to gravity on the moon differs from that on Earth; it is only a coincidence.

Notice that the above equation does not include the mass of the leg of the amplitude of the leg's swing as variables; it is true that neither affects the leg's period (swing time).

If ever needed, the moments of inertia of objects of many shapes in respect to various axes of rotation are listed in most college physics textbooks.

Active-ating the Physics InfoMall

Notice that this approximation for a leg is really a physical pendulum. You could search or browse the InfoMall for plenty of information on physical pendulums, as well as how the motion of a pendulum depends on gravity.

Planning for the Activity

Time Requirements

• One class period.

Materials Needed

For the class:

• *Active Physics Sports* content video (Segment: Bounding on the Moon)

• VCR & TV monitor

For each group:

• Calculator

• meter stick

• stopwatch

• hangible cylinders, lengths 0.1, 0.4, 0.9, 1.6 m

Advance Preparation and Setup

Reserve a TV monitor and VCR for viewing segments of astronauts bounding on the moon in the *Active Physics Sports* content video.

Optional: You may wish to set up a 1-meter solid cylinder able to swing about one end as a pendulum as a demonstration to match the model of the human leg used in the activity. A wooden post cut to a 1-meter length and drilled through one end accepting a nail or rod as a suspension would serve very well. The moment of inertia for a solid cylinder, $I = (1/3)(mL^2)$, used to develop the equation for the period assumes that the diameter of the cylinder is small compared to the length of the cylinder, so don't use a "fat" cylinder. Since some students may find it difficult to believe that the mass of the cylinder and the amplitude the swing does not affect the period, the amplitude could be varied to show that it does not affect the period, and a second cylinder of different mass could be used to show that mass does not affect the period.

Teaching Notes

Show video segments of astronauts "bounding" on the moon at the beginning of class, then direct students to the What Do You Think? questions.

Discuss steps 1, 2, and 3 with the entire class before having students proceed.

It is critically important for students to read and understand the basis for the cylinder as the model of the human leg before proceeding to step 6.

3

Activity Overview

In this activity students use a pendulum as a model of the human leg to predict how walking would be affected on the moon.

Student Objectives

Students will:

- understand and apply a solid cylinder as a model of a human leg acting as a pendulum during walking.

- measure the amount of time for a human leg to swing forward as a human walks on Earth.

- predict the amount of time for a human leg to swing forward as a human walks on the moon.

- explain why it is not possible to walk normally on the moon.

ANSWERS FOR THE TEACHER ONLY

What Do You Think?

Astronauts bound on the moon because they can't walk due to low friction, loose soil, and the effect of the moon's reduced gravity on human legs acting as pendulums during walking.

While running involves voluntarily bringing the trailing leg forward with each stride, not allowing the leg to swing as a pendulum, the great dependence on frictional force for running would seem to make running on the moon impossible. There would also be great need to adjust the timing of running strides on the moon because the body would fly for a much longer time after each pushoff.

SPORTS ON THE MOON

Activity Eight
Bounding on the Moon

WHAT DO YOU THINK?

Neil Armstrong was the first human to set foot on the moon.

- **Why do astronauts "bound" instead of walk or run on the moon?**
- **Would running events in track be able to be held on the moon, even indoors?**

Record your ideas about these questions in your *Active Physics log*. Be prepared to discuss your responses with your small group and the class.

FOR YOU TO DO

1. Observe the *Active Physics* video of astronauts "walking" on the moon. Record answers to the following questions in your log.

   a) Compare how the astronauts use their legs and feet to move across the surface of the moon to how legs and feet are used in normal walking.

   b) Do you think astronauts would be able to use their legs and feet to walk or run on the moon in the same way that they normally walk on Earth? Why or why not?

SPORTS S 150

ANSWERS

For You To Do

1. a) Students will observe that astronauts bound on the moon.

   b) Astronauts bound on the moon because they can't walk due to low friction, loose soil, and the effect of the moon's reduced gravity on human legs acting as pendulums during walking.

2.-9. Student observations will vary with the lengths of the pendulums used. Students will discover that the period of the pendulum depends on the length of the pendulum, as well as the acceleration due to gravity. Students should reach the following conclusions.

The time predicted by the model for the leg to swing forward on Earth is one-half of the period of a 1-meter cylinder acting as a pendulum:

$$T = (2\pi)/\sqrt{6}\sqrt{L/g} = (2\pi)/\sqrt{6}\sqrt{(1.0 \text{ m})/(10 \text{ m/s}^2)} = 0.82 \text{ s}$$

$$T/2 = 0.41 \text{ s}$$

2. Use a set of solid cylinders of various lengths as pendulums. Each cylinder has a hole at one end to allow it to be hung so that it can swing freely back and forth as shown in the diagram. Measure the length of each cylinder to the nearest 0.10 m.

&a) Record the lengths of the cylinders in a column in your log.

3. Suspend a cylinder from a hook or nail, pull it aside and allow it to swing as a pendulum. Use a stopwatch to measure the period of the pendulum (the period of a pendulum is the time, in seconds, for the pendulum to complete one full swing over and back). You may find it easiest to measure the time to complete 10 swings over and back, and then divide the measurement by 10 to get an accurate value for the period.

&a) Record the measurement of the period in a separate column in your log.

4. Using the same cylinder, pull it aside a distance different than you used in step 3. Measure the period of the pendulum.

&a) How did the period differ from your measurement in step 3?

5. Repeat the measurement of the period for all of the lengths of cylinders.

&a) Add a column in your log to record the period, in seconds, for each length.

&b) Plot a graph of Period versus Length for the cylindrical pendulums. Plot time, in seconds, on the vertical axis and length, in meters, on the horizontal axis. Enter the data points and sketch a line to connect the points. Observe carefully to decide whether the line should be sketched as straight or curved.

6. Observe a member of your group as he or she walks. Notice that immediately after one foot hits the ground in a step, the opposite leg, trailing behind, swings forward as a pendulum before it is used for the next step. Also notice that, except for the foot, a human leg is similar in shape to a cylinder and is suspended at the top from the hip joint. The person doesn't use muscular force to swing the leg forward because it's easier to let the force of gravity cause the leg to swing forward. Therefore, the forward swing of a human leg during walking can be modeled by the cylinders used above.

3

Answers

For You To Do (continued)

The time for the leg of a volunteer having 1-meter leg length to swing forward while walking normally should be reasonably close to 0.4 s.

The time predicted by the model for the leg to swing forward on the moon is one-half of the period of a 1-meter cylinder acting as a pendulum in the moon's g/6 gravity:

$$T = (2\pi)/\sqrt{6}\sqrt{L/(g/6)} \quad (2\pi)/\sqrt{6}\sqrt{6L/(g)} \quad (2\pi)/\sqrt{6}\sqrt{6(1.0 \text{ m})/(10 \text{ m/s}^2)}$$
$$= 2.0 \text{ s}$$

$$T/2 = 1.0 \text{ s}$$

Your foot would be at the bottom of the forward swing when your mind, according to your usual timing of walking, tells you your foot should be out in front, ready to step down on it.

7. See how good the model is by taking measurements for each member of your group. Measure, to the nearest 0.10 m, the length of the leg of each member of your group.

　a) Create a table in your log to record the name and the length of the leg of each person in your group.

　b) As each member of your group walks in a normal way, other members of the group should use stopwatches to measure the amount of time, in seconds, for the person's leg to swing forward during one stride. For accuracy, it may be desirable to take the average of several measurements of this short time interval. Since the forward swing of the leg is only $\frac{1}{2}$ of its period, double the measurement and record the period of the swing of each person's leg in the table in your log.

　c) Create a graph of Period versus Length to find out how well a cylindrical pendulum models the forward swing of your lower leg. Is a cylindrical pendulum a reasonably good model of a person's lower leg while walking?

8. Use the equation $T = 5.1 \sqrt{\frac{L}{g}}$ to calculate the periods of each cylindrical pendulum on which you made measurements. Be sure to include units of measurement when you do the calculations to be sure that the answer is in seconds. Divide the work among members of your group and share the results.

　a) Add a column to the data table in your log and compare the values predicted by the equation to the measured values. Comment on the comparisons.

　b) Use the same equation to calculate the period of the swing of your leg while walking. Share results within your group, enter the data in your log and compare the results to measured values. Comment on the comparisons.

9. The period of a pendulum on the moon should be 2.5 times greater than the period on Earth. (This is explained in Physics Talk.) Multiply the time for the forward swing of your lower leg (half of the period) by 2.5 to find how much time it would take your leg to swing forward if you tried to walk on the moon. With a member of your group providing signals separated by that amount of time, try to walk with the "swing time" that your leg would have when powered by the moon's gravity.

　a) What do you think now about why astronauts don't walk in a normal way on the moon? Write your answer in your log.

ANSWERS

For You To Do *(continued)*

NASA anticipated difficulties with walking on the moon.

The bounding style of astronauts is a "forward hop" which is directed both forward and upward; too much forward force (an amount exceeding the force of friction) would cause the astronaut to "break loose" from the ground, slip and fall forward.

Inability to walk or run on the moon presents serious problems for engaging in many sports on the moon. As pointed out in Activity Seven, astronauts could carry weights to bring the frictional force during walking up to the amount on Earth, but 5/6 of the length of legs would need to be "sawed off" to adjust the period of the leg acting as a pendulum during walking to the same period as on Earth.

PHYSICS TALK

Pendulums and Gravity

Physicists have developed equations to predict the period of many kinds of pendulums. For example, a "simple pendulum" such as a ball hanging on a string has a period:

$$T = 2\pi \sqrt{\frac{L}{g}}$$

where T is the period, in seconds, L is the distance from the point of suspension to the center of the ball and g is the acceleration due to gravity.

The equation for the period of a cylindrical pendulum of the kind you have been using in this activity is:

$$T = 2\pi \sqrt{\frac{2L}{3g}} = 5.1\sqrt{\frac{L}{g}}$$

Notice that the equations show that the periods of both kinds of pendulums are directly proportional to the square root of the length and are inversely proportional to the square root of the acceleration due to gravity. This explains why, for example, small children with short legs have such quick strides. The equations also predict that pendulums—and human legs swinging as pendulums— would behave differently on the moon than on Earth due to the reduced effect of gravity on the moon. Since the moon's gravity is known to be $\frac{1}{6}$ of Earth's gravity, the equations can be adjusted to predict the periods of pendulums on the moon by substituting $\frac{g}{6}$ for g. Therefore, the period of a cylindrical pendulum on the moon should be:

$$T = 5.1\sqrt{\frac{L}{\left(\frac{g}{6}\right)}} = 5.1\sqrt{\frac{6L}{g}} = 5.1(\sqrt{6})\sqrt{\frac{L}{g}} = 13\sqrt{\frac{L}{g}}$$

The above equation shows that the period of a cylindrical pendulum would be about (13÷5.1) 2.5 times greater on the moon than on Earth. Perhaps astronauts do not walk normally on the moon because they can't. The moon's gravity doesn't assist the swing of the leg enough to allow normal walking with normal rhythm on the moon.

REFLECTING ON THE ACTIVITY
AND THE CHALLENGE

There is a problem with walking on the moon, and perhaps the same problem would extend to running on the moon. Your legs will swing slower on the moon. The period of the natural swing will be 2.5 times longer on the moon. This could have serious implications for many sports on the moon, unless "bounding" like astronauts is an acceptable substitute for walking or running. It probably can't even be said that a good runner on Earth would necessarily be a good "bounder" on the moon because different muscles and skills are used; maybe Carl Lewis would finish last in the "100-meter Bound" on the moon! The time is nearing to write your proposal, so it's time to sort out the possibilities for sports on the moon.

INQUIRY INVESTIGATION

The equations for both simple and cylindrical pendulums presented in Physics Talk make no mention of mass or amplitude (swing distance) as variables which may affect the period. Do you think it really is true that such characteristics do not affect the period? Design experiments to test the effects of these and other properties of pendulums on the period and report your procedures and results.

ACTIVITY EIGHT: Bounding on the Moon

PHYSICS TO GO

1. The period of a "simple pendulum," a small massive object hanging from a string, is given by the equation $T = 2\pi \sqrt{\frac{L}{g}}$, where T is the time for the pendulum to swing once over and back, L is the distance from the point of suspension of the string to the center of mass of the object, and g is the acceleration due to gravity. Make a simple pendulum, let it swing, and see if the equation works

2. Describe how difficulty with walking or running on the moon would affect at least one sport.

3. How would walking and running be affected on a planet which has an acceleration due to gravity greater than g on Earth?

4. How long would a simple pendulum need to be to have a period of 1.0 second? Make one and see if it works. (Hint: Solve for the length in the equation given in question 1 above)

5. Pendulums were used as the mechanical basis for making the first accurate clocks. Why?

6. You also use your arms as pendulums when you walk. Do you think you use your arms for a reason? Why or why not?

7. Why do you "shorten" the length of your arms by bending at the elbows when you are running?

8. Obtain data for a small child's leg swing. Does it fit the data on your graph?

$\diamond$ 155

Physics To Go

1. The equation does work.

2. Student answers will vary but should include the slow swing of the legs..

3. You weigh more so your legs may not be able to support your body. In contrast the legs would swing very fast due to increased gravity.

4. $T = 2\pi \sqrt{L/g}$
   $L = T^2 g/4\pi^2$
   $= (1.00\text{ s})^2(9.81\text{m/s}^2)/4\pi^2$
   $= 0.249$ m

   The standard value of g expressed to three significant figures was used in the above calculation; Thomas Jefferson once proposed such a pendulum as a standard for one second of time.

5. Pendulums served as the basis for clocks because even as a pendulum "winds down" (decreases in amplitude, or swing distance) the period remains constant. Also, pendulum clocks which run too fast or too slow can have the period adjusted by slightly adjusting the position of the

3

---

center of mass of the pendulum, usually accomplished by designing the pendulum so that part of its mass can have its position adjusted.

6. Arms are used for balance while walking. As each leg is swung forward and back, the arm on the opposite side of the body is swung in unison but in the opposite direction, reducing the amount of shift in location of the body's center of mass caused by displacement of the legs.

7. Shortening the arms reduces the length of the "pendulum" that swings back and forth. This reduces the period of the pendulum (increases the frequency). When running you want to increase the frequency of the arm swings as you increase the frequency of the leg swings.

8. Students gather data.

ACTIVITY NINE
"Airy" Indoor Sports on the Moon

Background Information

The main phenomenon introduced in this activity is air resistance and its effect on moving objects.

It is recommended that you read the sections Air Resistance and Fly Like a Bird on the Moon? in the student text before proceeding in this section.

The dependence of the force of air resistance on projected area and speed presented in the student text is a simplified version of the governing equation for the resistive force acting on an object moving through a fluid:

$F = 0.5$
(drag coefficient) x (projected area) x (fluid density) x (speed squared)

Where:

- The drag coefficient depends on the object's shape and the so-called "Reynolds number." The Reynolds number, in turn, depends on the object's speed, diameter and the viscosity of the fluid through which the object is moving. A typical value of the drag coefficient for a sphere falling through air is 0.42.

- The projected area is the area of the object as viewed from ahead or behind as the object travels.

- The fluid density applies to the fluid through which the object is moving.

- The speed is the speed of the object relative to the air.

Obviously, air resistance is complicated to predict with high precision except for very simple shapes. That is why engineers who work with air resistance often use wind tunnels to measure air resistance directly. For the purposes of the activity performed by students here, the primary dependence of air resistance on projected area and speed as described in Physics Talk is sufficient.

Active-ating the Physics InfoMall

A search of the InfoMall for "terminal velocity" finds several good hits, including some classroom ideas. Check this out.

Notice that as we searched for various concepts above, air resistance and the range of a projectile have already been uncovered! For example, see Activities Four and Six.

Planning for the Activity

Time Requirements

- One class period.

Materials Needed

For the class:
- badminton racquet and shuttlecock, volunteer server
- golf club and Wiffle® practice golf ball, volunteer golfer
- Optional: Sonic ranger for monitoring fall of coffee filters (minimum of one for the class; one per group if available)

For each group:
- 21 coffee filters, paper basket-type

Advance Preparation and Setup

You must provide either real-time, or a video of a golfer driving a practice golf ball to maximum possible range (probably 30 to 40 yards). It would save class time to do this in advance via video; otherwise, a quick trip outdoors or to a gymnasium can be done. As a poor substitute, "canned" data can be reported to students.

A badminton shuttlecock probably can be served to maximum range in your classroom or in a hallway of the school. Have a large enough area in mind and available.

Teaching Notes

The dependence of air resistance on projected area and speed is established. Students may need help understanding what is meant by projected area.

NOTES

3

Activity Overview

In this activity students investigate the effect of air resistance on motion.

Student Objectives

Students will:

- observe and understand the dependence of air resistance on the speed of objects moving in air.
- observe and understand terminal velocity of falling objects.
- apply effects of air resistance to adapting sports to the moon.
- consider requirements for self-propelled human flight in an air-filled shelter on the moon.

ANSWERS FOR THE TEACHER ONLY

What Do You Think?

The acceleration due to gravity on the moon would be equal indoors and outdoors.

Air resistance would behave the same way inside an air-filled facility on the moon as on Earth. The horizontal component of a projectile's motion would be retarded (decelerated) in the same way as on Earth. Terminal velocity for falling objects would be reached at lower speeds than on Earth in an air-filled facility on the moon due to reduced weight of objects on the moon.

SPORTS ON THE MOON

Activity Nine
"Airy" Indoor Sports on the Moon

WHAT DO YOU THINK?

There is no atmosphere on the moon. If gas were released on the moon, the gas would escape and no atmosphere would form.

- **Is the acceleration due to gravity on the moon equal indoors and outdoors?**
- **Would air's opposition to the motion of objects moving through it (air resistance) be the same inside an air-filled structure on the moon as it is on Earth?**

Record your ideas about these questions in your *Active Physics log*. Be prepared to discuss your responses with your small group and the class.

ACTIVITY NINE: "Airy" Indoor Sports on the Moon

FOR YOU TO DO

1. What would the game of badminton be like on the Moon? Observe as a member of your class hits a badminton shuttlecock.

🖎 a) What is the range when the shuttlecock is hit in a direction approximately parallel to the ground using a hard-as-possible, tennis-like, overhand serve? Record the range in your log.

🖎 b) How is the range of the shuttlecock affected as the server "backs off" by hitting with less and less strength of serve? Describe in your log how the range is affected.

🖎 c) How do changes in the shuttlecock's speed during approximately the first one-half of its flight compare for hard and soft serves? Write your response in your log.

🖎 d) Explain in your log why hitting a shuttlecock harder-and-harder does not result in proportionately greater-and-greater ranges.

🖎 e) Including effects of $\frac{1}{6}g$, what aspects of the game of badminton would be the same as on Earth if the game were played indoors (in air) on the moon? What would be different? Within reason, would badminton be playable indoors on the moon? Write your answers in your log.

🖎 f) Including effects of $\frac{1}{6}g$, what aspects of the game of badminton would be the same as on Earth if the game were played outdoors, without air, on the moon? What would be different? Would it be playable? Write your answers in your log.

2. What would the game of golf be like on the Moon if regular golf balls were replaced by Whiffle™ practice golf balls? Observe a golfer drive a practice ball to maximum range.

🖎 a) Measure the range of the Whiffle™ ball and compare it to the golfer's estimate of what the range would have been if a regular golf ball had been used. By what factor does using the practice ball reduce the usual range of the golfer's drive? Why? Write your responses in your log.

🖎 b) Including effects of $\frac{1}{6}g$, would replacing regular golf balls with practice balls reduce the size of a golf course needed for outdoor golf, without air, on the moon? Indoor golf, in air?

🖎 c) Approximately what would need to be the size of an indoor facility for playing Whiffle™ Golf on the moon?

 157 ──── SPORTS

For You To Do

1. a) The maximum range of a badminton shuttlecock is in the neighborhood of 10 m.

b) The range does not decrease dramatically as the hits are made softer, this due to the fact that most of the speed is lost soon after launch when the air resistance is many times greater than later in the flight.

c) The speed during the first half or so of the range appears greater for hard serves and soft serves, but the speeds are about equal during the last part of the flight. Since the last part of both flights are about equal, the total ranges are not extremely different for had and soft hits.

d) The air resistance acting on the shuttlecock varies with the square of its speed; it's a limiting proposition to try to make a shuttlecock go farther by hitting it harder.

3

e) Indoors, in air on the moon, the horizontal component of a badminton shuttlecock's motion would be the same as on Earth; the vertical acceleration would be less on the moon. It would seem entirely possible to play badminton on the moon.

f) Badminton probably would not be playable without air on the moon because the shuttlecock would take off in the same way as a baseball hit by a bat.

2. a) A practice golf ball "limits out" at a range of about 30-40 m, about 10 times less than the usual range of a drive by a good golfer.

b) Golf in an indoor, air-filled facility on the moon may be feasible using a practice golf ball, but outdoors on the moon where there would be no air resistance, a practice golf ball would have tremendous range.

c) The minimum size for having several holes would seem to be in the neighborhood of 100 m x 100 m.

ANSWERS

For You To Do

(continued)

3. Student activity.

4. a) The filters will have fall times
    ranging in nonlinear order
    (not equally separated in time)
    from the heaviest set of six
    nested filters (least time,
    almost equal to free fall) to the
    single filter (which reaches a
    low terminal speed almost
    immediately after it begins
    to fall).

b) The greater the object's mass,
   the less time it takes it to fall.
   More precisely, the greater the
   net downward force acting on
   the object—the object's weight
   minus the force due to air
   resistance (which changes
   with the object's speed), the
   greater the object's acceleration
   at any time during its fall.

c) Certainly, the single filter
   seems by naked-eye observation to reach terminal speed,
   and so may some of the others.
   A sonic ranger, if used as an
   option would reveal which
   filters reach terminal speed; if

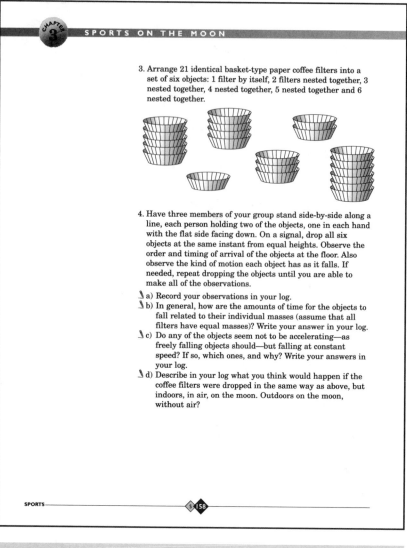

SPORTS ON THE MOON

3. Arrange 21 identical basket-type paper coffee filters into a
   set of six objects: 1 filter by itself, 2 filters nested together, 3
   nested together, 4 nested together, 5 nested together and 6
   nested together.

4. Have three members of your group stand side-by-side along a
   line, each person holding two of the objects, one in each hand
   with the flat side facing down. On a signal, drop all six
   objects at the same instant from equal heights. Observe the
   order and timing of arrival of the objects at the floor. Also
   observe the kind of motion each object has as it falls. If
   needed, repeat dropping the objects until you are able to
   make all of the observations.

   a) Record your observations in your log.
   b) In general, how are the amounts of time for the objects to
      fall related to their individual masses (assume that all
      filters have equal masses)? Write your answer in your log.
   c) Do any of the objects seem not to be accelerating—as
      freely falling objects should—but falling at constant
      speed? If so, which ones, and why? Write your answers in
      your log.
   d) Describe in your log what you think would happen if the
      coffee filters were dropped in the same way as above, but
      indoors, in air, on the moon. Outdoors on the moon,
      without air?

at all possible, use a sonic ranger to seek details of the motions of the filters.

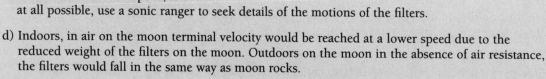

d) Indoors, in air on the moon terminal velocity would be reached at a lower speed due to the
   reduced weight of the filters on the moon. Outdoors on the moon in the absence of air resistance,
   the filters would fall in the same way as moon rocks.

One way to explore the possibility of self-propelled flight in an air-filled facility on the moon
would be to consider the power requirement needed to offset the power represented by a person
falling at terminal velocity. The equation Power = force x velocity may be useful for this purpose.

It appears that some sports which would not be playable outdoors on the moon would be playable
in an air-filled facility on the moon.

PHYSICS TALK

Air Resistance

Air resistance could have important implications for adapting sports—or even for inventing new sports—which could be played in an air-filled, indoor facility on the moon.

Air resistance exists because, as an object moves through air, it collides with air molecules in its path. Each collision with an air molecule is governed by Newton's Third Law and the Law of Conservation of Momentum. The air molecule is pushed in the direction of the object's motion, and, in reaction, the object experiences a tiny push by the air molecule in the direction opposite the object's motion. The result of steady collisions with many, many air molecules is that the object experiences a force and, therefore, an acceleration in the direction opposite its motion. The amount of force due to air resistance depends on the object's speed, size, and shape.

According to Newton's Second Law of Motion, $f = ma$, the effect of air resistance is to cause objects moving through air to decelerate, or slow down. Importantly, air resistance depends on the square of an object's speed, therefore tripling the speed of an object results in increasing the force of air resistance not by a factor of only three, but a factor of $3^2 = 9$.

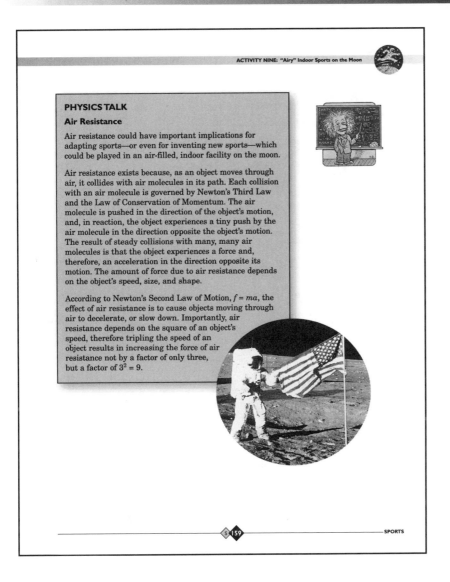

SPORTS ON THE MOON

FOR YOU TO READ

Fly Like a Bird on the Moon?

The effect of air resistance on a falling object is to cause the object's acceleration to decrease. If the object reaches great enough speed during its fall, it stops accelerating and continues its fall at a constant speed known as "terminal speed." This happens if and when the amount of the force of air resistance builds up enough to match the object's weight. Terminal speed is reached when:

Total force on object = (Weight) – (Force of air resistance)
= 0

Since the forces acting on the object are balanced, there no longer is any acceleration.

On Earth, a skydiver of average weight falling with an unopened parachute has a typical terminal speed of about 55 m/s (125 miles/hour); with parachute open, the terminal speed reduces to a safe landing speed of about 11 m/s (25 miles/hour).

On the moon, the skydiver's weight would be $\frac{1}{6}$ as much as on Earth, and so would the force of air resistance needed to balance the skydiver's weight be $\frac{1}{6}$ of the amount on Earth. Since the force of air resistance depends on the square of speed, the skydiver's terminal speed falling without a parachute through a moon atmosphere equal to Earth's would be less by a factor $\sqrt{6}$ = 2.45, or about 55 m/s ÷ 2.45 = 22 m/s (50 miles/hour). Therefore, a person would not be well advised to "free fall" very far through air on the moon, but it wouldn't take much air resistance to reduce the terminal speed to a safe landing speed.

This raises a possibility: people flying under their own power on the moon. Does a pedal-powered helicopter seem out of the question for an indoor activity on the moon? Could you equip people with gloves to create long, webbed fingers similar to bat's wings so that strokes similar to those used for swimming underwater and the breast stroke might be tried for "swimming" through indoor air on the moon?

REFLECTING ON THE ACTIVITY AND THE CHALLENGE

This activity has demonstrated that air resistance has profound effects on some sports on Earth and, if desired, also could have profound effects on indoor sports in an air-filled sports facility on the moon. Further, it seems entirely possible that the eternal human quest of self-propelled flight more easily could be realized in an Earth-like atmosphere combined with the moon's reduced gravity than so far has been the case on Earth.

PHYSICS TO GO

1. Think of a scheme which might work for people to engage in self-propelled flight in air on the moon.

2. Could a high-air-resistance replacement for a baseball serve to reduce flight distances enough to allow baseball as an indoor sport on the moon? How would the ball need to be altered?

3. What track and field events involving projectile motion could have equipment "fitted with feathers" (or other air-resisting devices) to reduce indoor flight distances on the moon?

4. How would outdoor table tennis be different on the moon than on Earth? How would it be similar?

5. How would indoor table tennis be different on the moon than on Earth? How would it be similar?

6. If you already have chosen the sport which you intend to propose to NASA for the moon, how, if at all, will air resistance affect your chosen sport?

7. Take a piece of crumpled paper and throw it horizontally. Compare the distance it travels with your expectation of how a thrown tennis ball would have travelled.

8. Have someone throw the crumpled paper horizontally so that you can see and record the path of the paper. How is it different from the parabolas you have seen as paths for other thrown objects?

S 161

Physics To Go

1. A pedal-powered helicopter.

2. A whiffle® version of a baseball might work for indoor moon baseball.

3. Low-inertia devices such as the javelin might be slowed down enough to be used indoors on the moon, but a shot put ball would have too much inertia to be slowed down even by adding "feathers."

4. Outdoor table tennis on the moon would have the ball travel faster due to lack of air resistance and fall slower due to reduced gravity. Spin applied to the ball would not affect its motion during flight.

5. Indoor table tennis would work about the same way as badminton in an air-filled moon sports facility — the greatest difference would be in the downward acceleration of the ball.

6. The response will depend on the student's choice of sport.

7. It will travel a much shorter distance.

8. The path of the paper will not be a parabola. After reaching its peak, it will fall almost straight down.

3

SPORTS ON THE MOON

PHYSICS AT WORK

Linda M. Godwin, Ph.D.

NASA ASTRONAUT

It was while working on her Doctorate in physics at the University of Missouri in the late 1970s that Linda first thought about becoming a NASA Astronaut. Prior to that time, there had been only male Astronauts. When NASA began putting the word out that they were interested in breaking down the gender barriers for their scientific mission specialists, Linda responded.

Now she is a veteran of three space flights and has logged over 633 hours in space, including a six-hour space walk. Her six-hour space walk has been one of her greatest and most thrilling challenges. During a space walk she leaves the shuttle attached only by a tether, and goes out into space wearing and depending solely on the pressurized life-support space suit. Dr. Godwin went up on Atlantis with the third docking mission to the Russian space station Mir. She performed the space walk in order to mount experiment packages on the Mir docking module to detect and assess debris and contamination in a space station environment. Setting up equipment under the micro gravity environment, while wearing the life support apparatus, was very difficult. "The suit is over 500 pounds, and even though you really don't weigh anything, you still have all the properties of mass. So, in terms of starting or stopping it's like you're moving 500 pounds—all the inertia is still there. You just have to learn to move a little differently. For example, it's harder to stay in one place, because you can get this motion build up."

"Seeing the earth go by while up in space is just an incredible experience," says Linda. "Earth is so beautiful and peaceful. In the space craft we are orbiting her every 90 minutes, so we see a sunrise and a sunset every 45 minutes. At nighttime you see all the city lights and during daylight you see the oceans, clouds, land masses and weather like storms and lighting flashes. You get a sense of how we're all tied together. You can look down and see that little surface of air that we have shining through the horizon during a sunrise or sunset and you realize there isn't much of that either. It really makes you think about how fragile and vulnerable our planet is."

SPORTS

S 162

Sports

Chapter 3 Assessment

You and your group have invented a definition of sport using the attributes of sports on Earth. You have seen what factors change on the moon which may influence the way in which a sport would be played. It is now time to use the information, knowledge, and understanding from this chapter to decide which sport you will adapt or invent that people on the moon will find interesting, exciting, and entertaining.

Write a proposal to NASA (National Aeronautics and Space Administration) which includes the following:

- **a description of your sport and its rules and how it meets the basic requirements for a sport.**
- **a comparison of factors affecting sports on Earth and on the moon in general.**
- **a comparison of the play of your sport on Earth and on the moon, including any changes to the size of the field, alterations to the equipment, or changes in the rules.**
- **a newspaper article for the sports section of your local paper at home describing a championship match of your sport on the moon.**

Review and remind yourself of the grading criteria that was agreed upon by the class at the beginning of the chapter. If you are going to submit one proposal as a group, remember that the grading criteria should satisfy every person's need for fairness and reward.

Physics You Learned

Factors on Earth and the moon

Acceleration due to gravity

Inertial mass and gravitational mass

Projectile motion

Collisions

Momentum

Effect of gravity on friction

Effect of gravity on pendulum motion

Air resistance

3

Alternative Chapter Assessment

Write your answers on a separate sheet of paper. Show your work if calculations are involved in your answer.

1. On the Earth, which object will hit the ground first? Explain each answer.
   a. A book and a marble dropped from the same height at the same instant.
   b. A book and a feather dropped from the same height at the same instant.

2. On the moon, which object will hit the ground first? Explain your answers.
   a. A book and a marble dropped from the same height at the same instant.
   b. A book and a feather dropped from the same height at the same instant.

3. If the same book is first dropped on Earth and later is dropped on the moon from the same height, in which location will the book hit the ground in the least amount of time? Explain your answer.

4. Compare each quantity on Earth to the quantity on the moon. If the quantity is the same amount in both locations, answer "Equal." If the quantity is a greater amount on either Earth or the moon, explain your answer.
   a. your mass
   b. your weight
   c. the amount of force necessary to lift a box.
   d. the amount of force necessary to swing a bat.
   e. the amount of time for a ball to fall 4 meters.
   f. the amount of time a person who jumps is off the ground.
   g. distance traveled by a long jumper.
   h. the amount of frictional force between your shoe and the ground.

5. The acceleration due to gravity on Jupiter is four times the acceleration due to gravity on Earth. Compare each quantity on Earth to the quantity on Jupiter. If the quantity is the same amount in both locations, answer "Equal." If the quantity is a greater amount on either Earth or Jupiter, explain your answer.
   a. your mass
   b. your weight
   c. the amount of force necessary to lift a box.
   d. the amount of force necessary to swing a bat.
   e. the amount of time for a ball to fall 4 meters.
   f. the amount of time a person who jumps is off the ground.
   g. distance traveled by a long jumper.
   h. the amount of frictional force between your shoe and ground.

6. It has been said, "Watching basketball on the moon would be like watching a game on Earth in slow motion." Explain this statement.

7. Choose from the following sports: basketball, baseball or football.
   Describe the changes that must be made to play the sport on Jupiter.

8. Assume that a swimming pool exists on the moon.
   a. Describe a diving meet.
   b. How would a 100-meter freestyle swimming race be different than on Earth?

3

Alternative Chapter Assessment Answers

1 a. They both hit at the same time because, in the absence of air resistance, all objects accelerate at 9.8 m/s/s.

  b. The book will hit the ground first because air resistance will reduce the acceleration of the feather.

2 a. They will hit at the same time because all objects accelerate at

  1.6 m/s/s on the moon (1/6 Earth's acceleration)

  b. Since there is no atmosphere on the moon, they will hit at the same time.

3. The book will strike the ground first on the Earth. It will take 6 times longer to fall on the moon.

4 a. Equal. Mass doesn't change on the moon.

  b. Your weight is 6 times greater on the Earth because the acceleration due to gravity is 6 times greater.

  c. Since the force necessary to lift a box is equal to its weight, it will require 6 times more force on the Earth than on the moon.

  d. Equal. Since the mass is the same, it has the same resistance to change in its motion.

  e. It takes less time (6x less) for a ball to fall on the Earth since acceleration due to gravity is 6 times greater.

  f. Same as e.

  g. The long jumper will travel 6 times farther on the moon.

  h. Since objects weigh less on the moon and frictional force is proportional to weight, the frictional force is 6 times greater on the Earth.

5 a. Equal

  b. 4x greater on Jupiter

  c. 4x greater on Jupiter

  d. Equal

  e. Greater on Earth

  f. Greater on Earth

  g. Greater on Earth

  h. Greater on Jupiter

  (See explanation in #4)

6. Since objects are in the air 6x longer on the moon due to the reduced acceleration due to gravity, it would be like watching a slow motion Earth game.

7. The ball (basketball, baseball or football) should be less massive so that it can be thrown with a greater initial velocity. The playing field should be smaller since the time in air will be reduced by a factor of 4. The goal posts and basket heights should also be lowered.

8. The diving board should be much lower since the height and time in air is increased. Also, the board should have less spring (elastic potential energy) so that the launching speed is lower. There will be more time to complete rotations, so possibly a new type of dive will be invented. As far as swimming, except that the water will be more buoyant, it should be the same as on the Earth because the force needed to propel the body through the water will be the same.

3

NOTES

NOTES

3

NOTES